911
MANAGEMENT

∎

JOSEPH J. BANNON

Sagamore Publishing, Inc.
www.sagamorepub.com

©1999 Sagamore Publishing, Inc.

Production Coordinators: Jennifer L. Polson and Anne E. Hall
Editor: Susan M. McKinney
Cover Design: Julie L. Denzer and Britt Johnson
Cover photo copyright ©PhotoEssentials
http://www.photoessentials.canon.com.au

ISBN: 1-57167-132-3
Library of Congress Catalog Card Number: 99-60712

Printed in the United States.

*To Kelsey, Keegan, Erin, Sophie, Doug, Leo,
and all future grandchildren*

CONTENTS

Foreword .... **vii**

Part One: General Management

1. Brainstorming 3
2. Business Ethics 9
3. Committees 17
4. Customer Service 27
5. Decision Making 37
6. Delegation 45
7. Developing Work Teams 51
8. Goal Setting 61
9. Grapevine 69
10. Managing Creativity 75
11. The Marketing Function 81
12. Meeting Planning 91
13. Negotiation 97
14. Organizational Charts 105
15. Organizational Communications 111
16. Problem Solving 121
17. Project Management 131
18. Public Relations 141
19. The Strategic Planning Function 151
20. Time Management 161

Part Two: Human Resources

21. Career Planning and Development 169
22. CEO-Board Relations 177

23.	Developing Personnel Policies	183
24.	Employee Appraisals	191
25.	Employee Fitness	197
26.	Employment Interviewing	203
27.	Hiring the Disabled	211
28.	In-Service Training	221
29.	Job Enrichment	233
30.	Job Satisfaction	243
31.	Motivation	251
32.	Participation Management	261
33.	Problem Employees	273
34.	Sexual Harassment	281
35.	Staff Development	291
36.	Substance Abuse	305
37.	Temporary Employees	313
38.	Terminating Employees	323
39.	Volunteerism	333
40.	Work-Time Options	343
41.	Writing a Résumé	353

Part Three: Executive Development

42.	Business Etiquette	365
43.	Effective Presentations	375
44.	The Importance of Writing Capability	381
45.	Leadership	387
46.	Listening	393
47.	Managing Stress	399
48.	Persuasion	411
49.	Procrastination	423
50.	Self-Esteem	429

Appendices **437**

Index **473**

FOREWORD

Park, recreation, and leisure service organizations need good management personnel in order to survive and thrive in a climate of downsizing, budget limitations, and reorganization. Executives must devote much of their time to understanding numerous new regulations and programs in managing the workforce of the next millenium. The environment in which they must operate is fraught with complexity and confusion. Errors of judgment have the most profound and resonant consequences. In short, the managers of the future, at virtually all levels, impose the most significant demands for administrative sophistication. Mangers are inundated with an enormous amount of information and advice from experts in the field. Information is also disseminated in bestselling management texts, magazines, scholarly journals, newsletters, newspapers, seminars, and instructional videos and tapes. The purpose of *911 Management* is to assist park, recreation, and leisure service managers to deal with the increased information needed to manage effectively as we approach the 21st century. It is also hoped that it will provide as complete a collection as possible of suggestions on how to meet the anticipated challenges of the future.

This book is primarily the result of my experience of over 30 years working with park, recreation, and leisure service organizations in the United States, Europe, Japan, China, Korea, and South Africa. During these three decades, I have conducted numerous workshops and seminars, programs evaluations, needs assessments, and organizational evaluations, and I have paid particular attention to the thoughts and concerns of the manager in the field. The topics addressed in this book reflect much of what they have told me. Many of my colleagues have encouraged me to put these thoughts and concerns in a book that would be an easy reference for the manager in the field. I have organized the book around three major categories: General Management, Human Resource Management, and Executive Development. However, the reader should not feel constrained to read a whole section before moving to another. Browse freely, reading whichever topic catches your interest or addresses an issue with which you are currently wrestling. After you read a topic, however, I suggest that you do three things. First, think of the last three situations in your own experience where you have observed similar situations. Second, roughly rate yourself as more of a manager or more of a leader with respect to the issue in question. You may be surprised to find that sometimes you manage, sometimes you lead, and sometimes you do a little of both. Third, think about your organization in terms of the issue presented in the topic of discussion. Over 1500 references have also been included so that each subject can be further studied. Suggestions for improvement for each of these subjects have also been included. This book represents a comprehensive encyclopedia of concise yet substantive information and advice on a variety of management issues.

911 Management presents information that will help the manager deal with peers, supervisors, subordinates, program participants, the general public, the press, and others with whom they may have contact. Information on such subjects as delegation, board-staff relationships, compensation and benefits, employee appraisal, decision making, communication, sexual harassment, marketing, employee fitness, public relations, and problem solving have been addressed. The reader will also find information on stress management, letter writing, leadership, career planning, goal setting, time management, temporary employees, negotiation, staff development, terminating employees, retirement, and employee commitment.

The book can be used as a point of departure on each of the subjects and as a guide for management personnel who want to update and keep current their knowledge of management issues, those who want suggestions to improve their basic management skills, or those who wish to improve their basic management function. Regardless of their position in the organization, managers will find the references at the end of each chapter useful. In addition, the book will serve as a valuable tool for students in leisure studies—the future managers.

Most importantly, have fun reading this book. Gaining new insights into yourself, your associates, and your organization can be exhilarating, surprising, uplifting, and even embarrassing. At times, most managers take themselves too seriously.

I am indebted to many individuals and organizations for sharing these thoughts and ideas. Hundreds of practitioners who provided me with input are also appreciated. The book also comes as a result of serving as president of Management Learning Laboratories and responding to hundreds of requests from park and recreation agencies. To each of them I extend my sincere appreciation. They all provided a reality check for the book. Of course, no acknowledgment would be complete without recognition of Carolina Zimdahl, Kathy Lankster, and Neva Summers for typing the initial manuscript. Appreciation is also extended to Jennifer Polson for designing and formatting the book, to Susan McKinney for her efforts in completing the final editing, and to Julie Denzer for designing the cover for the book. To all who participated in the development of this book, I am deeply appreciative.

Finally, I must thank the many people and organizations with whom I have worked as a consultant, for they have taught me much about management and leadership. I am deeply indebted and grateful to those who have inspired the many topics in this book. Their struggles, their victories, and even their setbacks have greatly influenced me in my quest to keep, in some small way, the men and women who manage and lead today's park, recreation, and leisure service organizations.

Part I

General Management

Chapter 1

BRAINSTORMING

Minds are like parachutes—they only function when open.

Brainstorming is a creative technique for problem solving that encourages development of radically new ideas and techniques. It does not initially require combining good judgment with good ideas. Brainstorming actually encourages "ransacking the brain" for ideas. Ideas are stimulated from the reservoirs of the subconscious mind, where most creativity and inspiration is believed to originate. These thoughts and ideas can begin a chain reaction, or a "brainstorm" of other ideas in a process of free association.

Brainstorming is a relatively simple procedure. It is basically a group of people tossing out ideas as fast as they can be recorded without outside or self-criticism. There is no discussion, elaboration, or selling of any particular idea. When brainstorming for ideas, everyone is encouraged to think in a free-ranging manner and to offer as many ideas and suggestions as possible. Unusual perspectives, innovative solutions, and wild shots should be encouraged. Participating in brainstorming should be stimulating and exciting.

Brainstorming may appear to be a substitute for more logical approaches to problem solving. It seems, at first glance, to demand little training or discipline, to be nothing more than an unplanned "bull" session. Nothing could be further from the truth. Brainstorming can be crucial to problem solving—a creative alternative to more traditional styles of problem solving.

ESTABLISHING A BRAINSTORMING GROUP

Although you can brainstorm alone, engaging a group of friends, coworkers, or others who are involved in or interested in the problem may really help. In terms of time and energy, more than one person brainstorming is very productive. Research shows improvement in both individual and group problem-solving skills when brainstorming is used.

A Few Basic Rules

It is important that everyone in the group understands and follows the four basic rules of brainstorming developed by Dr. Alex Osborn:

1. Criticism is ruled out. Judgment is suspended until later evaluation.
2. Freewheeling thinking is welcomed. The wilder the ideas, the better.
3. Quantity is encouraged. The greater the number of ideas, the greater the chance for good ones.
4. Combination and improvement of ideas is helpful. In addition to contributing ideas of their own, panel members should think how suggestions of others could be turned into better ideas, or how two or more ideas could be combined into an even better idea.

Who is included in a brainstorming session depends on the type of ideas you are trying to produce. If you have a problem in your family, at work, or in a community or political organization, all those affected or involved should be asked to participate or be represented.

The group should be bright and well informed. When the ideas you are trying to generate are work related, try to include participants from the same employment levels. Try not to mix people with vastly contrasting traits, if possible. For example, mixing shy people with outgoing people, or doers with thinkers, may cause conflict rather than stimulation. When there are too many differences among participants, participants might focus on those differences rather than on ideas to be generated. Bringing together people with greatly similar temperaments and experiences, however, may limit the group's vision and should be avoided as well. Groups composed of five to seven people are very effective. A leader or coordinator should be appointed to keep the session going no matter how small the group.

The location of a brainstorming session should be different from the usual place of work or home life—a nearby resort, motel, quiet picnic area, or a special room within your home or organization. A different location helps create a freewheeling exchange of ideas.

Leading a Brainstorming Session

Try to circulate your idea statement prior to the session and write it on a blackboard or flip chart immediately before your session begins. You may include a few examples of ideas that might be useful. However, be careful that when introducing "start-up" ideas you do not establish a mind set. Be prepared to offer more sample ideas in case participants get bogged down.

Brainstorming does not require total isolation to be successful. It can occur spontaneously at the breakfast table, in car pools, on airplanes, or during breaks in the day. Mini-sessions can be quite useful. Brainstorming should not be a marathon to drain people of ideas; during an eight-hour work day, no more than two hours should be devoted to brainstorming.

Research and experience show that when an authority figure attends a session, his or her presence usually puts a damper on participation. It is difficult for employees to be freewheeling and noncritical if a boss or some other authority figure is observing the process.

The seating arrangement, room layout, and decor should add to the "serious informality" of the occasion. Name tags or cards should be provided to participants if they do not already know each other. The group's leader or coordinator should be visible and reachable. There is no need to maintain any formality in seating, and allowing members to rearrange their seating literally offers them a new perspective, especially after a break. Whatever seating arrangement is decided upon, each member of the brainstorming group should be able to maintain eye contact with other individuals in the session.

Participants should raise their hands for recognition. If several participants have their hands raised, the coordinator should move quickly from one to another in order to maintain a fast tempo. The coordinator should also discourage anyone from reading lists of ideas prepared prior to the session, because that dampens the freewheeling atmosphere.

The coordinator should allow only one idea at a time from each person and should encourage ideas that are sparked by a previous idea. This chain reaction can be stimulated by asking participants to snap their fingers when raising their hands. It is useful to have someone keep track of ideas. Ideas can be jotted down, or better yet, recorded on tape. If a coordinator writes a few of the ideas on a blackboard or flip chart as the session goes on, it will stimulate more ideas, encouraging new combinations or additional ideas.

The coordinator's attitude is critical in ensuring that the atmosphere of the session is freewheeling, but not too casual or foolish. A knowledgeable coordinator is essential. This person should be careful not to expect miracles from brainstorming or to inflate its value to the group.

Consider starting your sessions with a brief warm-up, especially if you're working with a group that is not familiar with brainstorming. Using jokes is an easy way to learn the brainstorming process while preparing to tackle more serious problems. Some successful exercises that can be used for warm-up purposes include the following ideas:

- How many other uses can you think of for a paper clip?
- If you woke up tomorrow and were twice the size you are today, what difficulties would you encounter?
- You might also try bringing cartoons and asking the group to brainstorm captions for them.

There are some pitfalls to be avoided in introducing brainstorming. Again, avoid overselling the technique. Never attempt to use brainstorming as a substitute for individual thinking. Avoid this pitfall at all costs!

Brainstorming sessions often run smoothly. However, to prepare you fully for taking a position as a coordinator or leader of a session, refer to the list below that includes several of the difficulties we have discussed, as well as additional potential problems.

- Lack of adequate preparation.
- Lack of enthusiasm and support from participants.
- Putting too much faith in brainstorming before there are any results.

- Failure to clarify the problem, or to focus on the real problem.
- Failure to warn participants that many of the first ideas offered may not be creative, and not to become discouraged.
- Failure to keep participants focused on the real problem.
- Failure to seek more details when ideas are too general.
- Failure to ask provocative, idea-spurring questions when a session slows down.
- Failure to distinguish use of the technique from the use of individual or logical thinking.
- Failure to distinguish between an open atmosphere and one that has strayed from the problem or has gotten out of hand.
- Failure to evaluate and handle ideas effectively.
- Failure to put ideas into action.

After the session is over, try to shelve the problem for at least a day. Then, contact participants again to see if they have developed any additional ideas. Because participants will also have "slept" on the problem, you might get some of your best ideas at this time.

After you have contacted everyone, you may want to make a list of all the ideas, breaking them down into five or 10 categories. Some categories that may be useful are money, time, and space requirements.

SUGGESTIONS FOR IMPROVED BRAINSTORMING SESSIONS

- A typical brainstorming session should start with a brief warm-up to loosen up participants. Participants should receive a general statement of the problem at least a week before the session. This will allow them to begin to think of ideas before attending the session.
- A follow-up session a day or two later using the same participants is a good way to pick up on the ideas that are generated after the session has been completed.
- The brainstorming session should begin with the leader stating the principles of the brainstorming concept. If the session should reach a "lull," the leader should introduce his or her ideas or suggest ideas spurring questions to regenerate idea production by the group.

- Allow the ideas generated to stir around in the participants' minds. Do not be too quick to accept an idea before all aspects of its effects have been explored.
- Before the brainstorming session begins, be sure the problem is clearly stated and that you are focusing on a single issue. If not, the group will flounder and go off on tangents.
- It is important to remember the key to successful brainstorming, whether acting alone or with a group, is discipline and rigorously following the rules discussed here.
- It should be cautioned that brainstorming is not a substitute for problem solving or decision making. If the concept of brainstorming is presented as a cure-all to solving problems, your group will be disappointed.
- It is a good idea to include a few "self-starters" in your brainstorming group. These are people who can produce ideas on everything and anything and can assist in getting the ball rolling. Be careful, however, that these people are not allowed to "take over" the session.
- Do not allow participants to discuss, elaborate, or sell their ideas during the brainstorming session. Some session leaders sound a bell when a participant begins to violate this rule.
- The brainstorming leader should always inform session participants about the action taken on the ideas generated. Failure to do this will result in discouragement and apathy toward the brainstorming process and make future participation and cooperation difficult to obtain.

Chapter 2

BUSINESS ETHICS

Ethical judgments are most often clouded by the intrusion of personal factors. Decisions based on someone's irrational fears or prejudices, likes, and preferences are made daily.

"LEAKS PLAGUE BROKERAGE FIRM'S STOCK PICKS"
—*The State,* Columbia, South Carolina
"WAL-MART'S BUY AMERICA PROGRAM MISLEADS SHOPPERS"
—*USA Today*
"SEARS CHAIRMAN RESPONDS TO ALLEGATIONS OF AUTO REPAIR FRAUD"
—*Chicago Tribune*

Today it is difficult to read a newspaper or magazine that does not make some reference to business ethics. From Slippery Rock, Arkansas, to Washington, D.C., ethics has become the buzzword of the decade. Such diverse groups as professors, architects, political figures, and lawyers have discovered the importance of ethical conduct.

The ethical values of this country are a matter of debate and concern. Unfortunately, everyone cannot consistently agree on what is ethical behavior and what is not. To many people, the term "business ethics" is something of an oxymoron. The history of business in capitalist America makes it easy to be cynical, as we have witnessed breaches of ethics by companies all too often throughout history. Success and greed cause some to do whatever is necessary to get ahead. Businesses that deal unethically seem to have adopted the creed of "winning is everything," which makes dealings with these businesses seem comparable to warfare. These "win at all costs" corporations may experience short-term success, but may ultimately fail miserably.

Ethics can be defined as *standards or principles of conduct that govern the behavior of an individual or a group of individuals*. Ethics are generally concerned with moral duties or questions related to what is right and what is wrong. These moral standards have evolved over time. During the 19th century, commerce was pursued in a laissez-faire marketplace where social Darwinism reigned. Fortunately, most organizations seem to have moved away from this extremist view, but there are as many questions today about ethics as there were in the last century.

There are three levels of ethics by which everyone is guided. First is the law, which is how society codifies its ethical standards. But even the law can be subject to interpretation. Next, there are policies or guidelines that exist, in some form, in every institution. Unfortunately, these policies are not always emphasized and/or followed. Most people will recognize obvious unethical and illegal dealings (such as embezzlement), but overall, there is a lack of unanimity in many cases on what exactly is considered unethical.

While both the law and specific policies are critical to maintaining ethical standards, by themselves they are not enough. The most difficult, and least defined category, is the moral stance an individual takes when making judgment calls that are not specifically governed by written rules. Moral standards are subject to individual interpretation and application.

In his article "What's the Right Thing?" Andrew Grove is convinced that most ethical dilemmas do not arise because a person does not know the difference between right and wrong. Ethical problems arise because the solutions to these problems are unclear and decision making becomes difficult. Often, confronting an ethical issue is more difficult than ignoring it. In deciding how to handle an issue, Dreilinger and Rice, in their article "Five Common Ethical Dilemmas and How to Solve Them," believe that frequently the "right" answer is the one you can live with. This statement does not hold true for everyone.

For some people, unethical dealings are simply a means to an end. Being unethical may be an easier way to achieve a desired goal, or may even be an accepted practice. One particular salesperson admitted that "At times I don't feel entirely ethical. With anything involving sales or marketing, ethics don't really exist. I don't lose a lot of sleep thinking about ethics. Whatever I do I can rationalize in my mind."

Standards that guide people in these types of situations have been formulated throughout one's life. Not all people are morally corrupt, and when asking people what business ethics mean to them, answers range from "tell the truth, make decisions for the highest good, and treat people humanistically" to "screw them before they screw you!" Obviously, individual views on ethics vary widely.

Results of a recent survey on American ethics are staggering. When two in every three Americans today believe there is nothing wrong with telling a lie, and only 31% believe that honesty is the best policy, it is obvious that American ethics are questionable.

A study by Ruch and Newstrom produced an interesting ranking of some common ethical dilemmas that are worth mentioning. The items were ranked using a scale that included the following choices: (1) very unethical, (2) basically unethical, (3) somewhat unethical, (4) not particularly unethical, and (5) not at all unethical. The responses of the supervisors interviewed are listed below, in rank order from the most unethical to the least unethical:

1. Passing blame for errors to an innocent coworker.
2. Divulging confidential information.
3. Falsifying time/quality/quantity reports.
4. Claiming credit for someone else's work.
5. Padding an expense account more than 10%.
6. Pilfering company materials and supplies.
7. Accepting gifts/favors in exchange for preferential treatment.
8. Giving gifts/favors in exchange for preferential treatment.
9. Padding an expense account up to 10%.
10. Authorizing a subordinate to violate company rules.
11. Concealing one's errors.
13. Taking longer than necessary to do a job.
14. Using company services for personal use.
15. Doing personal business on company time.
16. Taking extra personal time
17. Not reporting others' violations of company policies and rules.

Although this study has some shortcomings in that the ranking of items was based on an average of 121 managers, the results are, nevertheless, interesting. One item that is especially curious is that "padding an expense account more than 10 %" (number 5) is viewed as noticeably less ethical than padding it less than 10 % (number 9). These responses are an example of the "moral relativism" of actions.

Nearly 20 years have passed since this study was conducted, but there has been no evidence that these findings have changed. In fact, ethical standards have been thought to have declined. According to Thomas Labrecque in his article "Good Ethics is Good Business," standards have significantly taken a turn for the worse over the past few decades.

The costs of this downward trend in ethics is enormous to businesses. When an employee feels he or she has been undermined or betrayed by unethical practices, that employee's contribution to the organization will not be what it could or should be. When a company is guilty of questionable behavior, much of a manager's time may be spent checking up on subordinates. Many people spend their time "covering their behinds," "putting it in writing," or "reading the small print," rather than getting their work done.

Labrecque believes that good ethics and profit are not contradictory. In fact, he says, they are inseparable. When examples of unethical companies are studied, two things surface consistently. First, the company derives only short-term advantages, and, secondly, over the longer term, skimping on quality or service and operating unethically does not pay.

So why does unethical behavior continue? One reason is that, as a society, we tend to envy the wrong people. We judge success from outward appearances and envy this success regardless of how it is achieved. Another problem that causes people to ignore unethical practices by themselves and by others is that they are not sure if they want to deal with the risk and the hassle involved with doing what may be right, or challenging something they feel is wrong. Chances are, when a person speaks up against unethical behavior, especially when it concerns a boss, the person's job may be in jeopardy. At the very least, life on the job may not be made any easier if one continually questions his or her morals and the morals of others.

Authors Soloman and Hanson offer another perspective on why unethical behavior continues. They believe that, sadly, unethical business, like crime, sometimes does pay. In any system, a few deceivers might prosper. Organizations that comply with the rules and care about their reputations set standards that allow them to provide high-quality services and, in the end, prosper. Organizations and individuals need to look past immediate gratification and recognize the importance of an untarnished image.

It is obvious that many people believe the ethics of businesses and individuals in America today to be of dubious quality. But there is some good news, however. According to a number of experts in business ethics, we seem to have reached rock bottom and perhaps more importantly, we know we have. The ethics issue is on the table at last, so while we may not have improved our ethical standards yet, at least we are paying attention to them. Companies of all sizes are recognizing that moral modes of operations may be beneficial and are taking steps to ensure a moral and ethical business environment.

Questions on ethics are not limited to executives in immense corporations. Managers in smaller organizations face smaller-scale, but equally important, ethical dilemmas every day. The way these situations are handled shapes the overall ethical environment of the organization. Actions that organizations take to maintain a high ethical standard may vary but often include three components. The first is a policy of hiring good-quality employees, especially managers, who set the ethical tone for the organization. Secondly, a written code of ethics is established, identifying critical areas requiring ethical conduct by employees. Lastly, some organizations employ the services of experts in the area, generally referred to as "ethicists."

Employees most often look to their supervisors for clues on what is or is not ethical and base their actions on the supervisor's behavior. Supervisors must be capable of setting the ethical tone of the organization. What a supervisor perceives as "right" affects his or her actions as well as the actions of employees. If a supervisor engages in questionable behavior, employees may either think this behavior is acceptable, or they may lose respect for the supervisor.

According to Lisa Newton, director of the Program in Applied Ethics at Fairfield University in Connecticut, "Good ethics start at the top and filter down. When the executive suite cuts corners, everybody down to the lowest jobs will follow suit." Critical areas requiring ethical conduct by supervisors have been noted by various individuals. Most of the literature identifies these critical areas as belonging to one of three categories, these being loyalty, human relations, and covert action.

Where a supervisor's loyalties lie is an important consideration when setting an example for employee behavior. If a supervisor is completely self-absorbed and constantly puts his own personal needs before those of the organization or fellow employees, he obviously is not a good moral model. A supervisor needs to maintain a balance between his own interests and those of fellow employees and the entire organization.

Another way a supervisor can work toward encouraging an ethical environment is to be especially cognizant of his treatment of other people, especially subordinates. A stable concept of fairness should prevail. A supervisor who is inconsistent, unfair, and uninterested in the careers of subordinates, or wants to make them look bad will undermine any ethical aspirations of employees. A manager must also realize that her personal actions reflect upon the organization. Not only must supervisors set an example internally, they must also ensure that they properly represent themselves, and thus the organization, well in the community.

Having good-quality people in managerial positions is mandatory for maintaining high ethical standards, but this precaution is no guarantee. Potential unethical situations must be identified, and guidelines and policies must be constructed and enforced. In response to complicated ethical questions, most organizations should have an established "code of ethics," consisting of written statements of principles that should be followed by all levels of employees in the conduct of business. The company's policy should never be subjected to an interpretation that would allow a supervisor to encourage an employee to "do as I say and not as I do."

A code of ethics eliminates guesswork by employees and molds the ethical environment of the organization. Many companies require employees to sign a form confirming that they understand and intend to follow the company's ethical codes.

Because there are obviously no clear-cut answers to many ethical questions, some organizations, usually larger ones, form an ethics committee, or employ a person to look at and interpret corporate policies, monitoring mechanisms, and training methods continually to enable people to wrestle more effectively with decisions requiring ethical judgment. It serves an organization well to monitor the standards that have been set constantly and to ensure that employees apply those standards in a complex, fast-changing marketplace.

Giving advice to corporations on how to behave ethically has become one of America's hottest growth industries. Professional ethicists conduct conferences or seminars, do research, or provide consulting services. A majority of ethicists are associated with universities or theological centers, but because there are no standardized credentials necessary in order to become an ethicist, corporations need to exercise extreme care when seeking these kinds of services.

When organizations believe that there is a strong correlation between integrity and success, ethical questions may be taken out of ambiguous context and analyzed as to what is and is not appropriate. Regardless of what measures are utilized, a clearly mapped-out ethical workplace is a more healthy environment. When people work for an organization that they believe is fair, where people are willing to give of themselves, and where loyalty and caring are hallmarks, people work at a higher level. The values around them become a part of them, and they think of the customer as someone to whom they owe the finest possible product and service. When a company builds a reputation of quality, it draws more new and repeat customers, which almost always ensures success.

SUGGESTIONS TO IMPROVE BUSINESS ETHICS

- Organizations should make the public and their employees aware that ethical behavior is important in all their transactions. When senior management communicates its support for ethical behavior, the public will accept it as practice and employees will follow with appropriate behavior.
- When employees are placed in a highly competitive situation, be aware of possible unethical behavior and take appropriate action to prevent it.
- Allow group decision making when moral judgment is required. Group decision making usually results in higher levels of moral reasoning than does individual decision making.
- When unethical behavior occurs in an organization, it should be punished immediately. The establishment of ethical behavior in an organization occurs when specific procedures for handling unethical behavior are in place.

- Every organization should develop a code of ethics and regularly communicate it to all employees. This code should describe the general value system of the organization, define the organization's objectives, and suggest guidelines for problem solving.
- An ethics committee should be established. The purpose of this committee should be (1) to serve as a sounding board and answer questions, (2) to provide directives in the interpretation of the code, (3) to investigate any noncompliance with the code, (4) to recommend appropriate changes, and (5) to serve as an appeal board for employees charged with unethical conduct.
- All ethical situations are not clear cut. Like many basic business situations, the organization should recognize there are ambiguous, gray areas where ethical trade-offs may be necessary.
- When it is necessary to send an ethical situation to the legal department, be careful. Ethics is not just complying with existing laws and regulations. Ethical behavior goes beyond simply complying with the law.
- Refuse to do business with organizations that are known to violate or seriously compromise laws or acceptable professional standards. By doing so, you support their efforts and encourage their unethical behavior.
- Every organization should have a procedure that will facilitate internal whistle blowing. If an organization is committed to ethical behavior, it must provide mechanisms to ensure that potential problems can be brought to the attention of management.

SUGGESTED READINGS

Carroll, A. B. (1987, March-April). In search of the moral manager. *Business Horizons.*

DeGeorge, R. T. (1982). *Business ethics.* New York: Macmillan.

Sims, R. R. (1991). The institutionalization of organizational ethics. *Journal of Business Ethics, 10,* 493-546.

Solomon, R. C., & Hanson, K. R. (1985). *It's good business.* New York: Atheneum.

White, B. J., & Montgomery, R. B. (1980). Corporate codes of conduct. *California Management Review, 23*(3), 80-86.

Chapter 3

COMMITTEES

When an individual becomes a member of a committee, he or she will become involved in discussion about problems and issues and their solutions, and as a result will become aware of why certain alternatives were rejected and others accepted.

The average executive spends about six hours per week in formal committee meetings and approximately 12 hours per week in informal consultation with other members of the organization. The significant amount of time spent in committee meetings by managers indicates that, even with its disadvantages, the committee process does have a use in the organizational environment. The committee process provides a forum for the discussion of many points of view on an issue that requires a decision. If the committee is to be successful, it should be given a specific area of responsibility. Most importantly, the committee members should be aware of the expectations of management. The committee members should know whether the committee is responsible for making recommendations or only an evaluation of some issue or activity.

Today, committees are used in organizations for a variety of reasons. Governmental organizations have made extensive use of committees in advising in such areas as long-range planning, land-use development, recreation and leisure services, library services, and special projects. The function of these committees has been the following: (a) to give employees and constituencies the opportunity to exchange views and information, (b) to make recommendations concerning issues and problems facing the organization, (c) to generate ideas for solving problems, and (d) to assist in making policy decisions for the organization.

TYPES OF COMMITTEES

There are basically four types of committee structures. These include the following: task forces, standing committees, boards and commissions, and the group management committee. The following is a brief description of each.

Task Force Committees

The task force committee is formed to deal with specific issues or special assignments. For example, the mayor and council in a community may form a special task force to study the long-range land need for the community. This committee may recommend how much land is needed, where it should be located, what development would be appropriate, and how best it might be paid for. These task forces are usually made up of key decision makers within the city administration as well as citizens in the community. They may include technical experts in the area being studied.

Standing Committees

Standing or permanent committees remain in existence to meet the continuing needs of the organization. These committees usually include finance committees, nomination committees, membership committees, long-range planning committees, and research and development committees. Usually such committees either make formal recommendations to higher levels of management or have the power to make their own decisions.

Boards and Commissions

Boards are usually made up of individuals who are appointed or elected to oversee the operations of public or private organizations. Park and recreation board members, for example, are often elected by their communities to set policy, levy taxes, hire directors, and to oversee the delivery of program services. Library boards are usually appointed by the elective governing body but carry out similar responsibilities. The board of directors of a corporation is selected by the stockholders. Their responsibilities include the management of the company assets, establishing goals and objectives, hiring a CEO, and reviewing the progress of the company toward those goals.

Multiple Management Committees

This group is a committee that is formally established to oversee the operation and management of a company or organization. Members of the multiple management committee are responsible for the overall management of the organization. In a sense, they carry out the functions of a chief executive officer. This committee functions in a democratic mode in that its members have equal authority; no single member can overrule the majority decision of the group. This management committee is usually responsible to an executive committee or a board of directors. Most management experts do not recommend this kind of operation because it leads to power struggles, personality conflicts, and competition for influence and authority.

Committees usually do not appear on the formal organizational chart, but they are almost always used to enhance the operations of the organization. Committees, when properly handled, can be very productive and helpful. The criticisms of committees are probably due more to their misuse than to the committee principle itself, because the manner in which committees are formed and operated has a significant effect on the results obtained.

When is it best to appoint a committee? Experts agree it is desirable to appoint a committee when it is expected that the recommendations of a group will be better than those of a single individual. When specific expertise or knowledge is required, the committee approach may be the best. Also, a committee is useful and helpful when continuing coordination with different units in the organization is needed. Committees can be effective in settling jurisdictional questions, because participants can express differing points of view. A good example of this practice is the use of a coordination committee to settle a dispute between a supervisor of production and the director of sales.

ADVANTAGES AND DISADVANTAGES OF COMMITTEES

For many years, the strengths and weaknesses of committees have been a matter of discussion. In spite of their weaknesses, the general feeling among administrators is that committees are essential in the operation of large organizations and are often useful in managing smaller groups. Proponents of the committee structure give a number of reasons for their use. The committee can be a major force in pulling together different abilities and knowledge of its members. No two individuals approach a solution of a complex problem in exactly the same way, and varied analytical abilities may, through committee deliberation, be brought to bear upon the same problem or issue. This approach implies that the committee is more than a rubber stamp, and that members can speak out on issues under consideration.

It also implies that the committee chairperson is sufficiently honest and alert to recognize personal bias when a challenge is voiced in the committee meeting and is willing to act accordingly. In the course of committee discussion, members become increasingly aware of the extent to which the activities of their units affect the workings of other units. From this awareness comes a new willingness to coordinate the work of all units in order to achieve the overall goals of the organization. Also, there is a better chance that the decision made by a committee will be accepted more readily. When committee members participate in the decision-making process, they are more likely to commit themselves and their subunits to implementing committee decisions. The increased likelihood that a committee decision will be accepted means that these decisions will probably be more effective.

The development of personnel is one of the most important responsibilities of management. By serving as a committee member, participants are exposed to ideas and knowledge outside the usual areas of responsibility. In addition, committee members may engage in the study of the process of preparing for committee participation, particularly if they are given some special responsibilities in connection with committee projects and performance. The greatest training value comes from the opportunity to participate within a group, to express ideas, and to defend a point of view.

COMMITTEES

Advantages	Disadvantages
Group study produces better decisions	Wastes time and money
Coordination of work activities	Vulnerable to compromise
More effective application of ideas	Dominated by one individual
Serves as educational laboratory for participants	No one has direct responsibility
	Delays decision making

Still, many feel that the establishment of committees has its disadvantages. The major criticism is that committees waste management's time and, as a result, cost money. Committees are often too large, meet too frequently, or possibly not enough, and they operate inefficiently.

Committees often spend too much time on unimportant issues that could be handled with little discussion. Another weakness of committees is that they may compromise decisions. Some experts feel that committees lack the desire or assertiveness to come to a conclusion that might be arrived at by an individual acting as a decision maker. Compromise is often the result of conformity. An individual who disagrees with the majority may comply and vote with the majority rather than risk taking issue with other members of the committee (see discussion of Irving Janis' "Groupthink" concept on pages 39–41).

Moreover, because no committee member feels responsible for the committee's final decision, members may act less carefully than they would if they assumed individual responsibility. In addition, if the committee's decision should falter in application, members might not work as hard to overcome the difficulty as they would in situations where they would be individually held accountable. Another criticism is that committee decisions take time. It is sometimes difficult to convene busy committee members. Business as well as personal schedules are difficult to coordinate. Business decisions must be made quickly and with dispatch. Opportunities may be lost if action is not taken quickly.

Management experts caution that committees sometimes are operated under the domination of one individual. This dominant person may or may not be the chairperson. If the chairperson expresses disapproval, whether it be verbal or by body language, he or she may convey a negative attitude to other committee members. Also, if the chairperson has an aggressive, hard-driving, autocratic personality, these characteristics will influence committee meetings. Once the advantages and disadvantages have been reviewed and the decision made to appoint a committee, consideration should be given to the following questions:

1. Is there really a need for a committee?
2. Is the idea or cause to be served by the committee one that can be given practical implementation?
3. Is the work proposed by the committee duplicating that of any other existing group?
4. Will the committee promote the democratic process?
5. Will the committee serve to develop an organized procedure for the advancement of the cause?
6. Are there sufficient resources to finance the formation and objectives of the committee?

Selecting a Chairperson

Once a committee has been established, one of the most important next steps is the appointment or election of a committee chairperson. It is estimated that at least two million persons in the United States carry the title of chairperson in nonprofit organizations alone. Add this number to all those who hold this title in government and business, and it would appear much of the nation's direction rests in the hands of the chairperson. The person who assumes the position of chairperson must possess an outstanding leadership ability, a keen appreciation of human relations, and a willingness to work as part of a group.

The two major functions of the chairperson are to assume an attitude of open mindedness and to display a willingness to listen. Preparedness is another major requirement for a chairperson. Certainly an agenda for every meeting should be prepared. This step involves the chair doing some inquiring for himself and not being satisfied by just a briefing from the staff or other committee members. The position of chair should be clearly established so that committee members do not waste time wondering who is in a leadership position.

The chair must understand the committee purpose along with important constraints, such as time deadlines and responsibility. The chair has the responsibility of using the knowledge and experience of all members and to use their participation in attaining the objectives of the committee. The chair should encourage discussion, mediate disagreements, avoid digression, and keep the discussion on the right track. A successful chair will guide the committee in making efficient use of time. Inefficiency leads to apathy, which will demotivate committee members.

Selecting a Committee

An invitation to become a member of a committee raises the question, "Why me?" Therefore, it is good practice to have a list of reasons for joining. People serve on committees for a variety of reasons, such as improving the environment or the construction of a new hospital. Some use committees for their own personal advancement in the organization or for political gain. Whatever the reason, the chair should be fully aware why committee members have chosen to participate. Some participate because they have a strong desire to contribute their knowledge and expertise for the public good. Those who do it for personal gain will make decisions that will not be in the interest of the committee objectives.

Who invites whom to join the committee? People respond to recognized leaders and care should be taken that invitations to prospective committee members come from persons the prospective members respect. It would appear that the process of recruiting a committee should do the following:

- Present an attractive case for accepting membership.
- Demonstrate fully the purposes and end results sought by the committee.

- Let the invitation to serve come from a person the prospective member respects and with whom he would like to work.
- Make it clear how much time and service the potential committee member will be asked to give.
- Point out the benefits that accrue to the organization or agency and the individual through service on the committee.
- Give assurance that the committee member will work in a group that will have the facilities and assistance needed to make the task a practicable one.

Committee Size

The ideal size of the committee is not and cannot be fixed, but some general limitations can be suggested. The size of the committee must be large enough to represent varying opinions, but not so large as to create a general debating body. It must be large enough to carry out the work required of it, but not so large that the division of responsibility becomes confusing. It must be large enough to divide its authority when necessary, but not so small as to become a "one-person committee." There are varying recommendations about the size of committees, ranging from three to five members and from three to seven members.

Regardless of size, the committee must have balance. Most committees work in areas where the professional staff are co-workers. It is necessary to provide the committee with enough professional assistance to make certain that the "lay membership" makes no moves that are in conflict with good professional service.

SUGGESTIONS FOR IMPROVING THE MANAGEMENT OF COMMITTEES

- The committee's goals should be clearly defined in writing. This practice will focus the committee's activities and reduce the time devoted to discussing what the committee is supposed to do.
- The committee's authority should be specific. Can the committee merely investigate, advise, and recommend, or is it authorized to implement decisions?
- A friendly climate rather than a pecking-order atmosphere should be established. The chairperson should try to help each individual member identify his or her role on the committee.

- Each member of the committee should be encouraged to speak and bring up ideas and suggestions.
- The agenda and all supporting material for the meeting should be distributed before the meeting. When members study each item beforehand, they are more likely to stick to the point and be ready with informal contributions.
- Committee members should not yield on any point for the sake of harmony. A solution should be selected based on sound logic.
- Begin all committee meetings on time and adjourn them at the stated time and with a record of decisive action.
- As a committee member, announce the amount of time you can give and let it be known before accepting membership on the committee.
- All committee members should be punctual for meetings and work in a manner conducive to rational discussion, never proving a point by wearing down other members.
- The committee should periodically evaluate its progress toward its goals and objectives.

SUGGESTED READINGS

Carver, J. (1990). *Boards that make a difference*. San Francisco: Jossey-Bass.

Fram, E. H. (1986). Nonprofit boards: They're going corporate. *Nonprofit World, 4*(6), 20-36.

O'Connell, B. (1985). *The board member's book*. New York: The Foundation Center.

O'Donnell, C. (1987). Ground rules for using committees. *Management Review, 50*(10), 63-67.

Witt, J. A. (1987). *Building a better hospital board*. Ann Arbor, MI: Health Administration Press.

Chapter 4

CUSTOMER SERVICE

Producing high-quality products and services benefits everyone. It pleases the buyer, improves the organization's services, sales, and profits, allowing it to expand and grow . . . and provides steady-paying jobs.

The purpose of an enterprise is to receive and keep a viable customer base. It is obvious that without customers there is no need for the service or product to be provided. Also, there will be no business. Peter Drucker, the renowned management consultant, states that

> the purpose of business is to create and keep customers. Organizations should constantly seek to offer better or preferred products or services, through the different combinations of means, places, and processes, so that customers prefer to do business with them.

No enterprise can exist successfully without a clear understanding of what the customer wants. Clear, usable information is essential in tracking customer satisfaction. This information is important in monitoring performance of key marketing assumptions of the enterprise.

Service is valuable for organizations hard-pressed by competition. The most successful organizations emphasize good-quality service, not price. Competitive pricing attracts shoppers but not necessarily customers. Give a customer something worthwhile, such as service that treats him or her in a personal, individual, sincerely concerned manner, and that person will gladly pay the asking price and return to buy again and again.

A formal survey of the Homewood/Flossmoor Racquet and Fitness Center found that 72% of the members considered "speed of service and atmosphere the staff creates" the most important aspect of the club. Value for the money rated third, supporting the idea that club members want service first and are willing to pay for it.

How important is customer service? The Technical Assistance Programs Institute in Washington, D.C., found that one out of every four Americans is upset enough to stop doing business with an organization, and 95% of these unhappy customers will switch their allegiance from an organization rather than complain for their rights. The Office of Consumer Affairs reports the average business never hears from 96% of its unhappy customers, and for every complaint received, 26 other customers also have problems that are not reported.

It is further reported that 44% to 70% of complaining customers will do business with an organization again if the complaint is resolved, with the percentage rising to 95% if the complaint is handled promptly. A study conducted by the National Family Opinion for the Consumer Research Centers of the Conference Board asked consumers in 6,000 households to assess the value received for money spent on almost 40 different products and services. Its conclusion was "the vast majority of consumers believe that they receive good value for their dollar when they purchase products. But there is a rather pervasive discontent with what they get for the money they pay for services."

When *Cambridge Reports* of Cambridge, Massachusetts, asked 1,500 people, "How well do service companies meet your needs and concerns as a consumer?" only 8% rated them *excellent*, 50% rated them *good*, but 42% said *fair, poor,* or *depends on service*. More than one in three agreed with the statement that service industries care less than they did a few years ago. A Gallup Poll asked 1,045 people what makes them decide not to return to a given restaurant. Number one on the list of reasons, by 83% of the respondents, was poor service—not food quality, not ambiance, not price, but poor service. All enterprises should constantly evaluate their customer relations program so that they know how it is perceived by the individuals who buy their product and/or service.

Service Industry Rated Low

Why is customer service rated so low in so many surveys? A number of interesting reasons have been proposed and there is probably a great deal of truth in many of them. There are several obstacles to customer service that one time or another affect most organizations:

- The attitude that customers are replaceable. Taking customers for granted is long past. Customers are considered to be valuable assets secured through hard work and high dollar cost. Unfortunately, employees are often relieved when unhappy customers leave. Organizations should never act as if they have exclusive domain over their customers or as if there is an unlimited supply of new customers.
- Managers are not alert or sensitive to customer needs. Managers must continuously respond to changing situations. They must be able to make quick and accurate decisions in a workplace where skills, perceptions, values, and prejudices impact customers' needs. Given many of the human frailties, managers must monitor, train, motivate, evaluate, and retain employees so that they can accommodate and be of assistance to customers.
- Limited financial resources sometimes cause customer relations programs to fail. In many organizations, the financial resources required to meet the customer needs exist, but they are being directed to other areas in the organization. Lack of funding lessens the importance of customer service in the minds of frontline employees.
- In many organizations, there is not a strong commitment to customer service. Some provide only "lip service." Unless the hierarchy of the organization gives its full support, the success of customer relations programs will probably be in doubt.
- Customers are oversold or promised unrealistic levels of satisfaction. Forget the person who informs the customers of this false advertising and concentrate on the service involved and the organization making the claim. Organizations must strive for consistency of service, as well as honesty and accuracy of information to the customers.
- Organizations that are concerned with only the future will not listen to today's customers. Management must be available to

employees as well as customers. Most importantly, they must be open to discussion and should not be apprehensive about being proven incorrect.
- Organizations that are inundated with policy, procedures, and red tape create unhappy and dissatisfied customers. Most people do not want to hear why they cannot do something, they want problems solved. Organizations that provide successful customer service programs focus on the customer and do not let bureaucracy interfere with what is best for the customer.

Failure to realize the high cost of losing customers results in too much emphasis on just selling and marketing. Many organizations consider the work involved in retaining customers dull and tedious. Some organizations view unhappy customers as chronic complainers who are not worth the effort to satisfy them. Admittedly, some customers are habitual complainers, but too often organizations apply this label to every customer who complains. If this attitude is consistent throughout the organization, dissatisfied customers will probably receive poor treatment and their problems will remain unaddressed.

Most executives believe they know what their customers need and want. However, if the perception is slightly inaccurate, it may cost the loss of some customers. Being only slightly inaccurate can cause an organization to spend a great deal of time, money, and human resources effort on an activity that does not help retain a strong customer base. In spite of a strong commitment and genuine desire to produce high-quality customer service, many organizations fail in their customer relations. This failure is often the result of the organization being internally directed rather than externally directed. An internally directed organization assumes that it knows what customers should want and delivers that. This focus can lead to providing products and services that do not meet the expectations or desires of the customer. There are some common gaps between customers' expectations and the products or services offered by the organization. These include

- Inadequate communication between the organization and its customers
- A weak market-research program
- Lack of evaluation of the customer service program

- Lack of communication between management and frontline personnel.

Figure 4.1 describes the activity in maintaining a viable customer service.

Figure 4.1: Maintaining Customer Service

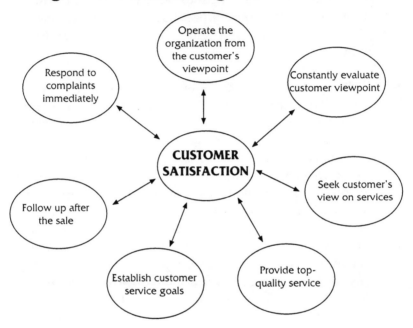

In organizations, an important priority should be listening to customers and trying to understand what they are saying. According to award-winning author Tom Peters, "listening to customers must become everyone's business. With most competitors moving ever faster, the race will go to those who listen and respond most intently."

In *Service America*, it was reported that successful service-focused organizations:

- Are obsessive about listening to, understanding, and responding swiftly to changing customer wants, needs, and expectations.
- Create and communicate a well-defined customer-inspired service strategy.

- Develop and maintain customer-friendly service delivery systems.
- Hire, inspire, and develop customer-oriented frontline people.

Obtaining Customer Information

There are many techniques that can be used to collect information about your customers. When selecting a technique it is important to consider costs, resources, time frame, and the organization's capabilities. Listed below is a brief discussion of some of these techniques.

- Telephone surveys. Phone surveys are an excellent method to use if quickness in obtaining the information is important, the questions are not too intrusive, and the survey is not too long. Training of the interviewer needs to be thorough, so questions are asked in a consistent manner. Efforts must be made to verify the phone numbers with respect to the demographics profile being targeted. Although phone surveys are excellent listening tools, they also have some limitations. Such as the amount of time passing since the complaint was made, so that recalling important details of the complaint may be difficult. Telephone surveys have a tendency to be fast paced—this fast pace may result in short, inaccurate answers. The validity of the survey can be affected if the interviewers do not follow directions precisely. Often people find telephone surveys intrusive and therefore give a negative response.
- Mail surveys. Mail surveys can take longer than phone surveys because respondents can complete the questions at their convenience. Mail surveys allow for the collection of a large amount of information. They also allow the customer to respond anonymously. However, their response rate is usually much lower than phone surveys. Sometimes the questionnaires get lost in the mail. As with phone surveys, it is important that a random sample of your customers is drawn, which will ensure the accuracy of your answers.
- Personal interviews. One of the simplest and best ways to get information from your customers is to ask them personally. This technique allows the interviewer to observe firsthand customer reactions to the questions. While personal interviews are very effective, their primary limitation is that they

require a significant number of interviews in order to complete the project within a reasonable length of time.
- Focus groups. This technique was used recently by the U.S. Postal Service to evaluate complaints about mail service. The Clinton administration also conducted numerous focus groups on citizen complaints concerning the health care issue. Focus groups consist of individuals brought together to participate in discussions concerning a series of topics, questions, products, or services. A moderator directs the questions and discussions and asks probing follow-up questions. Focus groups are particularly effective in assessing the validity of customer complaints. Factors that can affect the validity of focus group feedback are: one person influencing the opinions of the entire group; an ineffective moderator; and not selecting proper focus-group participants. Overall, this method may be the quickest way of getting customer feedback.
- The 800 number. Allowing customers to use a toll-free number may resolve customer complaints in the early stages. Toll-free numbers are also very effective in securing information from customers about new ideas, trends, and suggestions for new products and services. It is important for an organization to make its 800 number widely known and available to customers. It is important that a knowledgeable person answer the phone and have the information to respond to customers. Caution: your organization may actually increase customer frustration and dissatisfaction if an individual calls the 800 number with a problem or complaint and an uninformed person answers the phone.
- Comment cards. You see them in restaurants and in hotels. These cards can be very useful if well designed and part of an overall comprehensive customer relations program. When combined with surveys, the comment cards can provide a continuing profile of customer experience and attitudes.
- Frontline contact. Persons who are in daily contact with the customer should be given the opportunity to interact with the client. This contact provides an excellent forum to get new ideas about improvements and new services and products.

Measuring Customer Service Effectiveness

Assessing your organization's customer service program is critical to growth and survival, because most organizations are dependent on the success of these efforts. It is important to determine if the customer service program of your organization contributes to your organization's success and if so, how?

An assessment of customer service is helpful in guiding management to allocate resources efficiently to those areas with the greatest opportunity or greatest need. It helps to identify relative strengths and weaknesses by putting them in proper perspective. Many managers expend considerable effort supervising and/or performing many customer service-related tasks. But seldom is such intensity directed toward determining how well these tasks are performed. Without such an evaluation, an organization may not know why some programs are effective and others are not. A number of market research specialists have been trying to develop measurement systems that will be applicable to service quality and customer satisfaction from focus groups and survey research in a variety of industries. They offer a list of five statistically derived factors that they contend can be used to assess the effectiveness of a customer service program. The five component factors include the following:

1. Reliability. The ability to provide what was promised, dependably and accurately.
2. Assurance. The knowledge and courtesy of employees and their ability to convey trust and confidence.
3. Empathy. The degree of caring and individual attention provided to customers.
4. Responsiveness. The willingness to help customers and provide service.
5. Tangibles. The physical facilities, equipment, and appearance of personnel.

SUGGESTIONS TO IMPROVE YOUR CUSTOMER SERVICE

- Being nice to people is just 20% of providing good customer service. It is important to design systems that allow you to do the job right the first time. All the smiles in the world are not going to help you if your product and service are not what the customer wants.

- Educate your customers. Customers must be taught both how to use and how not to use a product or service. Through appropriate training programs, organizations can reduce the chances of calls of highly trained personnel to solve simple problems.
- Evaluate your customer service programs. Whether the customer service operation is treated as a cost center or a profit center, quantitative performance standards should be set for each element of the service performances.
- Be efficient first and nice second. Given the choice, most customers would rather have efficient resolutions of their problem than a smiling face. Efficiency is not intended to minimize the importance of a smile, however.
- Educate your employees. In many organizations, employees view the customer with a problem as an annoyance rather than as a source of information. A well-planned customer relations program is often needed to change such negative attitudes and to convince employees not only that customers are the ultimate judge of quality but also that their criticism should be respected and acted upon immediately.
- Get key customers involved in creating the service they want. Be sure that they see this involvement as a bonus for them.
- Limit the number of rules and regulations that affect your customers. For the rules you do have, your employees must be able to explain to the customers how each rule benefits that customer.
- To guide your organization into the next century, you must examine the external environment in which your customer relationships exist. Consider the following: technological changes, transportation and communications, legislation and the consumer movement, political environment, social environment, and economic environment.
- Ask customer-based questions. Tap both the customers' experiences (what happened to you?) and the customers' perception (how do you feel about what happened to you?). The customers' specific personal experience and interpretations enlighten far more than open-ended general questions such as "on a whole, how was your stay?"
- Thank customers when they give you comments and suggestions. Thanking them for their feedback says that you heard what they had to say and value their opinion.

SUGGESTED READINGS

Albrecht, K., & Zemke, R. (1985). *Service America.* New York: Dow-Jones.

Barry, L. L., & Zeitband, S. (1990, Summer). Five imperatives for improving service quality. *Sloan Management Review*, 31.

Desatnick, R. L. (1987). *Managing to keep customers.* San Francisco: Jossey-Bass Publishers.

Fine, S., & Dreyback, J. (1986). *Customers: How to get them, how to serve them.* New York: Dartnell.

U.S. Office of Consumer Affairs. (1986, April). *Complaint handling in America: An update study, part II.* Washington, DC: Technical Assistance Research Programs Institute.

Chapter 5

DECISION MAKING

Remember that choosing among alternatives often demands courage and moral judgment as well as intelligence—one alternative you should always consider is that you could be wrong.

Decision making involves more risk and responsibility than any other managerial activity. Decision making often requires that we try to predict results of events that have not occurred. The decision maker must accept full responsibility for a chosen solution. The work of problem analysis and evaluation can be delegated, but the responsibility of decision making is ultimately assigned to one individual. Some decisions are made by custom and tradition. Others are made in response to pressure, to escape other current difficulties, to emulate others, or for less certain reasons, both relevant and irrelevant. It should also be noted that the decision-making process has probably received more attention from management scientists than any other subject. Yet in most organizations, it is still slow, cumbersome, and sometimes nonexistent, even with the involvement of computer technology.

Effective decision making is vital to the growth of any enterprise. Making decisions or participating in the process by which they are made is an important and essential part of every manager's workday. Decisions can range from the profound to the trivial, from the complex to the very simple, with questions such as "Should we expand our markets now or later?" "Should we lease our office or build?" "Should we trim our payroll by letting personnel go based on seniority or by job effectiveness?" "Should you give your employees a 5% raise or a 9% increase?" Employing consistent methods of decision making has value beyond simply expediting the decision itself.

With many decisions, it is not possible to foresee all possible consequences of each alternative. Each involves risks and uncertainties. A bond issue may fail, a buyout may occur, funding from a federal contract may terminate, or a new product could turn out to be a failure. Such possibilities must be considered in decision making.

How should a manager generate and integrate all of the relevant information in a manner that responsibly addresses inherent complexities and help reach a decision that she feels comfortable with and can defend? One thing is certain, there is no completely objective process for decision making. There are, however, questions every manager should consider when making a decision. Do you know exactly what decision needs to be made? Is there a precise time when the decision must be made? Who will decide what the decision will be—the manager or a group of individuals? Will others need to be involved in the decision-making process? Can your decision be approved or disapproved by shareholders or a board of directors? Once the decision has been made, who will have to be informed of your actions? Answers to these questions will demand that the decision maker collect information that will require a rational approach to arriving at a decision.

One of the worst decision-making mistakes is to announce a decision impulsively before all the information has been collected. Victor Vroom, professor of psychology at Yale University, suggests we first ask ourselves a series of questions about what he calls "situation variables."

1. **Rational Quality Requirement.** Does it make a difference which course of action is adopted?
2. **Adequacy of Information.** Does the decision maker have the adequate information to make a quality analysis?
3. **Structure of Situation.** Does the decision maker know exactly what information is missing and how to get the information?
4. **Commitment Requirement.** Is commitment to the solution by others critical to effective implementation?
5. **Commitment without Participation.** Will they commit to an action made by the decision maker without their active participation?
6. **Goal Congruence.** Is there goal congruence between all those affected by the decision?

7. **Conflict about Alternatives.** Is there likely to be conflict about alternative solutions among those concerned?
8. **Subordinate Competency.** Do the people involved have the skill and know-how to implement the idea suggested?

A number of authors who study the decision-making process have suggested numerous causes of careless decision making. Below is a list of the most common responses:

- lack of clear objectives
- laziness
- complacency
- prejudice
- overreliance on past experiences
- copying other people's decisions
- impulsive reactions to events
- pursuit of private or irrelevant objectives
- uncritical pursuits of the obvious, and
- taking the easy way out

You can learn to avoid them by reexamining bad decisions and identifying the items on the list that may have had an effect on your decision-making process.

At some point in our experience, all of us have complained about groups, committees, and task forces becoming involved in the decision-making process. Yet we continue to spend considerable time in group decision making that might be better handled by one individual. Many CEOs and researchers believe that group decision making increases commitment to the decision on the part of those involved in the process. Others believe that group decision making may be preferred because "all of us know more than any one of us knows." Though it is not entirely supportable, this belief supposes that the group's multiple perspectives, talents, and areas of expertise applied in solving problems, setting goals, establishing policies, and carrying out projects result in a superior decision.

Periodically, we all have been part of a group whose decisions were poorer than those an individual might have reached. The major barriers to effective group decision making are those conditions that prevent the free expression of ideas.

Irving Janis, in his book *Groupthink,* warns that when groups get together to make a decision, they should guard against being caught in the "Groupthink Syndrome." According to this concept, members of any small cohesive group tend to maintain *esprit de corps* by unconsciously developing a number of shared illusions and related norms that interfere with critical thinking.

An important symptom of groupthink is accepting the illusion of being invulnerable to dangers that might arise from risky action in which the group is strongly tempted to engage. Essentially, the notion is that if the leader and everyone else in our group decides it's OK, our solution to the problem will succeed. Symptoms of groupthink include the following:

1. An illusion of invulnerability, shared by most or all of the members, which creates excessive optimism and encourages taking extreme risks.
2. An unquestioned belief in the group's inherent morality, inclining the members to ignore the ethical or moral consequences of their decisions.
3. Collective efforts to rationalize in order to discount warnings or other information that might lead the members to reconsider their assumptions before they recommit themselves to their past policy decisions.
4. Stereotyped views of enemy leaders as too evil to warrant genuine attempts to negotiate, or as too weak and stupid to counter whatever risky attempts are made to defeat their purposes.
5. Self-censorship of deviations from the apparent group consensus, reflecting each member's inclination to minimize to himself the importance of his doubts and counterarguments.
6. A shared illusion of unanimity concerning judgments conforming to the majority view (partly resulting from self-censorship of deviations, augmented by the false assumption that silence means consent).
7. Direct pressure on any member who expresses strong arguments against any of the group's stereotypes, illusions, or commitments, making clear that this type of dissent is contrary to what is expected of all loyal members.
8. The emergence of self-appointed mind guards—members who protect the group from adverse information that might shatter their complacency about the effectiveness and morality of their decisions.

A review of these eight symptoms may remind you of situations that you have encountered. The groupthink syndrome is evidenced in city hall council chambers, state legislatures, Congress, clubs, and organizations to which we belong, and places we work.

Computers now play a major role in the manager's decision-making process. The development of computer technology as we know it today has impacted decision making more than any other single influence. Any individual who writes a check or pays taxes is fully aware of the influence of computers on hi daily activities. Studies have shown that computers are used by management for three basic reasons: (1) they enable management to increase the quality and quantity of output of many clerical operations, (2) they either *make* or *aid* in making decisions, and (3) they assist in providing a total system approach to planning and operating an organization. Imagine the mountains of clerical work involved in reading, sorting, posting, and routing the billions of checks that circulate in the United States each year. Military personnel during the Gulf War in Iraq were able to make complex decisions regarding enemy movements, tracking scud missiles, and maintaining accurate information concerning Allied forces as a result of the high-tech equipment. The computer will never replace the human element in the decision-making process, but it will certainly improve, refine, clarify, and focus information on the problem. This will be valuable to managers in their roles as decision makers.

SUGGESTIONS TO IMPROVE YOUR DECISION-MAKING ABILITY

- Remember, not all decisions will prove to be successful even though you have based them on the best information available. No decision is irrevocable; other alternatives can be attempted. The success of the entire organization is what counts.
- Once you have decided what your decision will be, announce it so that everyone in the organization is aware of it. This will quiet the grapevine and the rumor mill. Procrastination can be costly.

- Take into account those involved in the decision. When employees know you do this, their response to unpopular decisions are often softened and their support is gained.
- Before you make a decision, try to find out what others have done in similar situations. If their decisions have resulted in success, then do likewise. There is no need to "reinvent the wheel" when others have found the right solutions.
- If all the necessary facts are available, make your decision. If they are not, make no decisions until the facts are available. Although getting more facts can become a delay tactic for some managers who do not want to make a decision, it usually is a valuable use of the decision maker's time.
- A sense of timing requires that we know when *not* to make a decision. Research shows that when we are depressed, or feeling low, our actions tend to be aggressive and sometimes destructive; when in good spirits, our behavior becomes more tolerant and balanced.
- There is much value in considering your intuitive feelings about a decision that is to be made; however, it is imperative that we be aware and place appropriate emphasis on our feelings, especially if they contradict the indications of all the data.
- It should be remembered that group decision making has its merits. Participation by others often increases the acceptance of the decision by group members and decreases the problem of persuading the group to accept the decision.
- Be careful not to become so emotionally attracted to a particular decision that not even the most accurate and up-to-date information will change your mind.
- Keep your information channels clean and shining. Early warning systems from this network will provide varied views on decisions to be made.

SUGGESTED READINGS

Einhoon, H. J., & Hogarth, R. M. (1987, Jan-Feb). Decision making: Going forward in reverse. *Harvard Business Review, 1,* 66-70.

Kernan, G. (1987). Decision making under stress: Scanning of alternatives under controllable and uncontrollable threats. *Journal of Personality and Social Psychology, 52*(3), 639-644.

Powers, R. P., & Powers, M. F. (1983). *Making participating management work.* Washington, DC: Jossey-Bass.

Thompson, V. A. (1971). *Decision theory: Pure and applied.* Morristown, NJ: General Learning Press.

Wild, R. (Ed.). (1985). *How to manage.* New York: Fact on File Publications.

Chapter 6

DELEGATION

Good delegation does not just happen; it demands time, effort, and persistence from the outset to develop and maintain the technique.

The modern manager is one of America's great assets: always busy, hard to get an appointment to see, and harder still to keep her attention. The telephone frequently interrupts. Secretaries run in and out with papers and perplexities. Her attention is usually submerged in numerous minor details. Subordinates invariably refer problems to her. Frustrated staff demand endless justification of what is taking place in the organization. Only a little time is spent on long-range plans and the future of the enterprise.

Not surprisingly, the "boss" sees herself as an expert in organizational theory: one who is well aware of the importance of delegation. Many managers who supposedly recognize that delegation of responsibility and authority is essential for organizational success attempt to prove it by displaying elaborate organizational charts and colorful administrative handbooks. Yet, in too many cases, they do not recognize the difference between delegation on paper and true delegation.

Scholars in organizational theory admit that although many managers make sincere attempts to put delegation into practice, few meet with success. Thus, delegation is one of the most complicated and least understood management principles.

Good delegation does not just happen; it demands time, effort, and persistence from the outset to develop and maintain the technique. A manager must face the challenge of effective delegation continually. Successful delegation takes thought, careful planning, knowledge of subordinates' areas of competence, effective personal communication, and a willingness to take risks.

WHAT IS DELEGATION?

When a manager assigns work to a subordinate, she is delegating responsibility and authority. In a recent survey, managers most frequently cited the following problems as obstacles to effective delegation:

1. Lack of agreement among supervisors and subordinates on the specifics of delegation. Lack of standards and guidelines.
2. Lack of training to accomplish delegated tasks.
3. Lack of understanding or organized objectives.
4. Lack of confidence by supervisors in subordinates.
5. Lack of confidence by supervisors in themselves. Unwillingness to take risks.
6. Supervisors' fear that subordinates will outshine them.
7. Fear of punitive action by supervisors.
8. Failure at all levels to understand the advantages of successful delegation.
9. Unwillingness of supervisors to delegate jobs they enjoy.
10. A desire for "nothing short of perfection."
11. A belief that things are going well enough as they are.

The problems cited illustrate that there are no easy answers to the questions that are at the heart of effective delegation: How much authority should be delegated? How much responsibility? How much and what kind of supervision should be exercised?

These are hard questions. The answers to them depend on particular circumstances and the people involved. There are no hard-and-fast rules for achieving or measuring success.

The manager who does not delegate effectively becomes more of a worker than a manager. She works harder, yet produces less than the manager who delegates effectively. By limiting her effectiveness as a manager, she limits her organization's success.

Delegation places decision making close to the point of implementation. The director of a park and recreation department knows that delegation to his neighborhood center supervisors brings them closer to program participants, encouraging the rapport that is a basic ingredient of program success. Delegation encourages subordinate responsibility and builds self-esteem, both vital to the health of an organization.

SUCCESSFUL DELEGATION

Successful delegation requires careful planning that takes into account the special character of the organization in which the delegation is carried out. The following guidelines will assist the manager in establishing an organizational climate favorable to effective delegation:

- **Set job standards that are fair and attainable.** It is essential for the manager and her subordinates to agree on standards for evaluation of subordinate performance. Subordinates should assist in developing organizational goals that are specific, yet general enough to allow for individual initiative on the part of the person accountable for achieving them. The standards agreed upon should be well understood before starting a delegated task.
- **Understand the concept behind delegation.** Successful delegation requires that both manager and subordinates recognize their respective roles. Delegation is more than just desirable; it is necessary for a successful organization. A manager should understand that subordinates do things their own way. Delegation is not a technique for ridding oneself of responsibility, but rather for dividing it up. It is a continuing process in which the manager is involved as planner, coordinator, and allocator of responsibility. She must understand that whatever is accomplished is done by working with her subordinates according to mutually acceptable guidelines.
- **Know your subordinates' capabilities.** The manager who knows the characteristics and capacities of her staff, as well as the facilities and equipment they use, can delegate tasks more realistically and more flexibly, thus more effectively. Selecting the right person for a job is an important aspect of

delegation. Delegating for the sake of delegating is always a mistake.
- **Develop goals and objectives.** Unless subordinates know not only what is to be done, but also why, how well, when, with what resources, by whom, and according to what priority, delegation is likely not to work. Many park and recreation organizations have given scant attention to statements of goals and objectives, resulting in time and energy wasted in endless clarification. New approaches—systems planning, participatory management, management by objectives—are now being applied to this problem.
- **Correct errors with tact.** The manager must use tact and discretion in correcting subordinate errors. Organizations that have employed joint goal setting, management by objectives, and similar systems of management have a distinct advantage over those that have not. Where the emphasis is on "setting the task" rather than on criticizing subordinate error, much of the correction is self-correction.
- **Reward subordinates for good work.** Subordinates who do good work should be rewarded. Managers must not lose sight of this principle. The reward may be no more than an increase in the subordinate's self-esteem. Motivation theorists agree that more authority and responsibility are particularly meaningful rewards.
- **Be a concerned manager.** By showing an interest in what a subordinate does, the manager backs up words with action. This can be done in a variety of ways: personal interest in the subordinate's work problems, open discussion concerning these problems, willingness to give support and guidance, and willingness to accept mistakes as a learning experience. How a subordinate sees a show of interest by the manager will depend on the manager's attitude. The manager who "snoops" will be resented, but one who shows a genuine interest in what is going on will be accepted and appreciated.
- **Evaluate performance.** Subordinates expect an evaluation of their work, even want it. Yet they have their own ideas of how the evaluation should be done, objecting to those that seem pointless, unnecessary, or haphazard. A variety of evaluation systems are presently in use that require evaluation by both manager and subordinate.

- **Be aware of areas of "No Delegation."** There are areas in which the manager will want things done precisely as she specifies. She should make clear what these areas are and why such a position is taken. When a subordinate understands the answers to these questions, there is less chance of a problem.
- **Provide for in-service training and development.** Delegation doesn't mean simply handing out tasks that a subordinate has not done before. The manager must provide appropriate training so that the subordinate has a reasonable chance of success. Training can be specific in nature, particularly when a special skill is involved. From time to time, the manager needs to assess what her subordinates know about their jobs. In most cases this does not require formal training, but rather a systematic plan to find out how well subordinates are doing and what their strong and weak points are. The manager can then plan an appropriate training program.
- **Don't be quick to take back delegated authority.** Making mistakes and finding and correcting them is a useful form of self-training. The good manager does not take away delegated authority the first time a subordinate makes a mistake.

SUGGESTIONS FOR SUCCESSFUL DELEGATION

- Relate your delegated task to the overall goals and objectives of the organization.
- Assign delegated tasks on the basis of an employee's needs, interests, and abilities.
- Allow the employee the appropriate authority to accomplish the delegated task and let him do his job in his own way.
- Delegation cannot be successful without agreement at all levels on the objectives to be achieved.
- Delegation always works best when superiors and subordinates respect and trust one another.
- Arrangements for reporting by subordinates to the superior should be worked out carefully and as far in advance as possible.
- The growth and complexities of organizations has necessitated managers to implement authority to carry out the organization's objectives.

- Delegation should not take place unless there is a joint meeting between the superior and subordinate with respect to its meaning and requirements.
- It is just as important to agree on areas of limited or no delegation as it is on those in which a subordinate's initiative and responsibility are sought.
- Authority must be specific enough for the subordinate to proceed without fear of exceeding her authority or having her actions reversed.

SUGGESTED READINGS

Bannon, J. (1978, March). Delegation—a misunderstood concept. *Parks and Recreation, B*(3), 38-41.

Engle, H. M. (1983). How to delegate: A guide to getting things done. Houston: Gulf Publishing.

Le Boeuf, M. (1979). Working smart: How to accomplish more in half the time. New York: Warner Books.

Smith, J. C. (1984, Summer). Management by delegation. *The Bureaucrat*, 31.

Vinton, D. (1987, January). Delegation for employee development. *Training and Development Journal*, 65-67.

Chapter 7

DEVELOPING WORK TEAMS

Team management flourishes in a supportive environment in which the manager effectively delegates and the team members effectively communicate in an atmosphere of mutual trust and respect. It's a common-sense approval of letting people improve their performance by improving the process they use.

It is time for park and recreation agencies to rethink their organizational and management styles. The present organizational environment is forcing us to look more carefully at ways we operate. Organizations are finding that they can no longer rely on former management techniques to guarantee a competitive advantage. Traditional methods of implementing improvements take a top-down approach. That is, top organization officials decide on changes to be made, assign personnel to specific tasks, determine when actions will be taken, and dictate expectation. Because management style does not encourage input or suggestions for improvement from employees, it greatly stifles employee involvement.

Employees must feel empowered to make decisions that help their organization. As they gain a vested interest in day-to-day operations, they will begin to work together toward improving quality. A technique that can assist an organization in attaining this goal is the self-managing work team.

What Is a Work Team?

Teams are small groups that are organized according to talent and to tasks that might be accomplished. A team is two or more people working together to achieve a common goal. That goal may be the mission of the entire organization, or it may be an achievement of one project or task. Teams are part of the search to find new ways to accomplish an organization's goals and objectives. Teams believe they share a common vision, both with other team members and with the entire organization. They have cooperative goals, they complement each other, discuss problems, and strengthen their work relationships.

A work-team concept builds cooperation on the employee level. A self-managing work team is a group of individuals with varying backgrounds, such as park planners, recreation programmers, marketing personnel, accountants, and computer technicians. Its members work together with a common goal—to arrive at decisions that create competitive advantages. For example, team members could share expertise in the development of a long-range park and recreation plan for a community. The park planner could determine the kind and amount of land that should be acquired for future development. The recreation programmer could suggest types of future recreation programs, marketing personnel could provide information relative to who the participants might be and how best to reach them, and the accountants could provide information as to revenues and expenditures.

The synergistic effect of people working together provides benefits to the organization. Each team member's skills are complemented by those of other members; weaknesses are identified and remedied. In today's constantly changing organizational climate, one person, the manager, cannot be expected to have all the tools necessary to carry out an entire task effectively. Harnessing the strengths of individuals enhances the output of the whole organization. The benefits of the team approach include the following: better discussions, quicker solutions, greater ownership by team members, and a more positive work environment. Most importantly, employees will witness the effect of their individual impact on the organization. As a result, the team will be driven by the desire to achieve rather than by the fear of failure.

Criteria for an effective team are:

1. Unity in and respect for the primary task.
2. Open communications—expression of ideas and opinions.
3. Mutual trust—particularly revealed in actions.
4. Support—the presence of care, concern, and active help.
5. Management of individual differences—an active process that involves the personal contributions of all members.

6. Selective use of the team—leaders or trainers should know which groups are appropriate for certain tasks; group effectiveness differs by project.
7. Variety of skills—each member brings certain strengths to the team.
8. Leadership—to manage and integrate the previous seven characteristics into the norm for team behavior.

INSTITUTING THE TEAM CONCEPT

The self-managing team cannot be implemented quickly without planning. In most organizations, this change will mean switching from an organization with sharply delineated functional departments to cross-functional teams. As with any organizational change, there are various stages of development. To create an atmosphere of openness and trust, a thorough discussion of the team concept should be held with all involved in the team-building process. Areas that should be addressed include the following: the philosophy underlying the creation of the team, the synergistic effects of the team, the importance of trust, and the breaking down of traditional functional barriers.

For the team concept to be successful, make sure it is carefully planned and supported by management. A strategic long-term plan must be developed. Management must commit itself to supporting the plan. Sound changes in organizational structure cannot be affected by shuffling names on an organizational chart.

Who Should Be Team Members?

Team members should include individuals with the type of expertise needed in the situation. Various teams will have different needs and opportunities. For example, to design and lay out a park, the team may include an architect, engineer, soil expert, forester, and maintenance specialist. A team to plan future recreation programs may include a survey researcher, a recreation programmer, a computer analyst, interested citizens, and program participants. It should be emphasized that team members have crucial interpersonal skills such as listening, leadership, brainstorming, communications, and consensus building. They should also be aware of and respect the other members' areas of expertise. As the team works together, each person becomes increasingly sensitive to the others' input. In time, team members will become more aware and knowledgeable in areas outside their specific functional responsibility. As the team makes more organization-wide decisions, members will increasingly understand the impact of their decisions on their constituents, as well as others in the organization.

Management must handle the change to the team concept carefully. The team can encounter problems and pitfalls. The most common mistake is organizing a team without the necessary training in teamwork. Even though individuals' functional skills may be adequate, their interpersonal skills and behaviors may not work in the team setting. A successful team requires careful nurturing and guidance. When people are simply thrown together and told to be a team without proper training and leadership, frustration quickly sets in. This type of situation lends to infighting and a return to the old, comfortable, individualistic approach. Teamwork should instill members with a desire to achieve.

Fostering Cooperation

To build the desire to achieve, crucial areas should be addressed. These include the following: common goals, trust of team members, willingness to cooperate, interdependence, and adequate feedback. Each of these areas must be addressed before the team can achieve maximum output. An important factor is providing common goals and accountability. If team members' goals are not aligned with the group, or if members are held accountable to people outside the team, they will suffer conflicts of priority, and the team's efficiency is reduced. In addition to knowing individuals' roles and responsibilities, each team member must understand his or her dependence on the other members. The team members must feel that they are trusted to carry out their responsibilities. Trust can best be gained if members commit to team goals rather than to individual ones. Within this framework, they should be allowed to accomplish their individual tasks.

The willingness to cooperate in the team setting is required to ensure a harmonious relationship among team members. Dissension can breed infighting in the group, and significant energies will be directed at merely trying to get along with others. Resolutions of this sort of dissension can eat up valuable time that could be better spent focusing on the team's mission. Team members must realize their interdependence. The team relies on the skills of each of its players, and they rely on each other to get the job done. If one of the links is missing, the whole chain falls apart.

Timely feedback is important to provide an open forum for discussion among the team members. Feedback dealing with the team's output, as well as its daily operation, creates the opportunity for continuous improvement.

The value of feedback is reflected in the attitudes of the employees. The ability of each team member to contribute is reflected in the attitudes of the employees. The ability of each team member to contribute to the end result and know *how he or she contributed* makes the work more rewarding.

Improved communication is a benefit of using the team approach. When applying teamwork to an organization, the advantages are reflected in an environment that promotes effective communication. Communication is a two-way action that means the free flow or exchange of ideas, information, instructions, and reactions that result in common understanding.

The reciprocal relationship between teams and the productivity of an agency is also reflected in the culture of an organization. Using teams, whether on a task basis or a permanent status, *enhances the culture of the entire organization,* which, in turn, serves the greater mission of the organization in its delivery of services.

Disadvantages of the Team Concept

While the introduction of the team concept to an organization holds these obvious benefits, there are disadvantages to team building. The most common disadvantage is known as *groupthink,* a concept previously discussed in Chapter 5. The pressure for unanimity can interfere with the development of suggestions and alternative solutions to issues faced by the team.

Training those working in teams in decision-making skills, as well as educating them to the possible failings in teamwork decision-making processes will assist in discouraging groupthink.

A less significant disadvantage to teamwork is burnout. A team can become complacent and exhaust its ideas without fresh stimulation. Emphasis should be placed on *individuals* working together to make good-quality decisions rather than on the team itself. The importance of the team leader lies in confronting and dealing with these disadvantages to team building.

IMPORTANCE OF FOCUSING ON THE GOAL

The key to achieving team success is persistence. Stay focused on the goal, keep a positive and open atmosphere, and encourage frequent communication. In time, the team will attain its goals.

A newly formed team cannot be expected to perform as well as teams that have been in existence for some time. Developing interpersonal relationships is difficult and time-consuming. At the same time, the organization may not provide much moral or financial support to the team. During difficult periods, the team will be forced to become more self-reliant. The leader must coach and guide the team to resolve and work through these situations. Self-reliance and internal focus provide the climate for repeated success.

To maintain the excitement and enthusiasm of the team concept, frequent reinforcement of the team's guiding principles and ultimate goals should be emphasized. Be careful not to focus on the process itself rather than on the outcome; too much attention to the process can lead to inefficient use of time and reduced team enthusiasm.

Management must be prepared to handle criticism from others in the organization. A frequently voiced complaint will likely be "What's wrong with the way we operated in the past? We're still here, aren't we?" Explain to the critics that traditional methods of operation cannot effectively handle the challenges of the new organizational climate. Understanding that organizations must adapt in today's dynamic organizational environment will help answer some of these questions. Most importantly, team success requires team spirit. Be cautious of individual celebrations and short-term thinking. While there may be certain times for individual recognition, an individual's achievements should not steal the stage from team objectives. As the team experiences success, small or large, celebrations can strongly reinforce the cooperative process. Though the celebrations need not be elaborate, they must be sincere.

With emphasis on continuous improvement in the team-building process and the commitment of management to allow it to work, individual capabilities will be harnessed into a coordinated team. Patience and persistence—these are the keys to successfully implementing the self-managing team.

The Manager's Role in Team Building

Managers can facilitate the growth of teams by assessing the necessary conditions for constructive team development. After recognizing the symptoms of an organization that require team building—low productivity, unresolved conflicts, unclear decisions, and inappropriate use of resources—managers can study a team's process and develop effective operations by meeting the conditions necessary for sound team development. These conditions include the following:

1. Team managers committed to team building.
2. Team managers who examine their own roles in the team.
3. Team members committed to objectives and willing to take responsibility.
4. Team members dedicated to examining and understanding the work process.
5. Frequent team meetings. Subordinates must be able to communicate with each other as well as with the leader.
6. Team building should not be limited to special occasions.
7. Team building must be acknowledged as a dynamic process, as a "process of continuous diagnosis, action planning, implementation, and evaluation."

A team manager who ensures the existence of these conditions will foster the healthy development of teams, as well as positively influence the culture and organization of his or her organization.

Leading the team through the various stages of development is done by the manager as leader and facilitator by guaranteeing that these conditions do exist. The manager is also responsible for getting the team past the first stage of dependence. In this stage, the manager must provide the group with structure as well as identity. During the awareness stage, the manager must achieve these goals, familiarizing the group with its goal and its individuals with their respective roles in the team.

Social activities cannot be overlooked when beginning a new team. Becoming familiar with each other as individuals allows the members of the team to then recognize their similarities in purposes and goals. This unity in objective is not only a characteristic of an effective team, but also of a thriving agency.

Team building not only serves the task at hand, but also furthers the organization's objectives, goals, and mission. An understanding of methods of team building and the theories that lie behind the application of team building are important to a manager's organizational role. The manager should be aware of which approach to team building is more appropriate to his or her situation: task, group, or individual. Each type of team building serves a different purpose that depends on the reasons the agency has for forming and building teams. Managers should also be aware of the conditions needed to manage the stages of a team's development successfully.

By recognizing the role teams play in an organization, acknowledging the benefits that team building contributes to the agency and its culture, and by fostering team development, managers can bring their organization beyond line hierarchies into the future. More and more organizations are considering the importance of group and individual influence on decision and policy making. Newer management techniques are evident to the creative value of team building. While establishing and developing teams, a manager not only creates a lasting impression of the way the organization functions internally, but also improves the manner in which individuals are valued.

SUGGESTIONS FOR SUCCESSFUL TEAM BUILDING

- Self-directed work teams should have substantial control over what they produce and how they produce it. Many organizations have seen significant increases in productivity when they have committed themselves to self-directed work teams.
- Management backing is essential to effective development of a self-directed work group. Everyone should understand that management is behind the team concept and will support it throughout the process.
- If management sees team building as a temporary balm to soothe workers' low morale, its sincerity may be questioned. Management's attitude must be clear to employees. Team building should not come across as a gimmick, but as a solid, long-term approach for building a better organization.

- Evaluate the effectiveness of the team-building process. Various strategies are available for evaluating the process. Have participants fill out standard written and oral evaluations after individual team-building sessions.
- Set team goals. Team goals tell team members where they are going. The goals, along with the rationale for choosing them, must be clearly understood by all team members.
- Teams need to examine whether members get along with each other. To what extent are team members effectively relating to each other? To what extent can they communicate with each other? To what extent do they resolve conflicts in a constructive fashion?
- An effective team should develop constructive internal relationships with its broader environment. It understands its role in the organization and works to improve relations with other units of that organization. It has good diplomatic relationships with all kinds of individuals and groups.
- Do not assume that teams are all basically alike. Even teams that are playing the same sport should not necessarily be coached and developed in the same manner.
- Respect and cooperation are essential. Mutual respect within and between teams does not require agreement on every issue.
- It is important to provide timely feedback to all team members. Feedback dealing with both the team's output and its daily operation creates the opportunity and environment for continuous improvement.

SUGGESTED READINGS

Dyer, W. G. (1987). *Teambuilding issues and alternatives.* Reading, MA: Addison-Wesley.

Guest, R. H. (1986). Work teams and team building: *Highlight of the Literature.* New York: Pergamon Press.

Lewis, J. P. (1993). *How to build and manage a winning project team.* New York: AMACOM.

Tjosvold, D., & Tjosvold, M. M. (1991). Feeding the team organization: *How to create an enduring competitive advantage.* New York: Lexington Books—Macmillan, Inc.

Woodcock, M., & Francis, D. (1994). *Teambuilding strategy.* Brookfield, VT: Gower.

Chapter 8

GOAL SETTING

"Participation by management in the goal-setting process communicates to all the employees that management cares about their productivity and their performance. Without it, the process will fail."

Organizations today are making an extended effort to improve the productivity levels of their employees. Establishing a goal-setting program has proven to go a long way in achieving this idea. There are many articles and books written on the subject of goal setting. More than a few tend to oversimplify the process. The goal-setting concept has different meanings and, as a result, is confusing, especially when one attempts to introduce the concept into the organization. Research that has been done on goal setting reveals that employees' productivity has been increased by as much as 15–35%. These studies have examined such jobs as insurance salespersons, maintenance workers, government workers, marketing specialists, and production employees.

Setting goals in organizations is a strategy—a way of leading the organization so that purpose and action are connected. The absence of controlling goals causes an organization to drift. Even though goal setting is a powerful management technique, it must be applied with care and caution if it is to be beneficial to an organization. The experience of many organizations has provided illustrations of areas of cautions to be observed and pitfalls to be avoided.

First, there must be commitment from top management to making the system work. That commitment must also be continuing. When managers lose interest, signals will be quickly picked up down the line and the system will soon evaporate. The entire staff must be trained in both the purpose and the process of goal setting.

The emphasis throughout the process should be on the results that are produced . . . results in terms of improved services or increased productivity. It must be clear and logical that a goal will help meet an important need and that the action plan will accomplish the goal. Broad assumptions and vague statements must be avoided.

In some cases, the goal-setting system can create an overload of paperwork. For some, this is a problem. It should be remembered, however, that the purpose of goal setting is to identify important needs, to plan to meet those needs, and to determine the degree in which the need has been met. In this light, the additional paperwork is well worth the effort, because the information focuses on the results produced and establishes accountability.

Figure 8.1: Factors Affecting the Success of the Goal-Setting Process

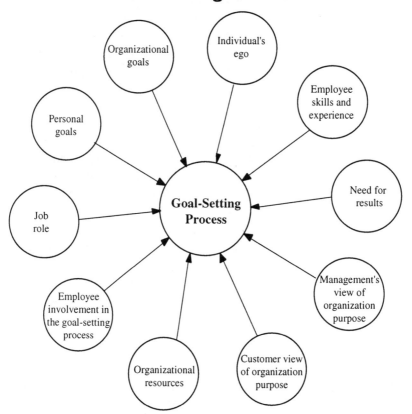

WHY ESTABLISH A GOAL-SETTING PROGRAM?

Establishing a goal-setting program in an organization can accomplish a number of objectives. It gives the organization a sense of purpose and direction. If people operated their automobiles the way that many organizations operate, they would never get out of the driveway, according to management consultant Bob Conklin. Fortunately, before they start driving, they anticipate where they want to go. They have a destination. If they do not know how to get to their destination, they consult a road map. If the car is never started, it will carry no one to a destination. These same principles apply to operating an organization.

A goal-setting program will encourage staff members to assume responsibility. The process also helps in judging performance of employees. A manager must evaluate the performance of every person in the organization in some way, either formally or informally. When individuals are working to achieve a full set of five to eight goals, their ability to get results on each goal can be a good, objective measure of performance. Traditional performance evaluation systems have been strongly criticized because they deal with subjective matters such as leadership qualities, rather than the more objective measure of results. Evaluating performance by using the goal-setting concept, while objective, is a complex task, which must be done with care by someone who fully understands the process. Failure to reach goals can be a result of setting the wrong objectives in the first place, the existence of organizational restrictions not taken into account, inadequate or improper measures of goal achievement, personal failure, or a combination of factors.

As employees become more disciplined and motivated to the task and committed to its success, the goal-setting process also spurs activity. Employees assume more aggressiveness as they approach their tasks. Completing a task often motivates them toward the next activity. Research has found that when employees announce their goals in a group setting, it leads to higher commitment and greater task persistence. The goal-setting process will assist in providing standards to judge strengths, weaknesses, risks, and opportunities. They can be matched against trends.

Goals comprise the criteria by which decisions are made among alternative strategies. Many organizations use the goal-setting process as a basis to make profit-planning decisions, cost-effectiveness decisions, and zero-base budget decisions. Goals can become a major factor in determining operating procedures on a day-to-day, year-to-year basis. Finally, if the goal-setting process is carried out properly, it will improve communications between a manager and his or her subordinates. Both the manager and the subordinate understand what the goal is, because they both participated in developing it and, as a result, both can more intelligently work toward its success.

KINDS OF GOALS

When an organization begins to set its goals, it is necessary to know what areas of operation are suitable for goal setting. What are the really important aspects of the organization rather than that part which is most visible? How can the organization be sure the goal-setting program is balanced for the long range rather than just reacting to immediate pressing problems? It might be useful to the organization to classify its goals that suggest opportunity. The following are suggested areas under which you might classify your goals. One or two goals in each of these areas would be appropriate:

1. Regular work goals.
2. Problem-solving goals.
3. Innovative goals.
4. Developmental goals.

Regular work goals refer to those activities that make up the major part of the manager's responsibilities. The head of production would primarily be concerned with the amount, quality, and efficiency of production. The head of marketing would primarily be concerned with developing and conducting market research and sales programs. Each manager should be able to find opportunities to operate more efficiently, to improve the quality of the product or service, and to expand the total amount produced or marketed.

Problem-solving goals will give an organization an opportunity to define its major problems; then it may set a goal to eliminate each one. Problems will always exist; new problems or new versions of old problems always seem to replace those that are overcome.

Innovative goals may be viewed the same way. A goal for innovation may apply to an actual problem; however, some innovation may not deal with a problem. For example, the head of building management may set a goal to invigorate the employees' suggestion program by putting five suggestions into effect during the next four months. There was no specific problem to be solved; management was just trying to improve its administrative efficiency.

Developmental goals may recognize how important the development of the organization's employees are to the company goals. Organizations should develop their employees, just as they should try to produce more effectively and efficiently. Every manager in the organization must be a teacher and coach; each manager must plan for the employee's continued growth.

> **Definitions**
>
> Mission—Described are organizations' reason for existence and its philosophical basis for operation.
>
> Goal—Designed to give an organization and its members direction and purpose.
>
> Objective—A statement of results to be achieved. The statement contains four elements: (1) an end result or accomplishment, (2) criteria for measurement, (3) a date and time for completion, and (4) amount of physical, human, and financial resources needed to achieve the objective.
>
> Task—An activity to be completed to accomplish the objective.

ESTABLISHING A GOAL-SETTING PROGRAM

When an organization decides to establish a goal-setting program, all supervisory personnel should be fully informed of the process, particularly if they have subordinates who will also participate. It cannot be successful if all concerned are not fully aware of the process and their roles in it. Early management concepts purported that goals were to be set by management and passed down the chain of command. Employee commitment is seldom achieved when goals are externally imposed. When they are, indifference and/or resistance are more likely to be a result. Joint goal setting by the employee and manager seems to be the most effective. By working together, this cooperation will result in the setting of more appropriate goals. Also, the employee's participation in the goal-setting process will increase his or her acceptance of the system.

Set Goals That Are Attainable

Goals should have challenge and difficulty, but they should be attainable. If they are not, subordinates will find it difficult to accept them. When accepted, they will become the bottom line of acceptable standards. There are a number of studies that indicate that the more difficult the goal, the higher the performance, so long as the goal is accepted. Goals that are challenging can encourage the need for achievement. Achievement-oriented behavior includes setting or accepting realistic, but moderately risky, goals and taking personal responsibility for solving problems and accomplishing results. If the goal is too easy, it is less likely to develop the urge to achieve. If it is too difficult, it leads to the belief that trying will not succeed. Only if the difficult goal is accepted and becomes the individual's level of aspiration does it tend to lift performances.

Why Goal Setting Fails

An organization may establish a goal-setting program but still have difficulty in achieving its goals. The reason why this happens has been studied by organizational experts, practitioners, researchers, and academicians. The trouble with the goal-setting concept is that it sounds simple in theory. Because it appears simple, organizations feel that it is a cure-all for an organization's failure. The goal-setting process can only be successful if it is implemented carefully and in a receptive environment. There are a number of reasons why the process fails. The following is a list of these:

1. Goals are set too high and are not attainable.
2. Lack of resources committed to the task that needs to be carried out to achieve the goal.
3. Insufficient planning.
4. Failure to follow the plan established to achieve the goal.
5. Unanticipated events beyond our control.
6. Insufficient motivation on the part of management and/or employees.
7. Lack of follow-up by managers and supervisors.
8. Management does not work effectively with staff.
9. High staff turnover and no trained replacement.

10. Insufficient coaching of subordinates.
11. Overemphasis on performance appraisals.
12. Resistance to change.

SUGGESTIONS THAT WILL ASSIST IN ESTABLISHING AND MAINTAINING A GOAL-SETTING PROGRAM

- Break down your goal into short-term objectives. If you break down your goal into specific objectives and consistently accomplish your goals, it will automatically lead you to your long-term goals.
- Staff training sessions should be conducted to ensure that each individual understands the purpose of implementing the goal-setting system and how he can be helped by the process, how the system will work, and how to write objectives and action plans to accomplish the objectives.
- Each individual must write out a separate action plan that will accomplish her specific goals. The action plan is a series of steps or tasks that must be accomplished in order to complete the goal fully.
- Conduct periodic performance reviews. Once a staff member has agreed on his goals, periodic performance reviews should be scheduled. These reviews should be informal in nature and conducted in a supportive climate. The purpose of the review should be twofold: (1) to determine how individuals progress toward their goals, and (2) to provide assistance if difficulties are being encountered.
- When an organization introduces the goal-setting process, it must encourage its managers to manage more and spend less time doing the work of those they supervise. When a manager and subordinate agree on goals and action plans, the subordinate should assume responsibility for accomplishing them.
- To initiate a goal-setting system throughout the organization may be more than you want to take on all at once. You may want to start with one major unit of the organization and expand to other units at a later time.

- Subordinates should receive feedback concerning their goals on a regular basis and in written form. Meetings discussing past performances and planning future performances should follow distribution of written reports.
- When management introduces the goal-setting process to the organization, it should anticipate a resistance from employees. Employees often fear that the system will increase their workload and paperwork and they may dislike any form of evaluation. When this occurs, management should deal directly with these concerns in an open discussion with the employees.
- Management must demonstrate a high level of commitment and involvement in the goal-setting process so that it communicates to employees that management cares not only about the achieving of organizational goals, but also about their job performance.
- Goals should be written on paper, not etched in stone. As situations change, so might our initial goals. This policy is not unusual, but the decision to change our goals should be a rational and conscious one.

SUGGESTED READINGS

Austin, J. T., & Bobko, P. Goal setting theory: Unexplored areas and future research needs. *Journal of Occupational Psychology,* 5(8), 289-308.

Bannon, J. J. (1992). *Take charge: How to solve everyday problems.* Champaign, IL: Sagamore Publishing.

Locke, E. A., Latham, G. P., & Eney, M. (1988, January). Determinants of goal commitments. *Academy of Management Review.*

Locke, E. A., & Latham, G. P. (1990). *A theory of goal setting and task performance.* Englewood Cliffs, NJ: Prentice Hall.

Vallacher, R. R., & Wegner, D. M. (1995). What do people think they're doing? Action identification in human behavior. *Psychological Review,* 5(15), 3-15.

Chapter 9

GRAPEVINE

The grapevine can slip by the tightest security screen and will carry anything, anytime, anywhere.

Grapevine is as American as mom and dad and apple pie. The grapevine is an integral part of any organization. The dictionary defines grapevine as "the informal transmission of information, gossip, or rumor from person to person." The grapevine is the informal and unsanctioned network within every organization. The network helps employees make sense of the world around them and consequently provides a release from emotional stress, as all informal information is undocumented. The grapevine is an expression of a healthy human motivation to communicate. As one organization expert states, if employees are so uninterested in their work that they do not engage in "shoptalk" about it, they are probably very job dissatisfied. Because the grapevine thrives outside the formal organization structure, it often causes problems for managers who see their influence decreased as they lose control of information flows. The grapevine operates without conscious thought; it will carry anything, anytime, anywhere. By gaining insight into how and why the grapevine flourishes, managers can become more efficient in directing employer/employee communication.

The term *grapevine* carries the connotation of inaccurate information; however, it can be a source of much factual data that gives it credability with employees. Messages concerning promotions and procedures have been found to be almost always true.

Moreover, it has been found that some organizational events were predictable over a month in advance of their official announcement.

Two differences exist between formal and informal (grapevine) communications networks. First, the informal network has no official standing; therefore, its leaders cannot be held responsible for harmful behavior or officially rewarded for positive contributions. Second, the informal organization is less permanent and less stable because its leaders and pattern of action change readily.

The grapevine serves several purposes. For example, during a period of retrenchment, workers may feel anxious about their jobs or status, but because their fear is unspecified, they cannot relate it to a particular event or decision. The grapevine provides structure by focusing workers' anxieties on a specific decision. The grapevine also helps make sense of a situation. Because people want a unified picture of a happening, they will use the grapevine to fill information gaps and to help clarify managerial decisions. Some people use the grapevine to organize a strategic positive—that is, to protect their decisions or influence others and to organize people into supporting factions.

HOW THE GRAPEVINE WORKS

Within the organization, a communication chain exists. The chain used by formal communication may be very rigid following the chain of command or authority. However, the chain used by the grapevine tends to be very flexible. Four different chains appear to dominate the grapevine network, according to the research by Keith Davis in his article "Communication within Management." They include:

1. **The Single Strand Chain:** This is a simple concept to follow. *A* tells *B*, who tells *C*, who tells *D*, and so on. Each person passes the information on to the next person. The longer the strand, the more distortion and filtering affects the information being passed, until the last person in the chain may find the information unrecognizable from the original message. Most inaccuracies occur in this chain.

2. **The Gossip Chain:** In this, *A* simply tells everyone with whom he or she comes in contact. This pattern is considered to be somewhat slow in moving the information.

3. **The Probability Chain:** In this case, *A* makes random contact with *F* and *C* and passes on information. They, in turn, randomly contact others in accordance with laws of probability. Some hear the information and some do not. In this structure, there is no definite pattern of communication. Information is randomly passed along to anyone willing to listen. The type of person who communicates in this manner might be a very outgoing and talkative individual.

4. **The Cluster Chain:** Here *A* tells contacts *B* and *F,* who may work with *A*. They may tell two or three other persons with whom they usually have close contact. The most predominant pattern is the cluster chain. Selectivity is the basis for this pattern. In any organization, individuals will generally feel more comfortable with some fellow employees than with others and therefore only relay information to those in their informal social groups. This pattern results in information missing some individuals completely.

The grapevine arises when formal communication channels are defined too rigidly or are adhered to too narrowly. When employees lack information, they feel they need and do not have access to the formal communication channel; the grapevine fills a basic human need to send or receive information. In this instance, employees will fill the information gaps with their own perceptions and conclusions. Employees tend to be more active on the grapevine during periods of excitement and insecurity. For example, the adoption of new technology may be perceived as a threat to workers who fear that it will replace them or that they will be unable to cope with its complex operations. At times like these, the grapevine will run rampant with activity and managers need to feed it with accurate information to keep it from getting out of hand. If they fail to do so, the fear and insecurity that set the grapevine in motion will result in diminished productivity. The grapevine thrives on the absence of news. In a healthy organization, there will be both formal and informal channels of communications. Managers should listen to and study the grapevine to learn who its leaders are, how it operates, and what information it carries.

Because grapevine activity increases during times of uncertainty, management must provide information through the formal system of communication about key issues and events that affect employees. Management should supply employees with a steady flow of accurate, timely information; in this way the potential damage caused by the grapevine can be minimized. Any attempt to soften or distort a rumor to make things look good is not the appropriate solution. The longer a rumor circulates, the more difficult it is to control. Facts should be released quickly. The grapevine can be controlled with prompt, clear, and accurate information on the issue important to the employees. Full facts must be presented. Formal communication lines must be kept open and the process as short as possible. Direct memos, large-group announcements, and intercom systems should be used. If employees perceive management is giving them the facts, they will be less anxious and less emotional when rumors are heard.

Should Managers Participate?

In many cases, lower and middle managers are already active participants. They should choose strategic positions in the communication channel, because they filter and block two-way communication between higher management and operating employees. Managers have three options when it comes to their participation in the grapevine:

1. Ignore the grapevine. This is difficult in most organizations, but can be accomplished. They do their job and let it operate unnoticed around them. In effect, they become isolated.
2. Participate only when it serves their purpose. In this case, they may seek out the grapevine and tap it to learn what is being said concerning a specific situation or issue.
3. Become an active full-time participant.

Reduction of Stress

Even though management does not always view it favorably, the grapevine has several positive aspects. One major advantage of the grapevine is that it is a release mechanism for stress. Bottled-up feelings have proven to have negative side effects for individuals, and the grapevine helps ease this type of situation. We know individuals like and need to talk about their work. The grapevine provides a forum for individuals to talk about this important facet of their lives. In talking about work, the grapevine gives employees the opportunity to convert official organization policies into their own language or jargon. In doing so, individuals are better able to understand the policies and better able to cope with their work environment. This open communication also enables employees to have empathy for those who are encountering stress outside the workplace. As stronger personnel bonds occur among workers, a greater spirit of team work exists within the organization.

In order for both of the formal and informal communication channels to complement each other, there must be an understanding of how the grapevine operates. Only the use of the organization's e-mail system gets information out faster than the grapevine. That news travels faster on the grapevine than through official channels can be attested to by employees who find out about their transfers before they are officially informed. Responses of 100 employees in a survey reveal that if management made an important change in the way the organization was run, most of them would expect to hear the news first through the grapevine rather than by any other method. Supervisors and official memos came in second and third, respectively.

Rumors on the grapevine are used to signal status or power. The explicit content of a rumor is its facts—the what, when, and where of a rumor. The implicit content is the relationship between the rumor's source and receiver ("I can put you in the know"); information is power.

Some studies have indicated the grapevine to be 80–95% accurate, with the inaccuracy normally taking the form of incompleteness rather than wrong information. In addition, some researchers believe that much of the information carried by the grapevine may be more accurate than that relayed by formal channels, particularly in situations where managers are less frank and honest than they should be. Regardless of its accuracy, it is important for management to realize that the grapevine is perceived as being accurate by many employees and may have a stronger impact on the recipient than other, more formalized methods of communications. Another finding from studies deserves special mention. Despite the generally held idea that females participate more actively in grapevine activity than males, this assumption is not true. Both men and women are equally active.

Chapter 10

MANAGING CREATIVITY

We ask our people to be creative, **but** *we ask them not to generate conflict.*

Any organization that does not come out with a new product or service within the next year will be out of business before this millennium is over. Furthermore, this product or service will have to be totally new, not just an old product or service with a new name or an old product designed in new packaging.

Most executives agree that creativity is the most profit-producing possession their organizations have. However, I have often heard executives say, "Sure I'd like to have more creative people, if I didn't have to put up with all the inconvenience they cause." Is the price of having real creativity too high? Is it worth putting up with the constructive discontent that characterizes the creative mind?

Most executives would be quick to say that no price is too high. Real creativity always brings change, and change can be unnerving and even shattering. The desire to cling to the familiar is human, and the higher a person rises in his or her profession, the greater the desire to secure the gains one has made. The more successful a person or organization becomes, the greater the desire to freeze things into a permanent mold. Although *progress* may not be inevitable, *change* is inevitable.

Speaking of this inevitability in "How to Be a More Creative Executive," Joseph Mason stated that over one-third of America's top-brand consumer products had lost their leadership in a 10-year period. Roughly 23% lost it because of competition from radically new products, 31% because of improved competitive products, and 25% because competing products provided new opportunities. So there is always a risk in venturing new ideas, but playing it safe is also fairly risky. Few people are thinking up to their full capacity.

According to estimates, nearly half of the new items in the $2.5 billion, highly imaginative retail toy market were originated by laymen—clerks, machinists, salesmen—very much like the men and women who work in your organization. While highly trained specialists sweat over costly equipment trying to solve the problem of color photography, a violinist by the name of Leopold Godowsky, aided by a colleague, tested formulas in a bathtub and came up with Kodachrome. Ideas can come from anyone, anywhere, anytime, but they are most likely to occur in a receptive, informal atmosphere and to the worker who feels appreciated.

Many managers fear creativity because they equate it with major changes, which they know can mean major adjustments. Sometimes pure originality is not needed. The average executive may seek and need nothing more than finding new ways to better existing conditions. As one executive summed it up, "Who needs an Einstein? What I need is a person who can help me save time and money in the production department." If this attitude is genuine, more creativity can be infused into each department with amazing results.

Innovation at Citicorp's credit card division suddenly dried up, according to former employees. The once vigorous and productive staff stopped coming up with new ideas. It was reportedly due to intense pressure from management to *deliver* as well as its indifference to the high rate of turnover. This encouraged a survival-of-the-fittest mentality that killed the incentive to take risks. The experience at Citicorp gives evidence that managing creativity is poorly done in most U.S. corporations.

Management often insists that it wants innovation from employees but then supports only the ideas that entail little or no risk. Creative types need a feeling of freedom and a sense of control over their own work. Teresa M. Amabile, associate professor of psychology at Brandeis University, states, "The model manager of a creative workforce is like an impresario who persuades and cajoles a highly self-motivated cast into working towards a common goal. Rather than set and enforce rules and regulations, they act as a teacher, coach, or cheerleader."

It is a maxim of management that if you want the job done, you must delegate authority along with responsibility. But if you want an *outstanding* job done, you should also delegate the freedom to apply imagination. There is a tremendous mass of sheer ingenuity available within the minds of most individuals if a way can be found to tap it. One thing is certain: the more rules, routines, and rigidity within an organization, the less creative its staff will be. To encourage creativity, management must believe there is more than one correct way of doing things. Rapidly coming into usage are training methods that combine instructions in standard procedures with training in creativity, aimed at encouraging the worker to look for ways to improve on the past.

What are the forces that affect the creativity of individuals, groups, and organizations? A growing number of studies are now available to help describe and understand some of these forces. There is considerable evidence that few of us use all of our creative potential. Creative people are not geniuses set apart from the rest. All of us have some amount of creative ability. The main difference between those who more fully utilize their capacities and those who neglect them lies in the way people allow their abilities and skills to develop and be nurtured.

THE CREATIVE INDIVIDUAL

Much new light has been shed on the question of what makes some people more creative than others. Research conducted suggests that creative people differ significantly from those who are less creative in a number of characteristics. Interestingly enough, most of these attributes are not directly related to intelligence, which, until recently, was assumed to be the key trait of creative people. Among the more pronounced attributes of so-called creative people are the following:

- *Sensitivity to Surroundings*—an ability to see things to which the average individual is blind.
- *Mental Flexibility*—an ability to adjust quickly to new developments and change.
- *Independence of Judgment*—internal strength to insist on evidence, while at the same time recognizing the importance of deeply felt, but more vaguely defined, feelings.

- *Tolerance for Ambiguity*—a continuing confidence that contradictions, complexities, and apparent disorder may generate richer types of experiences.
- *Ability to Abstract*—proficiency in breaking down problems into their component parts.
- *Ability to Synthesize*—the skills to combine several elements in a creative way to form a new whole.
- *A Restless Urge*—a special drive, or the motivation to look at problems as challenges and hurdles to be mastered.

THE CREATIVE ORGANIZATION

Organizations, like people, have personalities. Some permit wide divergencies in behavior, others reward their people for going by the book, for doing things the "organization's way." This creative activity is greatly influenced by the type of organizational atmosphere or climate in which an individual works. Free access to ideas from superiors as well as from peers is helpful. If an organization is lacking in creative vitality, it may be useful to look at the following factors:

- *Pressure for Non-Relevant Uniformity*—an organization should demand from its employees adherence to only those rules, regulations, or procedures that bear upon eventual productive performance. When superiors insist upon uniformity in behavior unrelated to eventual job success, they run the risk of introducing unnecessarily destructive pressures into the system.
- *Lack of Information Sharing*—people often tend to withhold information from those who need it. Although they may deny it, they frequently withhold it in order to satisfy subconscious personal needs for power and control.
- *The Climate for Failure*—some organizations put a premium on their people always being right. Their executives ascend the ladder to success as a reward for not making mistakes. Because people in general tend to be afraid to show themselves in a bad light, to make mistakes or to "goof," it is doubtful whether organizations that expect employees always to be right are likely to encourage much creative thinking.

- *Excessive Work Pressure*—creative effort cannot be legislated nor demanded with production control efficiency. Much creative work emerges from the leisurely and relaxed development of ideas, gathered in time of contemplation, stimulation over friendly lunches, at unscheduled stops over in neighboring offices, or on vacations designed to provide relief from the pressures of daily work.

Creative spirit is not something you can accomplish overnight in the organization, but it could be that your personal road to fame and success involves using your own imagination to analyze, develop, and solve the problems. If you do, you will begin to infuse your organization with both the necessity and the means of getting all-out imaginative creative thinking from everyone.

SUGGESTIONS FOR IMPROVING THE MANAGEMENT OF CREATIVE INDIVIDUALS

- Encourage laughter. A good sense of humor stimulates creativity. Often when people are trying to be funny, they come up with good ideas.
- Think long range. Be willing to spend money on ideas and projects from which you will not see immediate results.
- Prod employees to take risks. When they are successful, reward them with money and recognition.
- Keep up enthusiasm during nonproductive periods. No person is continuously creative. Individuals go through times when they cannot "think of a thing." If this happens, you should encourage him to go on trying without pressing to come up with specific ideas.
- Set occasional deadlines. Although we cannot and should not try to force a person to be creative at all times, occasional deadlines will spur new ideas. If you tell a person that you must have an idea by the next day, she will set her mind on arriving at a solution.
- Criticize constructively. When a person has a suggestion or a creative idea, it looks good to him regardless of its intrinsic merit. In explaining that the idea cannot be used, concentrate your criticism on the circumstances that make it impractical rather than on the idea itself.

- Listen and ask questions. Many persons grasp for an idea that does not seem to jell. Then when they try to explain it to someone else, the idea takes shape. If the listener asks questions, it often serves to further clarify a concept.
- Adopt a participative leadership style. Autocratic leadership will inhibit and, most probably, eliminate creativity. Creativity will flourish when employees feel secure and free of fear of failure.
- Support and protect creative employees. Creative people are often different and sometimes a little unpopular. Their ideas usually involve change that may threaten other employees.
- Permit more freedom in how tasks are carried out. Creative people dislike rigid step-by-step routines.

SUGGESTED READINGS

Amabile, T. M. (1983). *The social psychology*. New York: Sprintger-Verlag.

Bottger, P. C., & Yetton, P.W. (1987). Improving group performance by training in individual problem solving. *Journal of Applied Psychology, 72* (4), 651-657.

Kanter, R. M. (1986). Creating the creative environment. *Management Review, 75*(2), 11-12.

Pinchot, G. (1985). Entrepreneuring. New York: Harper & Row.

Rose, L. H., & Lin, H. (1985). A meta-analysis of long-term creativity training program. *Journal of Creative Behavior, 16*(1), 11-12.

Chapter 11

THE MARKETING FUNCTION

Managers do not need to dust off their Marketing 101 textbooks to understand marketing theory. You will have a strong grasp of it if you simply remember what the most important objective of any organization is: to satisfy its customers.

Any college student who took Marketing 101 learned (or should have learned) that every step of a successful marketing program revolves around one key element: the customer. No matter how revolutionary its product or how sound its financing, a business cannot succeed if it loses sight of the potential customer's needs, wants, and buying behaviors.

This important concept can get lost, though, once those college students establish their careers in management. Many middle- and top-level managers may begin to think of marketing in terms of sales and profits, or as something that is the responsibility of the company's advertising agency. Marketing actually plays a much broader role in the organization, and it is critical that managers understand exactly what marketing is and how its strategies can be used to enhance dealings with customers and competitors.

WHAT IS MARKETING?

One way to define marketing is to start with some common misconceptions. Marketing is not just the promoting and selling of a product or service, and it is not confined to profit-seeking organizations. Marketing is much more than simply advertising and public relations efforts. Perhaps the most important concept for management to understand is that marketing is not a function that can be kept separate from the daily performance of administrative and operational tasks.

Essentially, then, marketing is a part of all business activities that are not directly related to production. One widely used formal definition is that marketing is the process of planning and executing the conception, pricing, promotion, and distribution of ideas, goods, and services. *More specifically, the marketing function includes all these activities.*

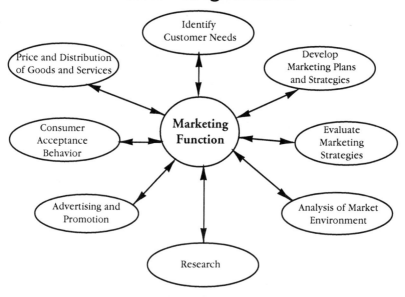

**Figure 11.1
The Marketing Function**

The purpose of marketing is to create exchanges that satisfy individuals and organizational objectives. Exchange occurs when there are at least two parties, and each has something of potential value to the other. When the two parties can communicate and have the ability to deliver the desired goods or services, exchange can take place.

Figure 11.2

> **Definitions**
>
> Advertising—Any paid form of nonpersonal presentation and promotion of ideas, goods, and services by an identified sponsor.
>
> Promotion—Any form of communication used by an organization or individual to inform, persuade, or remind people about goods, services, image, ideas, community involvement, or impact on society.
>
> Marketing—All individual and organizational activities directed to identifying and satisfying customers' needs and wants.

The Marketing Plan

Marketing should begin long before a service or product is offered or sold to the public. To be successful, marketing efforts must be selected to meet long- and short-term goals, and all activities must work together.

To develop a unified plan, you have to first do the homework—needs assessment, market analysis, customer research, product evaluation—and then use it to plan activities for a set period of time, typically one or two years. The plan itself should include:

- A list of objectives
- An overall approach (strategy) for meeting those objectives
- The specific marketing, advertising, promotional, and public-relations activities that will be used in the strategy
- A budget, broken down by market segment and marketing activity
- A weekly or monthly timetable
- A method for tracking results and evaluating the plan when it is complete.

One compelling reason for developing a long-range plan is that it saves time and money. By evaluating all possible activities in advance, management can select those that will do the best job of reaching all their customers and markets. The budget can be allocated for the year, so there will be no need to make major spending decisions until the plan is completed.

A STEP-BY-STEP APPROACH

1. Define Your Customers

The most important element in developing a successful marketing plan is a clear definition of your customers. This involves selecting and analyzing the groups of people who represent the greatest potential for increased sales, or good will, or whatever you want to accomplish. These "targeted" users should be defined and analyzed in relation to their needs and interests in specific products and services you offer, and they will be the focus of all your marketing activities.

Your "feel" for your customers' needs is important, but defining the customer—and the next two steps (evaluating your product and analyzing the market)—should be based heavily on your outside research. Your goal is to know everything you can about your potential buyers, which will require a good base of knowledge about their demographics (age, income, family size, education, etc.), psychographics (attitudes, interests, and lifestyles), and behavior patterns.

Once you've collected this data, the next step is to divide the total market into segments—groups of buyers who have similar characteristics or specialized needs. Markets are frequently segmented based on geographic, demographic, psychographic, or behavioristic variables. The number of segments and their sizes depends on your product and how homogeneous its potential buyers are. In practice, a segment should be large enough or important enough to support a decision to develop a marketing strategy specifically for that segment.

2. Evaluate Your Product or Service

The term *product* refers to everything you offer to your customers. To evaluate products as your potential users see them, think about each of your targeted customer groups. Then try to identify what you have to "sell" in terms of the benefits you are providing to each group. Keep in mind that no matter what type of product you are promoting or targeted customer you are considering, you have only four things to "sell" as benefits:

- *Product*—the physical features, quality, and selection of the products and services you offer.

- *Price*—what it costs your customers to buy or use your products.
- *Service*—the additional courtesies, considerations, and privileges you provide to your customers.
- *Location*—where your products are available or where your services can be accessed by your defined customers.

Even though your company may offer benefits in all four areas, it is unlikely that all of them are truly worth "selling." Marketing helps you evaluate them through your customer's eyes—it shows you what your real advantages are and how you should promote them.

3. Analyze the Market Environment

This involves making an objective assessment of all the external factors that may affect your marketing efforts. You can start by identifying and analyzing your direct competitors. Gather information on how they promote their products and try to determine which consumer groups they are targeting. Compare the four types of benefits they offer—product, price, service, and location—with yours, and decide where your strengths and weaknesses are.

You should also consider your indirect competitors, which include anything that affects your customer's desire or ability to buy your product. If you are an accounting firm specializing in personal income-tax returns, one of your indirect competitors may be the consumer's belief that he can save money by doing his own return. If you publish a newspaper or magazine, you are essentially competing against all other sources of the same information, including the growing popularity of the Internet.

Your indirect competitors can also include a wide range of uncontrollable factors in the marketing environment: economic forces, societal changes, new laws, changes in technology. Although these factors are difficult to predict and measure, they should be considered when designing the marketing strategy.

4. Identify Marketing Opportunities

The last element in the process of researching the market should be completed by management. Use your knowledge of your customers and your products to augment the research. Write down your observations about each of the three major research categories—customers, product, and marketing environment (competitors)—and try to develop your own "bottom line" conclusions about each.

To guide your thinking, ask yourself the following questions:

- Which types of customers find our products or services most beneficial? Which customer groups find our competitors more appealing?
- Which types of customers are our competitors "targeting"? Are they trying to reach the same or different types of customers than we are?
- Do individual groups of our customers patronize our competitors, too? If so, for what reasons and under what circumstances?
- Are there any indirect competitors, unfavorable trends, or environmental conditions that threaten our success? Is there a way for marketing activities to lessen their effect?
- Of the four benefits we have to sell (product, price, service, location), which one is our strongest/weakest? Which one do consumers perceive is our strongest/weakest?
- What is the ONE specific benefit we offer to our defined customers?

The answers to these and similar questions should help you identify the areas that you want to change, improve, or exploit. In most cases, you will want to think about these areas in terms of your individual groups of targeted customers. Each segment of your market will have different characteristics or motives for buying your product, and you will want to deliver marketing messages that are tailored to their wants, needs, and buying behaviors. Your overall marketing plan and the specific activities and strategies you select will cater to these differences.

5. Set Marketing Goals

Your organization should already have a clearly defined mission and a set of management goals. The next step is to develop specific goals for your marketing plan. This involves identifying the overall needs of your organization and using your research from the first three steps. Based on your market research and the opportunities you have identified, establish two sets of marketing goals:

1. Short-term goals that can be accomplished within the time frame of the marketing plan (usually one year). A short-term goal might be to increase sales by a certain percentage, to introduce a new product, or to add a specified number of new accounts.
2. Long-term goals that reflect intended accomplishments over a more extended period of time, usually two to five years. Long-term goals might include increasing overall consumer awareness of your organization, developing a reputation as an industry leader, or setting product standards.

6. Set the Marketing Budget

The size of the budget depends on how ambitious your goals are. One approach is to assign a set percentage of operating funds to the marketing budget. In practice, management's "what is available" figure and the marketing department's "what is needed" figure will not match. But if your marketing goals are realistic and based on solid research, both sides will be able to find a successful compromise.

The most important aspect of budgeting is for management to understand that marketing activities cannot be successfully planned or supported if they are paid for on a what-can-we-afford or a what-is-left-over basis from general operating funds. In order to effectively allocate funds and track success, the marketing budget should be separate from the operating and administrative budgets.

7. Design a Step-by-Step Plan

With clear goals and a budget, you are ready to identify and evaluate all the marketing activities that are available. These include media space and time, point-of-purchase materials, news releases and public-relations messages, trade shows, direct mail, product brochures, personal sales, and many others. The ones you choose will depend on which customer groups you are targeting, what your budget is, and your assessment of which will best help you meet your goals.

Your actual marketing plan will be a timetable of when each activity will be carried out, along with cost breakdowns and other details. The plan is more or less "locked in" for the year and serves as a guide for everyone in the organization. If you are targeting diverse customer groups, you may have separate budgets, plans, and timetables for each segment.

8. Track Your Results

As part of your overall marketing plan, you will want to build in a method for evaluating its success. This may be based on sales data, customer inquiries, ongoing research, or whatever you decide is the best measure of the effectiveness of your marketing activities.

You should also include a method for continuing to collect market research. The process of analyzing your customers, competition, and the marketplace does not end once the marketing plan is in place. It should be an ongoing process that develops new information as marketing activities are carried out.

The key is to build these methods into the plan and collect the data as you go. This will be the foundation for revising the plan to meet future needs and adapting your long-term marketing strategies to the needs of your customers.

THE LONG-TERM BENEFITS

Designing a good marketing plan is not a simple process. Many organizations rely heavily on professional marketing firms to handle the complex research, make recommendations, and carry out the activities. Even the most experienced professionals cannot guarantee a foolproof plan, though, and that is why it is critical that management has a good understanding of how the plan was formed and how it works. Management will be the best judge of whether a plan is working, and what strategies must be adopted to improve it.

Managers do not need to dust off their Marketing 101 textbooks to understand marketing theory. You will have a strong grasp of it if you simply remember what the most important objective of any organization is: to satisfy its customers. Management should never be so far from its customers or its competitors that it loses this focus. Every element of the marketing plan—and virtually every day-to-day decision that management makes—must be based on a devotion to meeting the customer's needs.

SUGGESTIONS FOR IMPROVING YOUR ORGANIZATION'S MARKETING EFFORTS

- Make your marketing plan a how-to manual for everyone in the organization. Provide details on who is to carry out each activity, exact dates for each stage of planning and execution, what outside agencies or contractors are required, what supplies are needed, and how costs should be allocated.
- Build a contingency fund into your budget—a certain portion of the marketing budget (typically 10%) that can be used to take advantage of opportunities that arise during the term of the plan.
- Marketing innovation requires a constant, free-flowing stream of ideas from every member of the organization. The marketing department does not have a corner on new ideas—other personnel should be actively encouraged to contribute their best thinking.
- Avoid "conference-room" market research. Some of the most notorious new-product "bombs" originated in the minds of executives as they sat around the conference table, remote from the real marketplace.
- Make sure the "boss" reads every horror story about organizations that failed because they neglected the marketplace.
- Make sure, as well, that management sees every success story of organizations that made marketing and the customer's needs their top priority.
- Design your marketing information system to give you early warnings, so you can do something about it before it is too late. The earlier you terminate an ineffective marketing scheme, the less it costs.
- Do not let yourself become so infatuated with one marketing approach that you neglect other possibilities. Have a backup ready to call up when it appears that a strategy is not working.
- One of your most important resources is brain power. Make sure your budget provides for salaries and incentives that will attract top talent in market planning and development.
- If you use an advertising agency, tell them everything. Rarely are agency personnel over-informed; most frequently, they do not have enough information on the organization, its products and services, its markets, its goals, and objectives. As a consequence, agency creativity is often misdirected, and time and money is wasted.

SUGGESTED READINGS

AMA Board Approves New Marketing Definition. (March 1, 1985). *Marketing News.*

Crompton, J. L., & Lamb, C. W. (1986). *Marketing government and social services.* New York: John Wiley & Sons.

Keegan, W., et al. (1992). *Marketing.* Englewood Cliffs, NJ: Prentice Hall.

McCarthy, E. J., & Perrecault, W. D. (1991). *Essentials of marketing.* Homewood, IL: Irwin.

McKenna, R. (1988, Sept.-Oct.). Marketing in the age of diversity. *Harvard Business Review*, 88-95.

Value Marketing. (1991, Nov. 11). *Business Week,* pp. 132-135.

Chapter 12

MEETING PLANNING

Meetings are inevitable. They can also be valuable. After all, the exchange of ideas and information can be one of the most stimulating processes in which humans engage.

Researchers estimate that there are 11 million internal meetings held daily by U.S. corporations and businesses. The cost of internal meetings is estimated at $2 billion annually, and the cost of out-of-office or external business meetings, including conventions and trade shows, is in excess of $5 billion dollars.

The average executive spends as much as 60% of her time in meetings, with the average external meeting costing as much as $1,000 per hour. What comes as a surprise is the minimal return in time and talent invested in many meetings. As a basic rule of thumb, meeting planners can figure the cost of a single meeting by determining the average salaries of the participants and multiplying by the number attending. Add 15% to this basic figure for general overhead, equipment, and fringe costs, as well as payroll costs for staff involved in planning the meeting. If travel is involved, these costs must also be added.

The answer is not to eliminate meetings but to develop workable systems of pricing and control to ensure that meetings accomplish their purpose. The most effective insurance is to call meetings only when necessary. It should be remembered that meetings are not the only methods for communications and problem solving. Alternatives include personal executive action, written communication, individual telephone calls, and conference calls. Added to this list are some of the newer communications techniques such as closed-circuit television, videotaping, computer networking, voice mail, and fax machines.

It should be pointed out, however, that nothing really takes the place of the personal interaction in each other's physical presence. It has been figured that the most effective meetings are those where there is a need to discuss and decide new policies and procedures and where the complexity, uncertainty, or importance of the subject matter necessitates direct participation by a number of people.

It is important to define what is considered a meeting. The accepted definition is a personal gathering of three or more people. There are six basic and frequent overlapping types: staff, information, fact finding, problem solving, decision making, and committee meetings. The two most abused types of meetings are staff meetings and committee meetings.

Staff meetings are sessions that bring the manager and the staff together. Because they are a gathering of people who have a common, continuing work relationship, they are excellent for making specific recommendations, securing approval of proposed projects, and communicating to another level of management. The meetings can be a waste of time when they are scheduled on an inflexible, recurring basis regardless of whether they are necessary, when they are not carefully rehearsed, or when they serve only as a platform for the manager to discipline the group when topics to be discussed are not of concern to more than one staff member at a time.

Committee meetings provide the greatest pitfalls. They are often used as vehicles for buck-passing and to avoid or postpone difficult decisions. They often have the effect of diluting distribution of management responsibility. The existence and number of committee meetings should be regularly reviewed and carefully controlled.

Meeting failure is a costly and deplorable situation that develops much more often than it should. Most frequently it comes not during the meeting itself, but in the early planning stages. A meeting is provided with a high failure probability when someone at the executive level commits one of the five pitfalls of meeting planning. There are, of course, many more than five mistakes that can be made in planning a meeting, but these five, all having to do with communications, have probably been responsible for the failure of more meetings than any other. They include the following points:

1. Being ambiguous about the purpose of the meeting.
2. Being unclear as to the format that will be followed.
3. Being enigmatic about the financial cost of the meeting.
4. Being indefinite about who is responsible for the meeting.
5. Being unclear about the decision-making authority.

In conducting meetings, the leader must arrange for lively participation. All of the participants should be urged to contribute actively at the meetings to keep their interest and to involve themselves personally in committee action. This participation can be done by listening actively to what others say. Skillful questioning can also be helpful—not to put a participant on the spot, but to bring out facts about which that participant is best informed.

It is also important to bring the meeting to a formal conclusion. Nothing frustrates participants more than no decisions. If it is a problem-solving meeting, make sure everyone is clear on the decision reached. If its an informational meeting, summarize all points made. Follow-up is essential. A meeting is only as effective as the results produced. Set specific deadlines for assignments. Keep minutes and send them to participants. It is also advisable to telephone participants after a few days to see that the work has begun.

We can do much to improve the effectiveness of our meetings. The following is a sequence of events that we should consider as we prepare for meetings where we will be the leader or where we may serve as a participant:

1. Determine why the meeting should take place.
2. Plan in detail what will happen at the meeting.
3. Prepare and disseminate an agenda in advance.
4. Determine the best place to hold the meeting.
5. Start at the announced time.
6. Explain to the group the meeting's purpose.
7. Be sure everyone in attendance knows one another.
8. Ask all in attendance to review the agenda before the meeting begins.
9. Do not allow the meeting to stray—establish time limits.
10. Stress openness and participation—avoid groupthink.
11. Do not allow interruptions.
12. Make sure everyone understands the decisions—summarize periodically.

13. Assign tasks to meeting participants.
14. Take a few minutes to evaluate the effectiveness of the meeting with the participants.
15. End the meeting on a high note.
16. The secretary should distribute minutes to all in attendance.
17. Design a follow-up program in all action items.
18. Look to the future—what next—who should do it, when should they do it, why should they do it, and how long should they do it.

Since the beginning of time, there have always been meetings, and they will probably increase in number rather than decrease. Meetings have been held by cavemen to determine which clubs to use in hunting, to present-day discussions on everything from space exploration to how to pick up trash in the community. Probably no medium can do a better job when handled correctly and, conversely, do such a bad job of boring participants, spreading misinformation, or just wasting time. Whole books have been written on meeting planning. The preceding essay, whether you are planning a foreman's meeting, producing a community planning board meeting, or a high-level meeting with your associates, should assist you in understanding the importance of planning the meeting, why meetings fail, and how you can improve your meeting planning skills.

SUGGESTIONS FOR SUCCESSFUL MEETING PLANNING

- Be very clear about the aims and objectives of the meeting. Decide what you are trying to accomplish. If there is no purpose, do not hold the meeting.
- Set clear time limits for your meetings. If some items require more time than you originally thought, ask the group for support in revising the agenda.
- Keep a follow-up file so that future meetings can be geared to what you have learned at the current meeting.
- Make a critical review of the meeting as soon as it is over. Talk to the participants to see whether or not they think you have accomplished the objectives.

- Keep control of the meeting. Keep checking to be sure that the overzealous participant does not take over, as well as seeing that the person might also have input.
- Pick the right audience for your meeting. If you are responsible for sending people to meetings, be sure that the subject matter meets their needs.
- Avoid overworking your volunteer participants. They usually have their own jobs to do and can donate only a certain amount of time to help you.
- Be careful your meeting does not take on too much formality. It is not necessary or even desirable because it may inhibit participation and reduce the amount of honest, open communication that may occur.
- When major issues and decisions are being made, action in the form of minutes or memoranda summarizing the meeting should be prepared promptly and distributed to all involved as well as concerned agency members.
- When a meeting is to be held, choose an appropriate time when both information and people will be available.

SUGGESTED READINGS

Ashenbrenner, G. (1988, July/August). Planning effective meetings. *Business Credit,* 43-46.

Half, R. (1985, December 16). This meeting will come to order. *Time,* 50.

Heckard, R. (1979). *Organizational development strategies and models.* Reading, MA: Addison Wesley.

Odom, R. Y., English, D. C., Mills, & Noe, R. M. (1990, October/December). Business and communication. *Business,* 52-55.

Williams, W. L., Buck, E., & Clark, M. P. (1987). Increased productivity through effective meetings. *Technical Communications, 4,* 264-269.

Chapter 13

NEGOTIATION

One secret to success is recognizing the difference between positions and interests. A party's position is what he wants; his interest is why he wants it, and that is the building block of a lasting agreement.

If you asked your company's managers and supervisors how often they negotiate as part of their jobs, a good number would probably say seldom or never. That is because many people think of the word "negotiation" in terms of win or lose scenarios that involve confrontations. What we do not often realize is that everything in business—and in life, for that matter—is negotiated, under all conditions, at all times.

In daily life, we negotiate every time we change lanes on a crowded freeway or try to convince our spouses to take out the garbage. In business, negotiation can be more structured and formal, but it is also used in most day-to-day work. Negotiation skills are required for everything from the simplest tasks—scheduling an appointment with a client, for example—to the most complex, such as working out terms of a union contract or securing approval of a budget proposal. According to Gerard I. Nierenberg, president of the Negotiation Institute, Inc., in New York, "Whenever people exchange ideas with the intention of changing relationships, whenever they confer for agreement, then they are negotiating."

WHAT MAKES A GOOD NEGOTIATOR?

Some people seem to be born negotiators—they have personalities that make it easy for them to influence people and get what they want. In general, there are some personal characteristics that give people a natural edge in negotiating. They include high self-esteem, the ability to think clearly under pressure, a flair for problem solving, a willingness to compromise, and a sincere enjoyment of dealing with people. One of the most important strengths for any negotiator is objectivity—the ability not to take a counterpart's strategies or comments personally.

Like any management skill, though, negotiation can be learned, practiced, and mastered. The first step is to go beyond thinking of it as a situation that involves conflict and confrontations. Negotiation is really just a resolution-seeking device designed to produce a solution that is satisfactory for both parties. It creates benefits by removing restrictions and providing a positive structure for the communication process.

Managing the Time Element

The success of any negotiation—from short, informal interactions to highly structured processes—involves an understanding of the roles that time, information, and power play in the outcome.

The element of time is one of the most difficult to manage and understand. One reason is that so many people think of negotiation as an event that has a definite beginning and end. In practice, though, most negotiations are a continuous process, with both parties basing their actions on events that occurred long before the discussions start. If you are asking for a raise in salary, for example, the negotiations actually began the day you started working for the company.

For negotiations that have a set deadline, the time element can favor either side, depending on the circumstances. To make the time element work to your benefit, it is helpful to remember that timed negotiations tend to follow the "80/20" rule: 80% of your results will generally be agreed upon in the first 20% of the time allotted. The most critical concessions and settlements, however, will occur in the final 20% of the time.

This phenomenon makes patience one of your most valuable assets. The best outcome seldom comes quickly. If negotiations do not seem to be going well in the early stages, remain levelheaded and wait for the right moment to act, which may be very late in the process.

Deadlines themselves are often negotiable, and in some cases, it may be to your benefit to change or eliminate them. Anticipate your time needs and if necessary, be ready to shift your discussions to changing the time frame. If you and your counterparts have different deadlines, try to find out what theirs are, but do not reveal yours. As you near their deadlines, their stress levels will increase and they will be more willing to make concessions.

Doing Your Homework

Most often, the side with the most information is the one that receives the best outcome. The reason many people fail is that they approach the negotiation as an event rather than a process. As a result, they do not think about the information they need until the face-to-face encounter begins, which is too late.

In the formal interaction, your counterparts will try to conceal their true interests, needs, and motivations. The best time to acquire this critical information is long before the discussions start. You can find useful material by researching facts and statistics, talking to someone who has negotiated with your counterpart in the past, and consulting people who have been involved in similar negotiations. You may even gain valuable information by having an informal, advance conversation with your counterpart.

It is also in your best interest to have clearly defined goals before the negotiation begins. These should be set-based on the information you collect. To plan your approach, make a list that summarizes the facts you have available, your needs as the negotiator, the negotiable issues, your position on each issue, and strategies and tactics you plan to use. Then use your research to construct the same list for your counterpart.

Some of your preparation will have to be based on intelligent guesses about the information you have collected. Try to anticipate more than just what you think the other side will want. More important will be why the other side wants it. It is equally important, and perhaps more difficult, to ask the same questions about your own positions.

Power Relationships

A third critical element in any negotiation is power, which is each participant's ability to influence people or situations. Most people have more power than they think, partly because they perceive power as coming mainly from a person's formal position in an organization or her ability to punish or reward.

However, there are several other types of power—some real and some perceived—that can give a negotiator the ability to influence people. Persuasive power can come from a person's knowledge or expertise, reputation, character, even gender. A person's behavior style and her ability to adapt it to a particular situation can also give her power over her counterpart.

A negotiator's innate self-esteem also affects her power and her counterpart's perception of it. People with high self-esteem generally feel they have more power and viable options in a negotiation. Those with low self-esteem are more likely to feel they lack power, even when their actual position offers them a clear advantage. A person who feels powerless will become apathetic, which gives her virtually no chance to prevail.

DEVELOPING NEGOTIATING SKILLS

Advance preparation and a clear understanding of negotiation principles are the foundation of the negotiating process. Once the actual encounter begins, a negotiator also needs some specialized skills, which can be learned and practiced.

Perhaps the most important skill is listening. To be a good negotiator, you must be able to understand the intentions behind what your counterpart is saying to you, and that requires active listening. Many negotiators fail because they think their job is primarily to persuade, which means talking. They see talking as active and listening as passive.

The first step to developing these skills is to become sincerely motivated to listen. You can find an incentive by reminding yourself that the person with the most information usually receives the better outcome in a negotiation. If you listen passively or impatiently, you will miss much of this information, and that can put you at a disadvantage. Being a good listener requires a concentrated effort, and it is not easy.

Successful negotiators rely on a number of different strategies and techniques to enhance their ability to listen and process what they hear. They include:

- **Create a good listening environment.** If the negotiation is on your "turf," make an effort to minimize distractions and interruptions. Clear your calendar, close your office, and shut off telephone calls.

- **Do not trust your memory.** When your counterpart says anything you think you might want to remember later, write it down. You'll gain credibility and power when you can accurately recount what your counterpart said in an earlier session.
- **Avoid overpreparation.** If you script your presentation too tightly, your mental focus will be on what you are going to say next, rather than on what is being said now. To listen actively, you have to relax and be patient. Wait until you have the whole picture before you decide how to respond.
- **Remember that it is impossible to speak and listen at the same time.** Your goal is to collect as much information as possible about your counterpart's position and motivations, and that process stops when you are speaking.
- **Let your counterpart tell his complete story first.** This strategy will put you one step ahead, since you will have more information about his position than he will have about yours.
- **Be patient.** Do not interrupt, even if the speaker is saying something totally inaccurate. Take quick notes if you need her to remember the points to which you want to respond.
- **Listen with a goal.** Decide what you want to learn from the person's statements on the current topic, and look for words and nonverbal clues that will give you the information.
- **Maintain eye contact.** Good eye contact shows you are paying attention, and that enhances your counterpart's perception of you as an honest, fair, and friendly person. A second benefit is that the eyes convey subtle, unspoken messages, and frequent eye contact gives you a chance to pick up on these clues.
- **Separate the words from the person.** No matter what your personal feelings are about the person speaking, try to focus your listening on the information he is presenting. React to his message, not to his personality.
- **Be alert to nonverbal clues.** A negotiator's entire message will not necessarily come out in her words. Gestures, facial expressions, tone of voice, and body language can tell you volumes about the attitudes and motives behind the words.
- **Gain your counterpart's respect by letting him know you have listened carefully.** When you are ready to respond, summarize what you have heard and agree on its substance. Then focus your first response on asking questions and gathering further information.

Reducing Conflict

By its very nature, negotiation involves some conflict. If both sides already agreed, there would be no need to negotiate. The goal of good negotiators is to keep a conflict of ideas from escalating into a confrontation.

One secret to success is recognizing the difference between positions and interests. A party's position is what she wants; her interest is why she wants it, and that is the building block of a lasting agreement. If you can figure out why your counterpart has taken a certain position—and why you have taken yours—you will be able to offer and accept more alternatives.

The classic story used to illustrate the difference between positions and interests involves two sisters fighting over the only orange in the family refrigerator. Each claims she must have the whole orange; neither will accept a half. A wise parent takes each girl aside and asks her why she wants the orange. One says she wants the juice; the other wants the rind to cook a pudding. What each sister wanted—the entire orange—was her position. Why she wanted it—and the source of the simple solution—was her interest.

Early in the negotiations, it is also helpful for all parties to make a commitment to fairness. This involves deciding on standards against which elements of the agreement can be measured. Neutral authorities or objective measures, such as lab tests, academic articles, and published industry standards or averages, are the most effective, but any standard is better than none. This can give each side a face-saving reason for making a concession, and it enhances the durability of the agreement. If all the parties view the process as fair, they are more likely to take it seriously and less likely to end up looking for excuses to get out of it later.

It is especially important to avoid confrontations with parties you deal with on a continuing basis. In these situations, the ongoing relationship will often take priority over a specific point being debated. Making a concession now can be a short-term cost that yields long-term gains. Active listening can also enhance the negotiation environment. When you listen patiently and carefully—and make it clear to the other party that you have understood what he has said—he may be so shocked that he, too, tries harder to listen to you.

One important element in active listening is silence, which can neutralize tension and give you an advantage. If your counterpart becomes emotional or confrontational, keeping quiet after he finishes speaking can be quite unsettling to him. Silence sometimes connotes disapproval, but since it has not been voiced, it cannot be treated as an attack. When met with an awkward silence, many people modify their previous statements to make it easier to agree with them.

Making Negotiation Work

Centuries ago, when two landowners had a disagreement, they would hire knights and wage war to reach a settlement. The civility of negotiation has improved significantly since then, but the process can still be stressful and costly if you are not properly armed.

The best ammunition today is a good knowledge of negotiating skills and techniques. In the business world, these are assets that can save time, reduce stress and improve productivity. The same is true for the negotiating we do in everyday life. In any setting, the ability to negotiate patiently, fairly, and effectively is an important element in personal development, and a skill that pays long-term dividends.

SUGGESTIONS FOR IMPROVING YOUR NEGOTIATIONS

- When preparing your goals, categorize them according to importance: essential, desirable, and traceable. Express them in quantified terms if possible.
- Do not arrive with a too-detailed package of positions. Make it clear to your counterparts that you are open to their ideas. If you need more time to react to their proposals, schedule another meeting.
- Being outnumbered is not necessarily a disadvantage, but watch out for "good cop/bad cop" routines. If your counterparts are presenting different positions on an issue, stop the discussion. Offer to leave the room and let them come to an agreement.
- Never give anything away; always receive a concession in return. Even in the early stages of a negotiation, when minor objectives are discussed, make every agreement a give-and-take proposition.

- Consider your BATNA—Best Alternative to a Negotiated Agreement. Decide in advance how little you will accept, and know when you should simply stop negotiating and handle matters yourself.
- Remember that in any relationship, the side with the least commitment generally holds the most power. Decide if that is you or your counterpart, and make your approach consistent with the relative power you hold.
- Be creative. Offer up new ideas, use brainstorming techniques, listen to outlandish proposals, act out different scenarios. Do not behave predictably.
- If you are asked questions in a delicate or risky area, do not commit yourself until you are ready. Try to change the emphasis of the question.

SUGGESTED READINGS

Cohen, H. (1994). *You can negotiate anything.* Carol Publishing Group.

Fisher, R., Ury, W., & Patton, B. (1994). *Getting to yes: Negotiating agreement without giving in.* Penguin.

Johnston, J. (1996, September 2). The deals people make. *The Boston Globe.*

The Negotiation Skills Company. (1994). *Pocket guide: The fine art of negotiating.* Prides Crossing, MA: Author.

Stark, P. B. (1994). *It's negotiable.* San Diego, CA: Pfeiffer & Co.

Chapter 14

ORGANIZATIONAL CHARTS

Organizational charts graphically portray an organizational structure. They show the skeleton of the organization's structure and depict basic relationships and groupings of positions and functions.

Executive business games have proven quite popular during the last few years. In management workshops and seminars, such games are used quite effectively to cover materials otherwise dull or unconvincingly presented. There are other games executives can play, however, which are time consuming, self-defeating, and illusory as to goals or objectives. One of these is the "game" of playing with organizational charts.

The organizational chart is widely used and appropriate for making organizational principles work. Every organizational structure can be charted, for a chart merely indicates how departments are tied together along the principle lines of authority. It is somewhat surprising to find top managers taking pride in not having an organizational chart or a belief that the charts should be kept secret. Some managers feel that organizational charts tend to make people overly conscious of being superiors or inferiors, tend to destroy team feeling, and gives the person occupying the box on the chart too great a feeling of "ownership." The manager who believes that team spirit can be endangered without clearly spelling out relationships is establishing an organizational environment for politics, vague policy, and uncertain decision making.

This is not to say that formal organizational charts have no purpose, or that time should not be given to considering the hierarchical relationships reflected in them. No management specialist would ever suggest these charts be discarded or ignored. An organizational chart, if properly designed, can clarify the organization's structure, simplify internal relationships, and prevent abuses of authority. However, many managers treat the organizational chart as if it were the index to a book—detailed, orderly, and inflexibly correct. In such cases, the chart is not analyzed for its relation to organizational reality, but rather the reverse—organizational reality is forced to resemble the chart.

It is a misguided manager who believes a formal organizational chart reflects, or should reflect, the *actual* operations of the organization. An organizational chart is a point of reference, providing an overview of how organizations should function, but in reality, *all* organizations deviate from it. Not only size, but the prerogative aspects of a bureaucracy—red tape, inefficiency, and corruption—cause the reality to differ from the chart.

Another difficulty with organizational charts is that they show only formal authority relationships and omit the many significant formal and informal relationships. They do not even illustrate how much authority exists at any point in the structure. Also, another limitation is managers who hesitate or neglect to redraft charts, forgetting that organizations are dynamic, and that the chart should be allowed to become obsolete.

One source of administrative failure is confusing the normative with the actual organizational structure. An organizational chart is comparable to a constitution, providing guidelines *and* limitations on organizational activity. A constitution is rarely taken literally, but is interpreted, extended, or amended according to specific challenges or to the reality at hand. The least flexible, unrealistic interpretation of an organizational chart is one that does not extend beyond what is in print.

The informal organizational structure, which is only partly revealed by an organizational chart, represents how things are actually accomplished in an organization. Any organization is far more varied and complex than abstract organizational theories or hierarchical charts suggest. Informal relationships, subgroups, and grapevines often provide more satisfaction and efficiency than the more formal structure permits. Furthermore, although the staff in an organization expect and appreciate an operational framework and directives, they are also quite capable of self-direction and imaginative problem solving.

Regardless of how rigidly structured an organization is or how iron-fisted or autocratic its leadership, all organizations have an informal structure or structures. In fact, one might argue that the more rigid an organization—as in the military, for example—the greater the number of informal substructures.

A manager preoccupied with the formal organizational structure, modifying or interpreting its relationships as depicted on an organizational chart, is not likely to be aware of the informal structure, or, if aware, may try to suppress the informal structure and force the more rigid organizational chart upon subordinates. Such a manager equates informality with insubordination, yet organizational informality generally tends to lean more toward creativity rather than insurrection.

Nevertheless, an organizational chart is not a meaningless sham to soothe board members or taxpayers. All organizations require some type of formal organization. What a formal statement of organization does is to encourage order, predictability, control, and fairly stable relationships. Without it, confusion, inefficiency, and irresponsibility may result.

Sociometric studies of organizations indicate that organizational charts *rarely* reflect how organizations actually operate. Yet an authoritarian administrator believes a formal organizational chart is built on the following seven assumptions:

1. Relations in groups are clearly defined by an organizational chart.
2. Behavior is governed by rational thinking.
3. Subordinates will always do as they are told.
4. An administrator knows best how to solve a problem.
5. The way to get things done is through use of authority.
6. People are merely instruments of production.
7. Man is isolated, unaffected by group pressures.

Other researchers describe these same hierarchical structures in model form. Douglas McGregor, in his Theory X, describes a formal organization showing subordinates as passive, rather unimaginative, requiring autocratic leaders to motivate them. For many administrators and workers, the formal, authoritarian style of organization is the only one conceivable. In fact, most organizations deviate from this conception, and in many instances, such deviation is natural and propitious. Perhaps with the exception of Weber, most organizational theorists describe, rather than defend, the formal order of organization.

Even though informal groupings or lines of communication such as cliques and grapevines are not shown on the organizational chart, an effective manager must acknowledge and deal with these often vibrant centers of activity. The wise manager is always opportunistic, alert to ways of getting things done effectively, rewarding those who do them, regardless of the dictates of formal organization. If necessary, he or she will change the organizational chart to reflect these more meaningful groupings rather than denying or suppressing what is perceived as insubordination.

Any manager who wishes to understand the function of the informal organization has to revise the concepts of traditional authority. McGregor's Theory Y describes the less autocratic organization. In it, workers are accepted as dynamic, self-motivated, independent, responsible, cooperative, creative, and imaginative. This more informal, participatory philosophy of organization openly involves employees and administrators in decision making and communication. Leaders encourage staff to be creative, flexible in work patterns and styles, and cooperative. Bringing the more informal structure to the surface decreases the time previously devoted by staff members to the "hidden organization," while they pretended to adhere to the formal structure as well.

Finally, it would be naive to suggest that all informal subgroups in an organization consist solely of subordinates. Many involve the organization's leadership as well. They may even be intentionally formed at the top for a variety of personal or political reasons.

Wherever informal structures exist, whether they are beneficial or harmful, an enlightened manager must be prepared to put aside the easier and more beguiling administrative toy an organizational chart can become, and recognize things as they actually are. Understanding the informal organization is not easy; it may take years of personal involvement and trust. Putting aside the organizational chart, except as a point of reference, however, can be a first step toward professional maturity. Otherwise, managers will continue to waste their time playing games, while the real life of the organization goes on beyond their reach or comprehension. Careers are not built upon such toys.

SUGGESTIONS FOR DEVELOPING AND USING ORGANIZATIONAL CHARTS

- When designing the organizational chart, clearly spell out authority relationships for all personnel and within each level of the organization.
- Be open and allow all individuals in the organization to see where they fit in the organizational chart and that they understand their job responsibilities.
- Be cognizant that the organizational chart shows inconsistencies and complexities and does not recognize the informal structure of the organization.
- Remember, when designing the organizational chart, make sure that it shows only formal authority and omits the many significant informal and informational relationships.
- Keep the organizational chart current. Be alert to changes and reorganization. Be sure all personnel are aware of any changes that are made.
- Do not try to eliminate the informal organizational structure. Recognize it and accept its existence. The wise manager will feed it accurate information because it is an effective way to disseminate information.
- Organizational charts can be effective insofar as accurate job descriptions exist. Their existence will spell out the individual's role in the organization as well as clarify lines of authority.
- Be careful of trying to maintain a rigid formal organizational structure. Experts agree that the more rigid the structure, as in the military, the greater the number of informal structures that will develop.
- Design the organizational chart so it shows functional relationships, chains of command, and interconnecting lines of subordinates and gives enough information to clarify the structure.
- The organizational chart should reveal the true relative status of the positions and describe the interaction of the various departments within the organizational structure.

SUGGESTED READINGS

Miner, J. B. (1985). *People problems.* New York: Random House.

Pattern, T. H. (1981). *Organizational development through team building.* New York: John Wiley & Sons.

Pfeffer, J. (1978). *Organizational design* (2nd ed.). Arlington Heights, IL: AHM.

Rahim, M. A. (1986). *Conflict in organizations.* New York: Praeger.

Schein, E. H. (1985). *Organizational culture and leadership.* San Francisco: Jossey-Bass.

Chapter 15

ORGANIZATIONAL COMMUNICATIONS

All messages are subject to the "Murphy's Law" of communications: "If a message can possibly be misunderstood, it will be."

"I just assumed that you meant..."
"Helen said she thought the department head said..."
"Nobody told me I was supposed to be responsible for that."
"But the last time you said 'right away,' you meant next week."

A lost sale, a costly equipment breakdown, the ultimate failure of a project—or sometimes an entire company—can begin with words as simple as these. They all signal a breakdown in communication and an interruption in the most critical of all organizational support systems.

Personal communication skills are important in the successful transmission of information, but a major source of problems is the overall structure of the organization itself. An organization's communication structure has been called its "lifeline" because it plays such a vital role in its very existence. It provides the information needed to promote the organization's goals and it sets up a means of channeling it among, between, and through the proper channels. Essentially, it ensures that accurate, complete messages reach the right people so they can make the best decisions.

DEFINING ORGANIZATIONAL COMMUNICATION

Organizational communication has many different definitions, but they all encompass the idea that the purpose is to facilitate the achievement of organizational objectives. It is management's way of letting each employee know what his role is, what his relationship is to others, and how all these roles work together to meet the organization's goals.

Communication is a continual process that changes as information is transmitted from one employee to the next. The exact way in which information is shared and understood is determined by the individuals who send and receive it, but management can benefit from providing a framework for channeling it. It is important to set up a structure that ensures that each employee gets the information he needs to do his job.

In designing this framework, management must consider both the formal and informal channels of communication. The formal communication structure is the one that management can best control—it defines who gets what, how messages are routed, and who is responsible for sending information to other employees. The idea of the formal structure is to give the organization and employee an outline of how communication in the organization should flow.

As many organizations have discovered, though, these approved patterns must be designed to coexist with the informal communication structure. Even the most comprehensive system cannot control the large amounts of informal communication that will occur inside and outside the workplace. A worker who has information she feels another employee should have will override the defined patterns and pass it along. This will not necessarily disturb the formal communications process, but it will change the number and type of employees who act on the information.

COMMUNICATION STRUCTURES

The impact of the formal and informal methods of transmission becomes clearer when we look at communication in its two main structures: vertical and horizontal. The vertical structure is the formal communication process. It involves the passing of information upward and downward between employee levels and is often referred to as a hierarchy or authority pyramid. The horizontal structure is informal communication that is shared by employees at the same status levels.

To examine each structure we must look at its effects on the organization. Within the vertical structure, the downward flow of information is the traditional view of organizational communication. As information passes down the organizational chart—from CEO to vice-president to supervisor to employee, for example—the messages are accepted with respect, attention, and little rebuttal.

Downward communication typically uses three channels to convey messages: oral, written, and a combination of the two. The various forms these channels can take include:

Vertical Communication

>*Oral*
>Personal instructions
>Lectures, conferences
>Committee meetings
>Interviews
>Counseling
>Telephone public address system
>Movies, slides, television
>Social affairs, union meetings
>Grapevine
>Gossip, rumor
>
>*Written*
>Instructions and orders
>Letters and memos
>House organs
>Bulletin boards
>Posters
>Handouts, information racks
>Handbooks, manuals, pamphlets
>Union publications
>Pay inserts
>Annual reports
>   —(Koehler 172)

Many organizations find a combination of oral and written channels is the best way to ensure that the message is received in the proper context and clearly understood.

As messages progress down the hierarchy and through the various channels, they are subject to change. Each level will interpret the messages in its own way, add to them, and transfer them downward. Ideally, this process will transform the message into a more specific, practical form at each level. For example, the message from the highest point of the organizational hierarchy may be a concise statement of a desired result. The level below may elaborate by including the means for achieving the result. At the next level, the receiver may add procedural details before he sends it to the operational level, which now has a fully detailed plan.

This flow of information is enhanced by the other vertical communication direction, which is upward. By requiring lower-level managers or subordinates to send information back to their superiors, the originator of the message can determine if the message was understood. Another advantage to this upward flow of communication is that employees have an outlet to express their ideas and attitudes about the organization.

Because the messages are subject to interpretation by people, problems are inevitable, especially in the upward flow of information. Frequently, lower-level managers will distort the message to conform to their idea of what the receiver wants to hear. As a result, the feedback they send to higher-level management may reflect only achievements and accomplishments, while suppressing failures or disapproval.

The relationships and rapport of superiors and subordinates can also affect the effectiveness of upward communication. If an employee feels inferior to her manager or lacks trust in him, the upward communication flow will be minimal. Employees can also be fearful about sending information upward if they lack an understanding of their roles in the organization's communication structure.

Horizontal Communication

The second structure of organizational communication is horizontal or informal. As the name suggests, it comprises messages that move horizontally across the organizational chart, such as information shared by co-workers.

Like vertical communication, horizontal communication is delivered in both oral and written forms. Types of horizontal communication include:

Oral
Lectures, conferences, committee meetings
Telephone, intercom systems
Movies, slides, television
Social affairs, union activities
Grapevine, rumor, gossip

Written
Letters, memos, reports
Bulletin boards, posters
Handbooks and manuals
Annual reports

Horizontal communication typically carries more information—on a much wider range of subjects—than vertical communication. It is a continuous exchange of information, which is critical in the development of group coordination and teamwork. Employees of the same status tend to be more open with each other than with their superiors, and they place fewer limits on the types of information they share.

Early studies of the organizational communication ignored the horizontal structure because it did not fit the traditional models. Today, managers are more aware of the importance of informal communication and the effect it can have on an organization.

A prime example is the formation of unions. These are classified as a horizontal, or informal, communication structure. If management does not acknowledge informal groups like these, they may work against the organization instead of for it.

Creating a Structure

The communications structure that will work best in an organization depends on its unique goals and on the types of people who work for it. The overriding concerns in designing a system, however, should be knowledge and control. The originator of a message—and, in some cases, his superiors—should know what type of information is being sent and to whom, and there must be processes in place to ensure its accuracy and control the flow.

It is also important to think of the twin goals of communications. In a business setting, the purpose of formal, downward communications is to (1) *provide information* that will (2) *guide employee behavior*. The root meaning of the word *communicate* is "to make common," so the process of sharing information should be understood as a means of achieving a common understanding by the sender and receiver, the supervisor and the employee.

The foundation of this process is the formal procedure of orienting new employees and periodically evaluating their performance. These communication tools have proven their worth over the years, and they deserve continuing attention. They constitute the main means of interacting with employees and provide an organized way to handle important, recurring communication needs.

To ensure that employees are adequately informed, supervisors and managers must regularly review and modify these procedures. The organization should establish a schedule for reprinting manuals and updating orientation classes, on-the-job training methods, and performance-review processes. It is also important to build upward communications into this process. Give employees a method for evaluating the information they receive and providing feedback about their training needs.

To set up an orderly process for other downward communications, start by classifying the types of messages that are routinely sent to employees. These might include policy changes, work schedules, instructions, production reports, announcements of changes in safety practices, and technical processes. For each type of communication, consider the intent of the message and the specific employee behavior that is desired.

The Medium Is the Message

The medium itself is a crucial part of each message, so the next step is to decide on the best way to deliver each type of communication. Oral or written? In a staff meeting or person-to-person discussion? In a memo, the company newsletter, a bulletin-board posting, a procedure manual? In general, a spoken message from the supervisor to employee is the strongest way to deliver information, but some types of information need to be communicated in several ways. The more important the content, the more it will benefit from repetition and the use of several media.

Some types of messages will go directly to all the people who need or will be affected by the information. Other messages—especially those that need to be expanded or modified before they reach their final destination—are best communicated through downward channels. For example, the highest point of the organizational hierarchy may send managers an announcement of a new production goal. The managers will add suggestions for how to increase output before passing the information to supervisors, who will add details about changes in procedures and work assignments. For these types of messages, the organization's communication structure should establish a clear routing method to assure that each level receives accurate, practical information.

Message Competition

Another important consideration is "competition" for your messages. Employees often receive messages from multiple sources, and some of these sources may be providing information that is different from that contained in your message. Many of these are delivered through horizontal communication channels. Employees may be acting on information they receive from co-workers, other departments, union representatives, trade journals, even their own personal experiences. Management cannot control the delivery of these outside messages, but it is important to be aware of them. By building repetition and multiple media in the communication structure, the organization can reinforce its messages and reduce any confusion caused by other sources of communication.

The most effective way to combat message competition is to make sure that all your communications are two-way. For every type of message and medium, incorporate a system for gathering—and adjusting to—employee feedback. Establish formal and informal methods that allow the receivers of messages to send their reactions back up through the channels—through staff meetings, written replies to memos, progress reports, informal discussions.

Modifying Your Structure

In any organization, it is important to create conditions that are conducive to an orderly flow of information, both vertical and horizontal. A clearly defined communication structure is the foundation of this process and the most critical element in the successful promotion of the organization's goals. It is important to recognize, though, that no communication system will ever be 100% effective in meeting goals or in influencing employee behavior. Even the best work units cannot be expected to have a communication efficiency of much more than 80%, so the communication structure needs constant attention. By continually reviewing and modifying the structure through which messages are sent and received, an organization can strengthen this "lifeline" and ensure its existence.

SUGGESTIONS FOR IMPROVING YOUR ORGANIZATION'S COMMUNICATION STRUCTURE

- Test every element of your communications structure with the "Five W's and an H" used by journalists. For every type of message, consider:
    Who will send and receive it?
    What will be communicated? What is the exact scope (and limits) of the content?
    When will the message be sent?
    Where will the recipients be when it's delivered?
    Why are you sending it? (What specific action or employee behavior do you want?)
    How will it be sent (the medium)?
  For all types of messages, make clarity the first priority. Recognize that all messages are subject to the "Murphy's Law" of communications: "If a message can possibly be misunderstood, it will be."
- To build clarity into your downward communications, it may be helpful to establish a format for some types of messages. Those that need immediate action, for example, should always end with a clear summary of who-does-what, when, where, and how.
- Never assume that the organization's formal messages are the employee's sole source of information. Establish a way to gauge and adjust to the information employees receive through the horizontal communications process.
- Make all your communications two-way. Establish formal and informal processes for information to travel up through the channels so you know your message has been understood and acted upon.
- Be ready to respond and adjust to feedback. If you cannot act on an employee's suggestions or complaints, explain why. If delays occur, send progress reports.

> • Always consider the state of the receiver. Employee attitudes, fears, and motivation levels will affect how information is interpreted, so message tone—and the medium used—may be crucial for the success of the communication.

SUGGESTED READINGS

Bittleston, J., & Shorter, B. (1981). *The book of business communications checklists.* London: Associated Business Press.

Connelly, J. C. (1967). *A manager's guide to speaking and listening.* New York: American Management Association.

Foltz, R. G. (1973). *Management by communication.* Philadelphia: Chilton Book Company.

Huseman, R. C. (1977). *Readings in interpersonal and organizational communication.* Boston: Holbrook Press.

King, S. S. (1994). *High-speed management and organizational communication in the 1990's.* Albany, NY: State University of New York Press.

Kochler, J. W., et al. (1976). *Organizational communication: Behavioral perspectives.* New York: Holt, Rinehart and Winston.

Lillico, T. M. (1972). *Managerial communication.* New York: Pergamon Press.

Murphy, J. F., et al. (1991). *Leisure systems: Critical concepts and applications.* Champaign, IL: Sagamore Publishing.

Rockey, E. H. (1977). *Communicating in organizations.* Cambridge, MA: Winthrop Publishers.

Chapter 16

PROBLEM SOLVING

Reluctant problem solvers may not realize that experience teaches, even if the experience is failure. We can learn much from our failures. It's important to remember that the process of trying is healthful, helpful, and life enhancing.

The most important function of a manager is to solve problems. To solve problems effectively requires using a systematic approach. The process of problem solving can range from one individual making decisions to a highly involved group effort taking long periods of time to reach a solution. This essay presents a problem-solving method that can be used by managers in their organization or by individuals who are faced with everyday situations for which they have no solutions.

This step-by-step approach simplifies problem solving. It may, however, produce an assumption that solutions come easily—even automatically. Nothing could be further from the truth.

The author, based on experience, has placed these nine steps in a common sense order. His intent is to develop an analytical and creative approach to difficult problems. In addition, if used consistently, the model will increase efficiency and improve the ability to make decisions.

Simple, well-structured problems require only logic, common sense, or intuition. However, complicated situations are easier to untangle if these successive steps are followed:

1. Determine the problem.
2. Define objectives for the solution.
3. Recognize barriers to potential solutions.

4. Observe changes needed and possible conflicts.
5. Identify factors that influence change.
6. Brainstorm for possible solutions.
7. Select a few alternatives.
8. Make a final decision.
9. Implement the solution.

These steps are all interrelated, and some take place simultaneously. Here they are shown as separate, visible identities, for ease of description and understanding. They show the problem-solving procedure in slow motion. In the application of this model, there should be an even, uninterrupted flow from start to finish.

The steps are examined in the following nine sections.

1. Determine the Problem

Problem definition, the first step, is the most critical stage of the model. It follows an analysis of what led up to the problem. Everything hereafter depends upon the accuracy of the problem statement.

Since there is a problem, there must be a deviation from an established expectation or a desired standard. Therefore, the end result of what is taking place must be known and understood. Seek answers to such questions as:

- What should be happening?
- What is the deviation?
- How serious is this departure from what was expected?
- What is the chance for improvement?
- What change will take place, if the problem is resolved?

Zero in on what caused the deviation from the desired result. Once this is isolated, it may be possible to prioritize and resolve some peripheral concerns, subproblems, or symptoms of other problems. But these side issues should not be confused with the real problem being defined.

Gather relevant facts related to the situation. Avoid speculation, bias, assumptions, and rumor. Ask the what, when, where, and who questions at this stage. However, fact finding can be a never-ending process unless the problem solver knows when to stop. At this point, decide what can be resolved and get on with the process of identifying and solving the problem.

No one directly concerned with the problem should be excluded from finding its solution. All will have a point of view that should be considered. For instance, a committee that sets policy affecting participation at the community level without citizen involvement can expect trouble.

A well-written problem statement will specify the anticipated outcome. It will clearly, concisely, and accurately reflect what is wrong in this particular situation and what can be done about it.

Check and recheck the problem statement with this series of questions:

- What was to be accomplished?
- What deviation has taken place?
- What is the objective for correcting the situation?
- Are all of the relevant facts gathered and recorded?
- Are subproblems and symptoms of other problems separated out; are they clearly identified and prioritized for possible resolution?
- Does the problem statement reflect all points of view?
- Does it get at the real issue?

If the final statement satisfies these searching questions, the first stage of the model is complete.

2. Define Objectives for the Solution

This step requires precision, candor, and realism. It is a complex part of the process. It is also closely related to the preparation of the problem statement and may move along simultaneously. Objectives set for the change desired (i.e., what ought to be) become the means by which alternate solutions may be measured. They are standards for indicating success or failure—check marks along the route. They are the inner framework of problem resolution.

Closeness to the difficulty may help during the information-gathering stage, but it can be a liability when objectives are set for the resolution of the problem.

Review stated objectives for the organization before considering new or modified policies. Have they kept pace with the needs and values of those who are served? Are social issues pressing for change? Is there conflict between stated and implied policies?

The problem solver must have a clear conception of the role of the organization, the procedures for carrying out these responsibilities, and the constraints of time or resources. There should be clarity but not rigidity. There is no room for guesswork.

Objectives can be modified as more is learned about the situation or a solution is selected from available alternatives. They are not meant to be final at this stage—although they may be. The better objectives are thought through here, the less need there will be for adjustment as the process progresses. There may be need for short-range objectives before a long-term goal can be achieved. However, setting interim objectives will require care if confusion with the real issue is to be avoided.

Objectives should be stated in both qualitative and quantitative terms. In recreation, qualitative statements are more difficult to prepare because they must deal with value-laden concerns and with personal feelings.

As various points of view are fed in, compromise may be necessary. The intent of the solution must have meaning and appeal for all who are involved. There is no room for ambiguity or misconception.

Formulate objectives that are precise in time, place, and quantity. When this task has been done, rank them by importance. Separate them, if necessary, from desires or wants. Peripheral desires only confuse the issue.

3. *Recognize Barriers to Potential Solutions*

Roadblocks to resolving a problem may emerge at any stage of this process, but they are particularly critical here where judgment must be applied. Mental blocks are common and spring from an individual's value system. They likely relate to how things are perceived, to a cultural background, or to emotion (fear, anger, hate) when under stress. Those who display emotional barriers to problem resolution frequently seek an "out" by attacking symptoms rather than the real issue.

Implicit or unwritten policies may stand in the way. The very structure of an organization may block the solution to a particular situation. Interpersonal relationships, false assumptions, or lack of acceptance for social change may prevent achieving the solution objective.

Attitudinal blocks will be signified by such comments as:

- It will not work.
- It costs money.
- It is against policy.
- We have tried that before.

Such barriers as these may be removed by a carefully planned educational program.

Lack of agreement by all who are involved will block action. Apprehension will be lessened if the proposal for change is simply and clearly stated.

4. Observe Changes Needed and Possible Conflicts

This is the stage to ask the "why" questions.

Change is the most important element within problem solving. What change is needed to remove the deviation from the desired result? Any shift from what actually exists toward what ought to be indicates a change in the right direction. A clear distinction between *what is* and *what ought to be* will narrow the search for change to reasonable limits.

As specific changes are identified and examined, a written change statement should be prepared. There is some room, during this step, for speculation or realistic hunches. But these should be regulated by ongoing problem analysis.

Personality conflict, both internal (within staff) and external (within community), may block problem resolution and should be noted here.

5. Identify Factors That Influence Change

If more information appears to be necessary at this stage, continue to gather facts.

These are some of the critical variables that need to be considered:

- cultural or ethnic factors
- ignorance, bias, false assumptions, or indifference
- emotional factors (fear, anger, hate)
- attitudinal barriers (inflexibility)
- mental blocks (limitations imposed by a personal value system)

- receptivity to change; entrenched resistance to change
- organizational restructuring to accommodate change (relinquishing power; broadening the base for decision making)
- unwritten or implied policies
- physical limitations of resources

Systematically examine all information gathered to this point. As possible causes of the problem are discovered and analyzed, weigh them for worth and place them in order of their probability.

6. Brainstorm for Possible Solutions

Try brainstorming for new or radically different ideas. This is one stage where imaginations should run freely. No judgment should be applied; no idea is too far out to be listed. However, this technique must be properly applied and should not replace the more scientific approaches to problem resolution, such as reviewing the literature for the resolution of similar problems.

Creative thinking about potential solutions enhances the process—freewheeling is welcomed. Here the role of the leader is critical. A brief warm-up session, using simple idea generators, helps to set the stage.

Limit spontaneous generation of ideas to 20 minutes at a time and the brainstorming session to two hours. Provide unusual surroundings for those involved.

The nature of the problem to be resolved will determine who should participate. The ideal group is six plus a leader, but this number may vary between five and 12. Avoid mixing radically different personalities; supervisors are out when staff is included.

The more specific the problem statement used, the more productive the brainstorming session will be.

7. Select a Few Alternatives

Here the stage is prepared for making a final decision.

Screen, classify, evaluate, and list all possible solutions, according to preset objectives. Combine similar ideas; eliminate those not producing the desired result. Edit the list, judging group relatedness, applicability, and the risks involved.

Establish boundaries within which consequences for change can be tolerated. Weigh each alternative and, as the selection is made, indicate its advantages and disadvantages. When nearing the completion of this stage, search for unknown or unsuspecting consequences.

Reduce the possible solutions to only the best workable alternatives.

8. Make a Final Decision

The problem has now been analyzed, clarified, and defined. An objective for the desired solution has been set. Potential barriers and conflicts have been considered. The change required has been stated. Alternatives have been identified, considered, and weighed. The most likely solutions have been narrowed down to a few. Even though decisions have been made throughout the process, this is the point where a final choice must be made—what will solve the problem with a minimum of risk?

A group of people may consider a problem and reach consensus as to what should be done, but one person must make the final decision. That person is the person responsible for solving the problem. He or she must accept the consequences of the choice made. Interpretation and application of gathered information can be sought, but making the final decision cannot be delegated.

For best results in problem solving, the individual must have a social conscience. He or she should be well informed, aware of the changes needed, and willing to adapt them. Personal bias and emotional involvement should not intrude. Then, and only then, will the best alternative be chosen.

Complex problems are usually resolved by striking a compromise between ideal and practical solutions. However, do not compromise on basic principles or long-range goals that have been set and accepted.

Selling the solution chosen comes after a rational decision has been made.

9. Implement the Solution

A plan for achieving change is useless unless it is put into action.

Change is usually resisted. Therefore, anticipate the difficulties that must be overcome so that all who are involved in the decision will respond positively to the idea.

This may require a variety of tactics and strategies to interpret, accurately and concisely, the change that is needed. Organizational or political pressures may help to achieve the objective set for the solution. Implementation requires personal enthusiasm—communication with conviction.

Before publicly announcing the decision reached, go back over all factors considered throughout the process so that potential or unsuspected objections can be identified and brought out into the open. This review will help with the preparation of interpretive material: graphs, statistics, pictures, or film. Timing of the presentation is important. Tact and good humor are necessary to get the cooperation of all who are concerned with the problem.

A rehearsal of the final presentation, with role playing, may be wise before that all-important public release of the solution chosen to resolve the problem.

SUGGESTIONS FOR EFFECTIVE PROBLEM SOLVING

- Find out if your problem has been solved before; if it has, find out if a similar solution will work for you.
- It is better to take additional time and ask if the problem that was so quickly identified might possibly be a symptom of a deeper problem.- solving process is to start looking for symptoms. If the problem has been solved, the symptoms will have disappeared. If the symptoms remain, you need to reexamine each step of the process.
- The degree of success one attains in implementing the selected solution to the problem depends on the amount of commitment, effort, and tenacity of the problem solver.
- Periodically look at the solutions or decisions to check if they are accomplishing their goal.
- Look at each of the potential solutions carefully to ensure that all are given equal consideration. Do not arbitrarily eliminate potential solutions because of personal prejudice.
- Do not base your solution exclusively on opinions; use the knowledge of experts who stay within their field of competence.
- Define the problem so that you and others can generate possible solutions.
- Remember, brainstorming is not a method to solve problems, but a technique for generating ideas that are potential solutions to problems.
- Do not get excited or emotional if someone resists your problem solution. Calm, reasoned responses will convince others of your confidence more than angry rebuttals.

SUGGESTED READINGS

Bannon, J. J., & Busser, J. A. (1992). *Problem solving in recreation and parks.* Champaign, IL: Sagamore Publishing.

Bannon, J. J. (1992). *Take charge: A "how-to" approach for solving everyday problems.* Champaign, IL: Sagamore Publishing.

Nadler, G. (1983). Human purposeful activities for classifying management problems. *Omega II,* 15-26.

Nickles, T. (1981). What is a problem that we may solve it. *Thythese, 47,* 85-118.

White, C. S. (1983, Spring). Problem solving: The neglected first step. *Management Review,* 52-55.

Chapter 17

PROJECT MANAGEMENT

In successful project management there is a complete and clear definition of project scope, objectives, and work to be done. Project responsibilities and requirements are clear and well understood by all involved.

In today's economy, most service and manufacturing organizations are involved in a number of special projects that provide work activities for all members of the organization. Usually, each project is in a different stage of completion. One project may be a feasibility study for a new product, another may be involved in research and development, and still another may be at some stage of production. All of these projects have their individual problems and require an individual to serve as project manager. Her responsibility is to keep all relevant personnel in the organization informed of the project's progress.

A project organization is made up of a combination of human and nonhuman resources on a temporary basis to achieve a specific purpose. Because projects have a temporary life, a method of managing and organizing is developed so that the parent organization structure is not disrupted and can maintain a degree of efficiency. Under the project management structure, individuals working on the project are officially assigned to the project from the parent organization, as well as experts from outside the parent organization. The major advantage of the project management structure is that the combination of people and resources used on the project can be readily changed to correspond to changing project needs. Other advantages include the emphasis placed on the project by establishing a project team and the relative ease with which projects' members can be absorbed back into the parent organization once the project has been completed. One serious potential disadvantage is that project organization can result in a violation of the principle of unity of command. A role conflict can develop if the authority of the project manager is not clearly delineated from that of the parent organization managers. Another problem can occur when the personnel assigned to a project are still evaluated by their parent organization manager, who has little opportunity to observe their work on the project.

Project management has a number of characteristics that are unique to it but do not exist in traditional management. First, project management requires that functional lines of the parent organization and organizational lines outside the organization be crossed to accomplish project goals. Project management is more concerned with the flow of work in horizontal and diagonal relationships than with the vertical chain of authority. When an organization introduces project management, two organizations will immediately exist: the formal structure of the parent organization and the project organization structure. The illustration on page 133 clarifies this concept. In this situation, the project manager will be concerned with the details of the project under way, particularly as they relate to costs, scheduling, personnel, and evaluation. Project management is concerned with a specific end product, and as such, the project manager is responsible for the following:

1. Bringing together project activities so that the recipient of the service or the user of the product is satisfied.
2. Creating the funding, scheduling, and evaluation criteria for the project.
3. Acting as the point person for the organization as it relates to the project under way.
4. Acting to settle any conflict that would endanger project activities.

Figure 17.1: Needs Assessment Flowchart

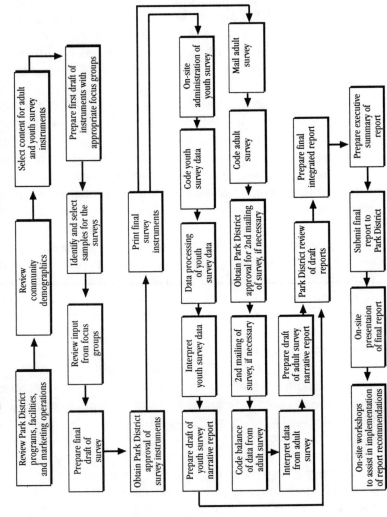

The manager in the parent organization provides facilitation to a project but also has responsibility for coordination of all ongoing projects. He also must provide resources for the operation of the project. Project managers are responsible for specific end results: a product, a specific service. The end result may be a prototype of a new product or the development of a new process or procedure. An example may be the discovery of a vaccine to reduce the potency of the AIDS virus or a new procedure to process welfare recipients. After the end result has been achieved, or determined that it cannot be accomplished, the project team may disband. The project manager's job will end following the evaluation of the overall project.

Project management is often employed for difficult and complicated projects that involve individuals outside the parent organization. Organizations will often seek the services of lawyers, engineers, financial experts, planners, and the like. These individuals may or may not be a part of the parent organization and, as a result, will need to be employed and become a member of the project team. Because many of these consultants are professional and often have creative skills, the project manager must use different management techniques from those used in the traditional superior-subordinate relationship. Personnel in the parent organization often do not understand this relationship and as a result become disenchanted and unhappy. A good example of this is NASA. Numerous specialists with diverse skills are needed to carry out the many special projects for NASA. The project manager must alter her style of leadership as it relates to motivation and persuasion.

For example, the space program will require exercise physiologists, engineers, nutritionists, mathematicians, statisticians, research design experts, language specialists, and many others. All of these individuals are brought together for one purpose—space exploration. Individuals with this high level of technical competency need special management direction. The project manager who does not recognize this need will surely fail.

The most effective project organization would be to secure the most qualified project manager while, at the same time, giving him the appropriate authority and responsibility. This may be the optimal situation but it is not very realistic. Therefore, the most competent manager in the organization should be assigned special projects. This is the responsibility of the parent organization management. Employing personnel to carry out a project requires that the project be broken down into units. The project manager should do so in collaboration with the parent organization management because it will provide support for the project. The major responsibility of the project team is the following:

1. Develop the parameters for the project as well as the overall plan to carry out the project recommendations.
2. Organize and direct the technical requirements of the project's product and service.
3. Secure equipment sources for items to be procured.
4. Develop a plan of costs, time lines, and all resources required to complete the project.
5. Create technical materials for the operation and implementation of the end product.
6. Provide technical leadership for the product or service under study.

Successful projects require top-management support, which affects the degree of acceptance or resistance to the project. Management can show its support by allocating necessary resources, backing the project manager in times of crisis, and granting the project manager sufficient authority and influence. In successful projects, the project manager is confident about top management's support and agrees with them about the level of responsibility and authority conferred to him.

In successful projects, the project manager is strongly committed to meeting time, costs, safety, and quality goals. Most importantly, successful project managers have above-average skills in administration, technology, communication and human relations.

The successful project team must be committed to the goals of the project and to the project management process. The project team participates in estimating, setting schedules and budgets, helping solve problems, and making decisions. This process develops positive attitudes about the project, builds commitment to project goals, and motivates the team.

In order to bring special projects to successful completion, goals and objectives must be set. In setting goals and objectives, a person accomplishes two things: (1) focuses oneself and the project team on the end result, and (2) creates a commitment and agreement about the project goal. Effective project managers keep their focus on the goal and make sure that everybody else on the project team does the same.

Once a clear statement of goals and objectives has been established, the project manager needs to identify team members, resources required, and input necessary to attain results. By identifying each objective with specific team members, you establish "ownership" on the part of the team member. Ownership leads people to take responsibility and feel committed to accomplish the stated goals and objectives.

In the very beginning, the project manager, along with team members, needs to estimate the time and resources necessary to complete the project. Specific tasks and activities need to be identified, which will help in planning the project in complete detail. Who needs to be involved? How long will it take to complete each task? What dollar cost and human resource cost will be required? By completing this exercise and answering these questions, you can be in a better position to estimate potential results of the project. It should be emphasized that this goal may not always be achieved because things often change during the process of conducting the project. Alan Randolph and Barry Posner, in their book *Effective Project Planning and Management,* suggest that three scenarios should be developed for every project: the optimistic scenario—the shortest possible time in which the project can be completed; the pessimistic scenario—the time it will take if things go wrong and you run into many difficulties; and the most likely time scenario—the amount of time required if all the planned activities are completed without difficulty. By looking at each of these scenarios carefully, you are in a better position to anticipate what could go wrong and what is involved in completing each activity. It will also require involving project team members and, as a result, build their confidence and commitment to the time estimates. Remembering that all three scenarios are still estimates, you can calculate a weighted average to get a better idea of how long a task or activity will take. Randolph and Posner suggest the following equation for this calculation.

Equation for Time Estimates

$$\text{Expected time} = \frac{\text{Optimistic time} + (4 \times \text{Most Likely time}) + \text{Pessimistic time}}{6}$$

Another method of estimating the amount of time for the completion of a project is the development of a flowchart. Some refer to this method as CPM (Critical Path Method), or PERT (Program Evaluation and Review Technique). The flow of activities is easily visible on flowcharts. Flowcharting frequently will uncover means in which a task can be preformed more easily or eliminate dead-end steps that might have been useful at one time but are no longer necessary. Each of the steps should be clearly stated and in proper sequence. A project manager can clearly see the specific activity that needs to take place. Each event can be established as a benchmark and crossed off when completed. The project manager can easily establish time sequences for each task. If something intervenes that impedes the flowchart, the flow chart can be modified without difficulty. Because of its graphic visual presentation, the flowchart assists with keeping all project team members informed as to the progress of the project.

SUGGESTIONS FOR EFFECTIVE PROJECT MANAGEMENT

- Successful projects have good control and reporting systems. These systems should provide monitoring and feedback at each stage of the project, allowing comparisons of schedules, budgets, and team performance with project goals.
- The project manager must be delegated the proper authority to carry out project objectives. The project manager, in effect, is the "general manager" of the project. No other person in the organization should be allowed to exercise control over the project.
- The project manager should have the ability to establish rapport with the project team, reach consensus, and demonstrate technical and managerial competence that will create a positive influence over the project.
- Project managers must strike a balance between the technical and administrative experts involved in the project. It is not always easy to get highly technical engineers and scientists to realize that costs must be kept within reasonable bounds. The project manager must rely on his interpersonal skills to accomplish this.

- Be careful. Project management-type organizational structures lend themselves to power struggle among managers because this type of organization allows and encourages dual reporting. The project managers who take into consideration the overall organization point of view should be rewarded.
- Because project management uses the team concept to achieve results, group decision making frequently occurs. As a result, groupthink can set in and become a problem. In many instances, project management decisions involve detailed matters requiring only one or two people. Each member of the project team should invite outside experts or qualified colleagues within the parent organization to decision-making meetings, and they should be encouraged to challenge the views of project team members.
- Project managers should establish specific criteria that will measure the success or failure of the projects. Areas where criteria should be developed are costs, schedules, quality of product or service, acceptance by upper management and customers, project team satisfaction, and safety features.
- Project team members must be strongly committed to maintaining and fulfilling project goals. All participants must be committed to the concept of project planning, control, and implementing that concept into practice. They must understand the project management process and be committed to the steps and procedures necessary to conduct the process.
- In successful projects there should be frequent and regular meetings to exchange information, data, and changes. Meetings should be open and everyone encouraged to attend.
- Project organization requires support from top management in the parent organization. This support may take financial, psychological, and human forms.

SUGGESTED READINGS

Alkins, W. (1980, October). Selecting a project manager. *Journal of Systems Management, 20,* p. 66.

Duffendack, S. C. (1977). *Effective management through work planning.* Schenectady, NY: Corporate Publications.

Kegner, H. (1987, February). In search of excellence in project management. *Journal of Systems Management, 16,* p. 96.

Lasden, M. (1980, March). Effective project management. *Journal of Systems Management, 23,* p. 80.

Chapter 18

PUBLIC RELATIONS

Public relations, if done well, presents an image that corresponds to the reality.

Every organization needs to secure and maintain the confidence of the public. Public understanding and the appreciation that comes as a result of an organized program of public relations is necessary for an organization's success. Public relations is no longer just something to talk about, it is something we must do.

Public relations has been with us for more than 50 years; however, today there appears to be a sudden discovery of its effectiveness in helping to market a product, service, or cause.

According to the U.S. Department of Commerce, a growth rate of 18-20% has been forecast for the public relations industry as compared to only 7% for advertising in the coming years. Public relations is the management function that evaluates public attitudes, identifies the policies and procedures of an individual or an organization with the public interest, and plans and executes a program of action to earn public understanding and acceptance.

Getting and keeping a good reputation is the primary purpose of public relations. As a profession, public relations is a 20th-century development, but the reason for it has been well understood for many centuries. A writer of the Biblical Proverbs stated that "a good name is to be desired more than great riches." The Greek philosopher Socrates was closer to understanding today's PR when he said, "The way to a good reputation is to endeavor to be what you desire to appear."

Over the last decade, public relations has steadily built its reputation, increased its prominence, and earned respect across a wide span of society. As today's institutions strive to understand more clearly the forces of change to adapt their activities to new pressure and aspirations and communicate more effectively, public relations has become more important.

VALUE OF PUBLIC RELATIONS

The value of public relations is usually not realized until it is absent. Public relations can be reflected in the morale of an organization, in the quality and quantity of goods and services produced, or by public support or lack of it. C. A. Schoenfield summed up the value of good public relations when he stated that we must have public support if effective management is to be practiced. He pointed out that a favorable climate of public opinion must precede management. Public support or tolerance and favorable public opinion are indeed the end product of good public relations. For private enterprises, this means the privilege of doing business and the freedom to do it profitably. For public agencies, it means the ability to exercise professional judgment in the management of the organization.

The value of good public relations is big. It includes being able to attract high-caliber workers both in the labor force and at the management level.

As pointed out by Fraser Seltel, the term *public relations* is really a misnomer. "Relations with the public" would be more accurate. Practitioners must communicate with many different groups beyond the general public, each with its own special needs and each requiring different types of communications. Figure 18.1 on page 141 describes in more detail these publics.

One reason that public relations has not been strongly established in many organizations is that it is often viewed with suspicion and distrust. To many, public relations is, at best, synonymous with "cover up" or "whitewash." In some cases, we have contributed to this image by shady acts or outright deceit. The principles upon which public relations is based completely preclude honesty.

Figure 18.1

Adapted from Fraser P. Seltel's *The Practice of Public Relations.* Charles Merrill Publishing Co. 1980.

THE PRACTICE OF PUBLIC RELATIONS

There are two distinct ways in which public relations is practiced. The first and most common is the constant day-to-day contact all employees have with the public. This includes such ordinary but important occurrences as answering the telephone, greeting office visitors, driving organization vehicles, answering questions, and generally conducting oneself in view of others. All of these functions should strive to influence positively public opinion toward the organization. The general daily practice of public relations should include all employees at all times. As the employees' contact with the general public grows, so grows the impression that an individual has of the entire organization. One rude or incompetent employee can significantly affect the organization's reputation or progress toward its overall public relations goals. No employee should be exempt from his or her responsibility of performing the general functions of good public relations. Subsequently, employers should be responsible for adequate training in these functions just as they undertake training in the more obvious or traditional duties of the employee.

The second type of public relations practice is the staff function. As the term implies, this is the more structured approach to public relations and uses the services of one or more individuals who specialize in the practice. This practice could range from the large public relations department to the assignment of one individual to carry out the public relations responsibility.

PUBLIC RELATIONS CONSULTANTS

In private enterprise, consultants or public relations firms are often used to provide at least part of the public relations staff function. There is much to be said for this approach, including a more accurate perception or few prejudices by someone outside the organization. Private consultants offer other advantages, including the following:

- Perspective into publics, markets, and products beyond those familiar to the client.
- Wider range of skills, talent, and experience that the consultants obtain by working for a variety of clients and different problems.
- Objectivity and independence of thinking derived by not being part of the client's organization and politics.

Another advantage is the use of outside services for such things as mailings, opinion polling, art, or photography that can often be obtained for much less expense that would be required to develop the necessary related staff positions within the organization.

Good public relations cannot be practiced in a vacuum. No matter what the size of the organization, a public relations program is only as good as its access to management. No matter how skilled the public relations techniques and technicians, they simply cannot succeed if top management is unaware of or sidesteps its responsibilities in describing its place in the community and defining its objectives.

Central to effective public relations is the ability to know how to communicate. In recent years, more and more students of organization have come to realize that communication is one of the crucial elements in administrative behavior. This knowledge sets the professionals apart. Before those in charge of the public relations program can earn the respect of management and become trusted advisors, they must demonstrate a mastery of skills in writing, speaking, listening, promoting, and counseling. These abilities are key to success, although expertise in management and marketing also contributes. Thus, it is essential that practitioners of public relations understand the theory behind interpersonal communications.

A Good Communicator's Dozen

1. Plan communications carefully, keeping the audience clearly in mind.
2. Know the purpose of the communication in advance of communicating.
3. Be flexible in adapting to the audience, content, purpose, and medium.
4. Don't expect complete understanding the first time around; always follow up.
5. Keep aware of nonverbal overtones, such as the mood of the audience, the flow of the presentation, graphics, and delivery of the message.
6. Be a good listener.
7. Remember that actions speak louder than words.
8. Neither overestimate nor underestimate the receiver's knowledge of the topic.
9. Avoid words that antagonize the audience.
10. Give reasons that are meaningful to the receiver, not the source.
11. Get feedback.
12. Work on building a long-term relationship with receivers.

Adapted from Fraser P. Seltel's *The Practice of Public Relations*. Charles Merrill Publishing Co., 1980.

Naturally, communication is less important than performance. Organizations must back up what they say with action. Slick brochures, thoughtful speeches, intelligent articles, and good press capture the public's attention, but in the final analysis, the only way to obtain continued constituency support is through proper performance.

OBJECTIVES OF PUBLIC RELATIONS PROGRAM

Public relations activities have five major objectives:

- *To obtain favorable publicity.* A large part of any public relations program is to get maximum "good new coverage" in newspapers, magazines, and other media. Press releases about new or modified products, business expansion plans, personnel promotions, and social contributions are sent to editors in an effort to obtain favorable mentions. Many public relations managers feel that a good news story is more effective than regular advertising because people are more inclined to believe statements that are not paid for.
- *To have the organization accepted as a "good citizen."* Many organizations consider it good public relations to develop community respect for the organization as a good citizen. To achieve this objective, they may "lend" executives to community fund-raising drives, operate programs for the unemployed, donate products to charity, and take the lead in various civic improvement projects.
- *To identify the organization with education.* Public relations-conscious organizations recognize that our society believes strongly in education. Hence, they often seek to develop goodwill in the educational community by providing scholarships, paying all or part of employees' education, offering films to schools, sponsoring open-house events for students, and providing speakers for special events.
- *To humanize the organization.* Too many large organizations seem aloof and too busy to care about the individual. To counteract this feeling, many organizations sponsor an assortment of activities designed to convey the impressions that they are just a large family of ordinary people. The name *Betty Crocker* used on some General Mills products is an example of a successful effort to give a business giant the human touch. Betty Crocker, actually a department of 50 people, receives several thousand pieces of mail daily, each of which is answered in a personal manner.
- *To counteract rumors of negative publicity.* On occasions, public relations campaigns are used to counteract rumors by clarifying the organization's positions on an issue or to win support for its stand in a public controversy. Public relations

work may also be needed to affect negative publicity about some of the organization's operations by emphasizing the organization's positive contributions to the community. Organizations that say nothing or issue denials when they are attacked by the media or the general public are often presumed to be guilty or at least have something to hide.

Contribution of Public Relations

It should be emphasized that public relations goes far beyond simple publicity. Public relations contribute to the following:

- Launches of new products and services.
- Assisting in the repositioning of products and services.
- Promoting interest in products and services.
- Influencing specific groups and target audiences.
- Defending services and products that have encountered constituency problems.
- Improving the organization's image in a way that favorably projects its services and products.

Implementation of Public Relations Program

It is generally accepted that those in charge of public relations have eight techniques they can use to implement an effective public relations program:

- *News*—a major responsibility of the public relations personnel is to create positive news about the organization's products and services. Often public relations persons can create events or activities that will make good news copy. Getting the various media outlets to use the press releases, attend press conferences, and write feature stories requires personal skills: making stories interesting, well written, timely, and attention-getting. It also requires the cultivation of those in the media, such as television news commentators and newspaper editors.
- *Public service activities*—organizations can improve goodwill by contributing money and time to good causes. Examples of this practice are the National Football League sponsoring advertisements on television for the United Way, beer companies that sponsor ads against drunk driving, or when Nike sponsors recreation programs in the inner city. All

of these programs help to maintain a good public image for the organization.
- *Events*—organizations can draw public awareness to new products or services by sponsoring special events. One local recreation and park department sponsored a fitness fair just before the beginning of a community-wide fitness program and the construction of a new recreation center. McDonald's sponsors a number of special events that aid the disabled. Many football bowl games are sponsored by such enterprises as Blockbuster Video, John Hancock Insurance, and UPS. All of these special events create a positive awareness for the sponsor.
- *Telephone information*—many organizations are providing direct contact from customers through special telephone numbers and the use of 800 numbers. The IRS has established a hotline to answer taxpayer questions. City park and recreation departments provide direct-line services for citizens who want information about ice skating, swimming registration, and other program services. A number of chambers of commerce provide hotlines that provide information about special events in their service areas.
- *Audiovisual material*—many organizations develop video and audio cassettes, movies, and slide presentations to promote a positive image of their agency. These materials usually cost significantly more than written material, but in most cases are much more effective. Audiovisual material can provide a firsthand observation of a new product or service. These materials should be developed with great care and professional advice. Poorly developed audiovisual materials can detract from the organization rather than project a positive image.
- *Identity materials*—all successful organizations make materials such as brochures, flyers, and signs that create positive images for the organization available to its public. Some of the most widely used materials are unique logos, special stationery and business forms, business cards, and even uniforms for organization's employees. The organization's identity becomes an effective marketing tool when they are properly designed and are attractive. The organization should employ a professional design expert to develop attractive and colorful organizational symbols. Organization administrators must play an active role in identifying the desired image.

- *Oral presentations*—speeches to clubs, organizations, and the general public can create a great deal of goodwill and result in a positive image for the organization. Oral presentations have created confidence and credibility for the Remington Shaver Company. Lee Iaccoca's convincing remarks influenced Congress to provide loans that saved the Chrysler Corporation. Public officials give talks to Lion's Clubs, Rotary Clubs, taxpayer groups, neighborhood associations, and government officials. All of these speeches assist in creating a positive image for the organization.
- *Written materials*—organizations depend on written materials to inform and influence their public. These materials include annual reports, organization newsletters, magazines, and articles in newspapers. Many private companies go to great lengths to develop materials for stockholders. Government agencies recently extended their effort in developing reports and informational materials to influence the citizens toward a positive image.

SUGGESTIONS TO IMPROVE YOUR PUBLIC RELATIONS PROGRAM

- Those in charge of the public relations program should start with gaining complete knowledge about the organization, its offerings, and market direction. They should undertake research to understand consumer behavior, demographics, and psychographics.
- When evaluating the public relations alternatives, the public relations director must seek answers to the following questions: Is the strategy aimed at the right market segment? Will it attract a substantial amount of media coverage? Will it be substantial over an extended period of time?
- It should be emphasized that public relations can and should work in conjunction with advertising, promotion, and personal selling to introduce new products and services, to promote product and service modifications, to maintain "cutting edge" products and services, and to help ward off trouble in times of crisis or emergency.
- Public relations personnel must be prepared to demonstrate a return on their public relations investment.

- Evaluation of effectiveness must be integrated into the public relations process.
- Agencies must increase public relations budgets so they achieve their desired impact. The public relations program cannot be funded as a "stepchild" of the advertising budget.
- Public relations personnel must learn to use marketing language and methods. They must undertake research to know about product and service positioning and future directions of markets.
- The most important factor in developing the success and growth of a public relations program is the acceptance by all personnel in the organization of its importance in meeting the agency's goals and objectives.
- In many cases, rapid and effective response to a negative situation can minimize the damage to an organization's image. Caution: an attack on the media may only keep the issue alive longer than it would otherwise. Aggressive retaliation may irritate the media to a point where they refuse to work with the organization.
- A simple way to gauge the effectiveness of the public relations program is to ask where the constituent heard about the program or service. Questions such as this one can help target the right public relations in today's highly competitive business environment.

SUGGESTED READINGS

Baskin, O. W., & Aronoff, C. E. (1988). *Public relations: The profession and the practice* (2nd ed.). Dubuque, IA: William C. Brown Publishers.

Corrado, F. (1984). *Media for managers*. Englewood Cliffs, NJ: Prentice Hall.

Culver, D. S. (1988, April). PR booming despite self-inflicted woes. *Marketing News*.

Gruing, J. E., & Hunt, T. (1984). *Managing public relations*. New York: Holt, Rinehart, and Winston.

Chapter 19

THE STRATEGIC PLANNING FUNCTION

Don't be misled by a fancy report in a colorful spiral binder and lose sight of the unmet needs that have been articulated on paper only—another impressive report to grace our bureaucratic bookshelves with only a printer's or consultant's bill demanding any real attention.

Strategic planning is becoming an increasingly important top-management preoccupation. This kind of planning, which some organizations call long-range planning, establishes the direction in which the organization will go. It is a formalized procedure that considers such decisions as whether to inaugurate a new program, restructure organizational responsibility, or embark on an expansion program.

IMPORTANCE OF STRATEGIC PLANNING FOR LEISURE SERVICES

Planning is directly related to action. It forces us to think beyond the present and to consider possible actions required in the near and distant future. Organizations need to develop the capability for long-range, or visionary, planning that is essential to the resolution of more complex, long-term problems. Planning accomplishes two vital tasks: (1) it provides foresight to anticipate future needs, and (2) it generates tools for more immediate decision making.

Planning is especially useful in making decisions because it requires us to learn from the past, consider the present, and prepare for the future. There are logical steps to planning such as:

1. Understanding the problem, need, or opportunity involved.
2. Determining what end result is desired.
3. Formulating and evaluating alternative ways of achieving the desired results.
4. Selecting and following the appropriate actions.
5. Evaluating the results.

Through such steps, an organization can comprehend the issues, alternatives, and implications of a particular decision. Through the planning process, people can learn about the social, physical, natural, and economic conditions and interrelationships in their community. Planning can also help an organization decide upon and obtain the best possible results given limited resources; it helps people allocate scarce resources and save money, energy, and natural resources. Planning also provides a common reference for evaluating progress the organization is making toward achieving its goals.

A recreation and park system has a good chance of being successful if certain "fundamentals" are observed when preparing and implementing a strategic plan.

The following are eight key fundamentals to be observed:

1. Make planning an indispensable tool for decision makers.
2. Involve participants.
3. Maximize public-private cooperation.
4. Set reasonable goals and objectives.
5. Develop practical techniques to implement recommendations.
6. Form a rational basis for priorities.
7. Provide sound management.
8. Evaluate effectiveness of recreation plans and services.

These fundamentals are part of a rational, comprehensive, and forward-looking approach to recreation, park, and leisure planning.

In simple terms, strategic planning is the recognition of an existing or anticipated need and the devising of specific steps for fulfilling that need. The primary motivation for planning is sensitivity to and awareness of a particular problem, dissatisfaction with present conditions, and a desire to change them. Planning should be initiated because of a real need for action in a given situation. We live in a time when change for change's sake is often considered a virtue. The future beckons us, so we indiscriminately topple the past in an effort to reach a millennium. An early stage of any plan is to determine through preliminary examination whether change is indeed needed. Once the organization is assured of the necessity for action, then it becomes essential to get everyone involved.

Planners determine in a formal and detailed way what actions are necessary to implement a particular leisure service or reach a certain goal. Short-range plans are used to determine what an organization's goal might be for leisure services, and a long-range plan operates to implement these services. Whatever the type of planning need, the process is ongoing, not completed when a written report is submitted or a model closely followed. A written plan is the framework for action, not the goal.

The Manager and the Planner

Many organizations now have individuals and, in some cases, departments responsible for developing strategic plans. These departments are staffed by specialists such as architects, market researchers, statisticians, systems analysts, demographers, geographers, and social scientists. Over varying periods of time, strategic plans are produced and supported by vast amounts of data that have been carefully documented and analyzed in the best scientific tradition. Despite this systematic approach, there is some evidence that management does not always use the plans its planners have developed. Yet, it is only fair to say that if some planners are dissatisfied because their plans are often contradicted or ignored, many managers are unhappy because they feel the plans frequently are impractical and irrelevant.

To a large extent, the dilemma is inevitable because of the way in which each job is conceived.

The traditional view of managerial work is contained in the writings of Luther Gubick who, in the 1930s, presented the acronym POSDCORB—planning, organizing, staffing, directing, coordinating, reporting, and budgeting. The manager is depicted as an all-powerful individual, continually pushing ahead with an ever-better organization strategy. It is obvious that the POSDCORB image is not very helpful. The manager's job is far more complex and far less glamorous than this acronym indicates.

Conversely, the planner's day differs considerably from that of the manager. The information, the telephone calls, the requests, and problems that constantly interrupt the manager are not a central part of the planner's concern. The planner has one long-range and well-defined task: to make plans. Thus, he is not disrupted from the main task. Because the planner's job is unrelated to the day-to-day operating problems of the organization, he is removed from the natural flow of information. The planner is not in touch with ongoing operations and, therefore, is unable to appreciate the daily pressures that act on the manager. As a result, the planner cannot understand why the manager treats planning so lightly. When the cry goes up that the organization needs to plan, the cause has inevitably been a conflict between decisions made at different points in time. For example, when management discovers that a new park is needed in one area of the community, someone in the planning area points out that facility planning would have taken into account the decision to embark on a new park.

The Difference between Managing and Planning

Factor	Manager	Planner
Work pressures	Many demands on time; no chance to delve	Much free time; one job—to plan
Decision-making method	Intuitive; implicit	Analytical; explicit programs
Authority	Formal and real; must choose among values	Formal authority only with factual questions
Information	Nerve center	Lacks effective access to channels, but has time to collect information
Feedback	Short run, adaptive	Long run, reflective
Timing	Needs to react to unanticipated problems, chance opportunities	Neglects timing factors
Strategy making	Strategy evolves as precedent-setting decisions are made in stepwise fashion	Integrated strategic plans created at one point in time

In contrast to the manager who must act quickly and intuitively, the planner has the time and training to use a programmed or systematic approach to the making of strategy. James Hekimian describes this in the previous illustration.

Include Line Managers

Strategic plans made solely by staff planners cannot be expected to have a high survival rate. Even if the plan is sound, it is not likely to be respected and followed. Effective planning goes through several phases: stock taking, conceptualizing, fact finding, weighing alternatives, decision making, implementation, feedback, and review. These phases require different combinations of staff and line skills. Staff management can provide the legwork, some of the analysis, and the mechanics of presenting the data. But the thought synthesis, judgment, and implementation steps require the practical experience of the general and line officers who not only must understand, but must use the resulting plans. Without the full participation and enthusiastic endorsement of line management, the plans are likely to be stillborn. Participation by line managers thus focuses attention and develops the commodity of thought and interest that encourages working together for organization success.

WHY PLANNING FAILS

Organizational planning has not been integrated into the organization's total management system. If planning is a process, it must mesh with the pressures of organizing, leading, and controlling. The effort managers devote to planning is determined by the importance they attach to planning as a means of managing. The lack of understanding of benefits of planning confidence, participation, resource allocation for planning, use of plans by executives, and control of what was planned seem to be an impediment to the effectiveness of planning systems in many organizations. An equally difficult area occurs in organizations in which managers plan but do not act. Planning is managing. It is one of line management's key functions. Without understanding this concept, one often develops paper plans only, and organization plans become nothing more than a "paper tiger."

Planning must be based on philosophy, inputs, process, outputs, and system. Often one of these dimensions is neglected for others. Inputs may be poorly assembled or incomplete, or planning may be insufficiently conceived as a process; the planners may not review the plans often enough or make the plans adaptable to changes in the environment.

In one large park and recreation organization, the planning dimensions were poorly understood and insufficient resources had been committed to the planning department. The plans were theoretically formed by the agency and then meant to be reviewed by upper management. Following this process, a synthesis was to occur. Based on this analysis, a total plan for the agency was to be established. The planning department, however, consisted of only two people plus a secretary. It could not do all the work expected. The agency realized that planning was not just a tool or technique by way of managerial living. Training sessions were organized to inform management personnel how to approach planning and to explain its many dimensions and the involvement necessary for it.

Management at different levels in the organization has not properly engaged in or contributed to the planning activities. In many organizations, lip service has been paid to the planning function. Top management has, in words, recognized the importance and desirability of formal planning, but the necessary action and involvement have not resulted. In many cases, top management has embraced planning as a panacea to all organization ills. There has been a false acceptance of the planning function, but appropriate steps to ensure its success have not been taken. The general tendency is to set up a formal planning staff to whom planning has been delegated. Because planning is such an integral part of management, it must be the ultimate responsibility of line management. Planning attempted by staff alone is doomed to failure.

The responsibility for planning is often wrongly vested solely in the planning department. Often, line management abdicates its responsibility to a planning department. The plan developed by the planning department is often removed from the reality of operations and is either too complicated or simply irrelevant. Line management then categorizes the plan as being an "ivory tower" and refuses to use it. The plan becomes meaningless and in vain.

Management often expects that plans will be implemented as they are initially developed. Planning must be a continuous, dynamic process, reflecting changes taking place, particularly outside the organization. A given plan is static. Reflecting the premises and assumptions, the forecasts, and the general inputs available at the time it was developed, the plan may be out of date fairly rapidly after development. There will be many changes before getting to a predetermined point in the future. Plans will change and we need to have flexibility in the planning process. It is not wise to do a plan every five years—planning is constant. The plan is only an indicator of what the agency should be doing and the direction in which the agency should be headed.

When beginning the formal planning process, we often attempt to do too much. Many organizations launch themselves wholeheartedly into their planning system, in many cases lacking adaptations to the unique aspects of the organizations.

Management fails to operate by the plan. As long as the plan developed is the best available and appears right under the circumstances, the plan should be used. In cases where planning as an overall management philosophy has not been fully absorbed, the tendency is to engage in short but hectic periods of plan development to meet certain organization requirements. After much frustration, a plan is finally put together and submitted. At that point, the tendency is for the executive to say, "I'm glad that's over! Now we can get back to business again."

The extrapolation and financial projections are sometimes confused with planning. To the extent that the future will be the same as the past, one can extrapolate past action and results into the future and argue that this scenario is what will happen. However, when the future differs significantly, such extrapolative thinking becomes inadequate. In areas of technology competition, the changes taking place are of such a fundamental nature that a new approach to planning may be required.

Much planning today can be described as extended budgeting. In many organizations this has meant planning, starting with today and going five years into the future. The concern has largely been with the financial and quantitative results with which one is concerned in one-year budgeting. However, the underlying causal factors that determine the financial results often have not been adequately covered. More fundamental, qualitative thinking is required.

Inadequate inputs are used in the planning process. Much of the information readily available in the organization is not suitable for planning purposes. The information usually generated is control oriented and has not been developed for decision-making purposes. The information needed in planning usually requires a substantial alteration of the organization's management information system. For strategic decision making, where the concern is with relating the organization to the environment, environmental surveillance is required.

SUGGESTIONS TO IMPROVE THE STRATEGIC PLANNING PROCESS

- The good planner should remember the useful elements of all the information that is available and is able to reject inappropriate or useless information.
- The good planner is flexible in her thinking.
- Remember, financial people have difficulty in thinking long term. They feel comfortable thinking a year or two ahead. They prefer to report what has happened—not what is going to happen.
- Planning cannot eliminate risk and uncertainty. However, it is the best way of selecting the right risks with the least uncertainty.
- Planning must result in something more than words on paper. It has to influence behavior.
- In order to develop a sound strategic plan, there needs to be a clear understanding of the plan on the part of line management with key executives participating in the process from the outset.
- Effective strategic planning does not stop when the strategy is identified. The work actually begins as action plans need to be developed and carried out.
- When developing strategic plans, ask these six important questions:

    1. What is the present nature of our business?
    2. What are our opportunities?
    3. Where do we want to go?
    4. What must we do to get there?
    5. What must we do right now?
    6. What progress have we made and have conditions changed?

> - Keep the strategic plan simple, concise, and non-bureaucratic. The plan should be extensively presented, reviewed and discussed. It should outline how the organization will proceed and focus much less on what it will do.
> - The strategic plan should produce a coordinated, conscious effort to achieve the goals and objectives of the organization.

We should not expect perfection the first time around. It is a process requiring experience and the mutual education of all management personnel. The plan's accuracy and practicality should improve from year to year.

SUGGESTED READINGS

Brodurin, D. R., & Bourgeois, L. J. III. (1984). Five steps to strategic action. *California Management Review, 26*(3), 176-190.

Donaldson, G., & Lorsch, J. W. (1983). *Decision-making at the top.* New York: Basic Books.

Hainesworth, B. E., & Wilson, L. J. (1992, Spring). *Strategic program planning public relations review, 18*(1), 9-15.

Halten, M. L. (1982). Strategic management in non-profit organizations. *Strategic Management Journal, 3*(2), 89-104.

Nutt, P. L. (1984). A strategic planning network for non-profit organizations. *Strategic Management Journal 5*(1), 57-75.

Sink, D. S., & Mize, J. H. (1981, May). The role of planning and its linkage to action in productivity management. *Spring Annual Institute of Industrial Engineering Conference Proceedings,* 11E, Norcross, Georgia.

Chapter 20

I have so much to do, there's just not enough time for me to do it all. If you don't manage your time, it will manage you.

Time, as a valuable resource, affects the way we use other personal resources, and unlike other resources, it cannot be stored or hoarded. Are you spending time the way you really wish or are you harried, concerned with tasks you do not really want to do? Take a few minutes to answer the following questions that will provide a perspective of your use of time:

- How many hours per week do you work at your job?
- Do you feel productive?
- Do you take work home evenings or weekends?
- Are you giving your family as much time as you would like?
- Do you have time to keep fit?
- Do you have time to take vacations and long weekends?
- Do you have time for a favorite hobby, sport, or a good book?
- Do you feel you must always be doing something productive?

Time management is, in essence, management of all activities engaged in during our life. Self-management is like managing anything else—it involves planning, organizing, implementing, and controlling. If the answers you gave to the questions lead you to be concerned about how you are spending your time and if you are serious about making improvements, then you must be committed to making changes. Revamping time-use patterns is like overcoming any bad habit.

In devising a time-use strategy, you must accept that it will take a while to learn how to use time effectively and that the principles of time use are not universally applicable. Although many time-use principles are simple commonsense precepts, "uncommon" sense is often required to make a plan that fits your needs. Learning to make better use of time involves pursuing three key questions:

- Where does my time go?
- Where should my time go?
- How can I use time better?

By answering these questions carefully, you will be better prepared to control time. Frequently, time-management problems arise from poor work habits. Therefore, you will need to analyze *how* you spend your time *and how* you waste time.

As a result of your analysis, you are likely to find three revealing facts. First, you are probably spending 80% of your time on activities that produce only 20% of actual benefit. Being involved with low-priority activities could explain your inability to accomplish more important tasks. Low-value tasks are generally easy to do and give a sense of accomplishment, but they are illusory and time-consuming as major goals.

Some managers find it helpful to use various techniques for establishing and revising priorities. For example, use index cards and jot down one task per card. Then sort the tasks according to your priority system; reorder if necessary. On priority schedules or "hot sheets," list the jobs to be done in order of importance. Or use dated tickler files to keep track of reminder slips, correspondence, and reports, among other things, and note an action date. Finally, record your schedule on display boards. You will have an overall view of your activities and be able to see them in relationship to each other.

A useful exercise to begin the process is to list "my five major goals for the next six months" and put them in priority of importance. If you have not done this activity before, you will probably find it difficult. Developing goals requires discussion with others in an organization. In the process, resources needed to attain a goal will be revealed.

It is helpful, at this point, if not before, to use various planning guides. Keep a *daily calendar* to schedule meetings, holidays, luncheons, social events, and conferences. Many managers keep

12-month calendars for such activities. Use a *to-do list* when faced with an unusual or complex project. List everything that must be done. It gives you a visual picture of what you have to do. Cross off each line as it is accomplished. When you have many things to do, you can save time and effort by writing them down, rather than trusting your memory. Make lists every day.

Delegating your work is one of the most effective ways of "expanding" time. Successful delegation does not just happen—it demands time, effort, and persistence to develop and maintain. Worthwhile delegation requires thought, careful planning, knowledge of staff members' talents and competence, and effective communications, as well as a willingness to take risks.

The prime reason for delegating is to involve others, because a manager cannot do everything himself. Another reason is that others often do a better job! Studies have shown that managers generally fail to delegate because they believe if a job is to be done properly, they must do it themselves.

If you believe that the only way to get something done is to do it yourself, then you are probably overwhelmed with work while your staff enjoys less harried work schedules. Try to break the "do-it-yourself" syndrome. Delegation does not mean dumping a task on another; rather, it involves responsibility for making sure that the person has the requisite skills and knowledge for the job.

Another problem that contributes to inefficient time management is constant interruptions that are often considered unavoidable. For example, we assume that every telephone call must be accepted, whether or not we are busy. You will spend less time on the phone if you plan to make all important calls during one period of time. Get all the numbers together, note the topics you need to discuss, and pull together files and correspondence that might be needed. Dial numbers yourself; to involve a secretary wastes two people's time. Immediately after hanging up, make a note of what was discussed and decisions made.

For some managers, the length of phone calls is as much a problem as the number of calls. Your job and the nature of your relationship to people will require a certain amount of informal conversation. However, if you spend five to 10 minutes discussing general topics, then you are wasting time. A telephone can be your biggest helper, as well as a major time-waster.

According to a recent survey, the average manager is interrupted about every eight minutes of the working day! A visitor drops by or someone buzzes on the intercom. Some managers reported that because of constant interruptions, only two hours of essential work is accomplished in an eight-hour day. Visitors are a troublesome interruption if they arrive when you are busy with an important matter, if the intrusion ruins concentration, and if they cause things to fall behind schedule.

You should develop a procedure that fits your needs for ordering your desk and handling paperwork. Give your secretary instructions on screening your mail—monitoring trade journals and magazines and disposing of those you do not read. Useless items should be discarded before they cross your desk. Some materials can be routed to others.

To stem the flood of paperwork, decide what can be streamlined or eliminated. Throw out junk mail, cancel unused subscriptions, and have mail routed directly to others. If possible, handle each piece of paper once; do not pick up a piece of paper unless you plan to do something with it. For example, a complaint does not go away simply because the letter has been put aside. Move the paperwork along instead of letting it stack up.

Getting a handle on time management may take some time—and much patience. There is a tendency to work on minor tasks first, with the idea of leading up to larger projects. What often happens is that the more difficult jobs simply do not get done, as too much time is devoted to unimportant activities. By the time a manager gets to the tougher jobs, she is too tired to work effectively. You need to reverse the process: start with the important work when your energy level is highest and follow your list of priorities.

Putting things off is easy. In fact, people generally do the things they enjoy first and procrastinate on those they do not. You need self-discipline to overcome procrastination. Break seemingly overwhelming jobs into bite-sized pieces that are more palatable. The project will seem less burdensome. Also, unfinished work is a better motivator than unstarted work. When you have started a job, you will be more likely to complete it because you have made an investment of time.

Finally, learn to say "no." Someone is always asking for your time. Instead of being honest and saying "no," the tendency is to hedge, to accept a responsibility you neither want nor have time for. Saying "no" requires some courage and tact, but you will be pleased with the positive results when you have learned to use that simple word effectively.

SUGGESTIONS FOR IMPROVING YOUR TIME MANAGEMENT

- Do not conduct a meeting without an agenda. There should be a written agenda for all scheduled meetings. This will ensure that people will come prepared for meaningful discussion.
- Delegate as many tasks as possible. Do not try to do everything yourself. You not only overload yourself with others' work, but also deny them the experiences.
- Do the most important and high-priority jobs first. Impose deadlines on yourself. When you achieve them, reward yourself.
- Single out the critical from the unimportant tasks. Take time to set priorities. Balance short-term demands and long-term objectives.
- Understand that everything takes longer than you think. Allow "slush" time for the unexpected and uncontrollable. Leave a minimum of 20% of your day unplanned.
- Recognize what you are presently working on is probably not the most important. Do not lose yourself in details and minor tasks and forget the big picture.
- Do not let perfectionism strangle you. If you put off the completion of the task until you have it perfect, you probably will never get anything accomplished.
- At the end of each day, make a list of the tasks you would like to accomplish the next day. Arrange them in the order of importance. Categorize *A* tasks as high value, *B* tasks as moderate value, and *C* tasks as low value.
- Take a few moments and make a list of the most frequent time wasters. Rank them on a sheet of paper, determine the ones you can control, and then develop a plan to do something about it.
- Select a portion of your day and make as many of your calls as you can at one time. Have material ready so you can discuss the specific concerns of each caller.

SUGGESTED READINGS

Bannon, J. J. (1982, November). Reduce stress and increase effectiveness through time management. *Park and Recreation, 17*(11), 38-43.

Ferner, J. D. (1980). *Successful time management.* New York: John Wiley & Sons.

Mackenzie, R. A. (1972). *The time trap.* Chicago: AMACOM.

Mackenzie, R. A. (1975). *New time management method for you and your staff.* Chicago: The Dartell Corporation.

Stalk, G., & Hout, T. M. (1990, November/December). How time-based management measures performance. *Planning Review, 18,* 26-29.

Part II

Human Resources

Chapter 21

CAREER PLANNING AND DEVELOPMENT

If you don't plan and manage your career, you won't have one.

The world of work has undergone a sea of change that can result in careers entering rough waters. Planning a career is far more fluid and transitory than ever before. During this decade, many will go from position to position, company to company, and career to career. There no longer will be long-term relationships with a fixed group of employers where you will be "taken care of."

In order to have a successful career, you must always be aware of what is going on in your field, keep an eye out for professional connections beyond your current employer, be persistent in selling yourself, and be flexible and self-confident.

When done thoroughly and wisely, personal career management can be used to launch your professional future, enhance your reputation in your field, and increase your visibility as a highly regarded expert. That, in turn, could lead to increased pay, a promotion, a higher position with another organization, or more importantly, it may allow you the flexibility to detour in and out of the workplace.

What is implicit in these changes is that careers must be planned. According to employment experts, changes that must be taken into account include the following:

- *Large versus small firms*—Small business will create many of the new jobs and opportunities. Large corporations will only have to replace employees who have resigned or retired, or to staff new departments.
- *Diversity*—The workforce in this decade will be increasingly female, composed of minorities and older adults.
- *Technology*—Computer literacy and technical know-how will be a requirement for every job, including unskilled ones.
- *Training*—Ninety percent of the jobs created in the 1990s have required education beyond high school. As job requirements keep changing, on-the-job training is necessary. Continuing education outside the workplace is a fact of life, and professions must upgrade licensing and certification requirements.

Individuals are now recognizing that planning their careers is as much their responsibility as it is the organization's. Organizational career development is a relatively new field. It is only in the last decade that organizations have shown an interest in providing career development programs for employees. Many companies recognize they must play an important role in their employees' career planning, and they also realize that people are a strategic and important resource. Employees are becoming more assertive in demanding that the organization recognize their needs for personal growth and career opportunities. Career planning and development is also becoming an accepted human-resource strategy among training and development administrators, personnel officers, and organizational consultants.

The primary purpose of career planning is to help employees analyze their abilities and interests to match personal needs better with the growth and development of organizational goals. Management also sees the objective of career development as a tool through which it can increase productivity, improve employee attitudes toward work, and develop greater worker satisfaction.

Career planning programs must focus on career development rather than just promotion and advancement. Promotion and advancement opportunities are often limited; there is simply not room for every employee at the top of an organization. However, all employees can further develop their skills and can increase their capacities to make better contributions in their present job.

Career development has also been defined as an organized, planned effort by the employee and management that can result in a logical career path for the employee. This process holds the employee responsible for career planning, and the organization is responsible for career management. Here career planning refers to activities that assist individuals in becoming aware of self, opportunities, constraints, choices, and consequences. The employee identifies career goals, work-related experiences, and continuing education that provides direction and timing that will allow the employee to attain career goals.

Career management is an ongoing process consisting of human resource activities, such as recruitment, selection, placement, and performance appraisal designed to facilitate career development.

In the past, an individual's career needs were often driven by corporate needs, but today many people are exercising more personal choice in making career decisions. They may turn down a promotion or resign from a job that is unsatisfactory or does not match personal goals and family needs. Values are changing and so are attitudes about work. People want more security and money, and they are speaking up. They want to be able to learn from work experience and expand their knowledge and skills. In fact, many people are taking on several careers in a lifetime. It is not unusual for individuals to make moves into unrelated fields. They may change directions not only because they are forced to or because they have failed in their jobs, but also because of their widening interests and heightened self-awareness of what they do best, what they enjoy most, and what direction they are heading.

As mentioned, organizations can do much to assist their employees in developing career plans. The organization can assist in the career planning process by providing an environment that is conducive to personal growth, training and development opportunities, and support for career planning programs. Many experts suggest the following for establishing a career development program:

- *Awareness.* Employee development, which should be available to all employees, must have company support. Human resource and other managers should be prepared to explain the program and make all employees aware of the information available.

- *Encouragement.* Every employee should be encouraged to do some basic thinking about his strengths, work preferences, develop needs, short-term/intermediate goals, and long-term goals. This evaluation may involve a structured self-assessment that uses any of the formal inventory tools or perhaps a less formal thought process.
- *Discussion.* A planning session should be held with the employee to discuss the self-assessment, determine strengths, and develop needs.
- *Understanding.* Depending on the employee's interests, capabilities, and the needs of the business, employees may move from one job area to another, either within a facility or to another branch, plant, laboratory, or division—or even to headquarters. Once a manager gains an understanding of the employee's work interests and objectives, it may be appropriate for the manager to provide information about different jobs in the organization.
- *Development Plan.* The preparation of a development plan should reflect activities that can be implemented by either the manager or employee. This plan should be flexible since it has to be subject to business conditions.
- *Follow-up.* In this fast-changing business environment, employee development must be an ongoing activity. The manager and the employee should periodically check progress against the plan. The plan should be reviewed and updated as appropriate. Also, the manager should encourage the employee to reassess goals and her strengths.
- *Employee development and equal opportunity program.* Employee development can be a key affirmative action tool. The manager should encourage minority, disabled, and female employees to participate in the process of self-assessment and the preparation of the development plan.

In the past, organizations have played a minor role in the planning and development of careers for employees. This is due to a number of reasons; however, some of the more apparent involve many managers and supervisors who are poorly equipped with the necessary skills to carry out the counseling required for career planning of subordinates. As a result, career planning decisions between managers and subordinates are meaningless. It becomes almost impossible to assist employees in resolving conflicts among the organization's personal and family goals. Also, organizations fail to provide incentives or rewards to managers who assist subordinates in career planning.

Managers are paid for achieving results; most organizations do not view planning employees' careers as a major goal and, as a result, place this activity as a low priority for managers. Until organizations view this activity as an important function and reward for it, it will remain a problem. Subordinates also do not give much thought or concern to their careers, nor do they have the skills and knowledge to manage them. When time and thought are devoted to career planning by subordinates, we tend to simplify the process.

We think of only immediate job opportunities or pursuing goals that make no sense in terms of long-range career planning.

A discussion of career planning would not be complete without discussing dual-career couples. In 1995, the U.S. Department of Labor predicted that 81% of all marriages would be dual-career partnerships. This is an issue employees cannot ignore because there are studies that indicate that fewer employees are willing to relocate without the support, agreement, and assistance from their spouses. Recruiting in tomorrow's labor market will require significant attention to the needs of dual-career employees. They will demand and get special benefits such as on-site child care, longer maternity leaves, flex-time, flexible benefits, assistance in career planning and management, and employment for spouses at the same location.

Traditionally, the male species has been the one whose career location has guided the family, but with women gaining more visibility in the work world, the "trailing spouse" is just as likely today to be the husband. Many of the benefits discussed here are presently being offered to upper-management workers, but in the future, they will be common in those companies that attract and keep top-quality employees.

Career counselors need to spend a great deal of time advising young couples in their career paths. Employees should be encouraged to set priorities and define as precisely as possible the direction they are headed. There are three sets of priorities to be considered: (1) employees' personal priorities, (2) their priorities as a couple, and (3) their priorities as a family. Couples should be encouraged to develop open communication and to discuss their careers honestly and openly, rather than wait for a crisis to begin the discussion.

Today, organizations are providing more advice and counseling to dual-career couples. It is good business to do so if they are going to retain top-quality personnel. Experts who advise dual-career couples suggest the following areas need to be considered:

1. *Commitment to the relationship.* Time demands of career and family can be a constant problem. Each member of the dual-career partnership must make a commitment to the relationship itself. Young people starting out in their careers may have difficulty understanding this. Practicing time-management principles is absolutely essential. It is suggested that specific time be set aside each day that is devoted to the relationship.
2. *Children of the dual-career couple.* If there is any one area that can be a major source of stress and tension for the dual-career couple, it is having children. In many organizations, parental leave and child care have become a major personnel policy issue. Recently, AT&T created a $5 million fund to develop community child-care centers as well as to provide services for employees' elderly dependents. Unpaid parental leave for both mothers and fathers is also being extended from six months to a full year. Careful planning, as well as agreement by the parents as to who is going to assume the major burden and under what conditions, is needed.
3. *Changing job locations.* Moving is traumatic under any circumstances. The sheer logistics of selling real estate and contending with the movers can be overwhelming. Leaving the familiar for the unknown can create fear, loss, and depression. Combine this stress with success-oriented couples and their egos, and there could be a potentially volatile situation. More and more companies are providing job-search assistance for the dual-career spouse. It is estimated that half of corporate transfer problems involving dual-career moves concern how to accommodate school-aged children. Parents are reluctant to disrupt their children's education, particularly during the school year. Their welfare must be taken into consideration in the relocation process.
4. *Who does what at home.* Dual-career couples need to decide how they will share the home responsibilities. In the past, these were often decided along gender lines; more recently, however, there is acceptance of these responsibilities by each

member of the dual-career couple. Some dual-career consultants report that both men and women feel considerable dissatisfaction with how these responsibilities are addressed on a day-to-day basis. While society now expects more from women in the workplace, women are still doing the bulk of the domestic work at home. Career couples should be advised that whatever arrangement is worked out, it should be agreed that it is fair. There is no right or wrong way to do it; it simply must be mutually agreed upon between the spouses.

5. *Competition between career couples.* Competition will become more apparent when career couples are in the professional ranks, more so than when each is in a blue-collar position. It is also more likely when both individuals work in the same industry, or even in the same company. To some degree, such competitiveness is natural and to be expected. More controversy can be generated by salaries or bonuses, a promotion, or perhaps one spouse's job being more demanding or more visible. It is important that the career couple acknowledges these competitive feelings and deals with them rather than deny them.

6. *Know what is important.* Dual-career couples should be encouraged to think in terms of setting priorities. They should be encouraged to discuss these in advance rather than wait for a crisis to occur. By doing so, they will know what is important to them. For dual-career couples, this prioritizing takes flexibility and communication. The important goal is for couples to create a win/win situation where each party comes away feeling good about him- or herself and their relationship.

In the next decade, organizations are going to need more employees—especially more highly skilled employees—than ones currently available. Competition for employees will increase and work environments designed with dual-career couples in mind will be necessary if organizations are to be competitive in attracting the best talent.

SUGGESTIONS FOR IMPROVING CAREER PLANNING & DEVELOPMENT

- The organization should advise and assist its managers and supervisors in the planning of training and career development opportunities for subordinates. Recommendations should be made relative to training programs and courses internally or externally that could enhance the employee's development at a particular point in his career.
- Each individual must assume the responsibility for her career planning. One of the biggest career blunders that workers make is assuming that once they have landed a job, career planning is finished.
- Take charge of your own career planning. Ask yourself what really makes you happy. Examine your interests, skills, strengths, values, work, and lifestyle preferences. Match prospective career paths to those professional and personal life choices.
- Career planning should be started long before college graduation and should be thorough enough to include at least the first 10 years out of school. The process should be repeated at progressive stages throughout your work life.
- Hands-on experience and a thorough understanding of your career field is essential. Find a trade group or professional society in the field of your interest and follow up with the people with whom you make contact.
- When an individual enters the organization, he should be informed as to his role in career planning and development as well that of the organization. This information will increase the likelihood that such action will occur.

Chapter 22

CEO–BOARD RELATIONS

The argument is not that the CEO should wrest control from the board, but that he or she be given a substantive say on policy. A close and cooperative relationship should exist between CEO and board.

Is the role of a CEO simply one of following the policy dictates of a board? It is commonly assumed so, yet most practitioners know that CEOs *create* policy as well as *implement* the policy decisions of boards. The CEO inevitably is drawn into policy deliberations, no matter how appealing a strict organizational separation might seem.

A board representing the wishes, beliefs, and aspirations of its constituents and working in tandem with a professional has been found most effective for meeting the needs of its constituency. Inherent in such a setup is the assumption that the board represents those served in establishing the goals, objectives, and purposes of the agency, and generally determines the scope of programs. The CEO, through his abilities and those of the staff, provides the professional competence and expertise necessary to assist the board with its policy-making functions—to implement the policies and directives of the board—and to administer the organization on a daily basis. The CEO assumes responsibility for all policies designated and approved by the board and does his best to apply them successfully. However, this situation rarely occurs because the relationship between boards and CEOs often deviates from such a pattern—in many instances to the benefit of all.

How the relationship between a board and the CEO is balanced has always been crucial; today it is even more so in view of the massive, and often confusing, social changes affecting society.

A board should encourage policy recommendations from the CEO or at least recognize that most CEOs are already "policy makers." The *administration* of these policies—the actual daily operations—is a task that requires professionals for effective implementation. Because of their professional knowledge of the situation and their complementary administrative skill, the CEO can be an important agent in the policy making as well.

The expansion of the CEO's function—from implementing policy to *co-designing* policy—already has occurred in many organizations. This new form of responsibility, or amalgam of previously divided functions, enables the CEO to offer stronger, clearer leadership rather than merely following board policy. Of course, this relationship will take time to surface fully in organizations where most semi-independent policy boards are established by law and not easily modified. Yet, we should recognize that even when not clearly stipulated, the CEO is often involved in proposing and formulating many of the policies.

When strict policy control is held by the board, the potential policy-making role of the CEO becomes that of wisely employing strategy to make the organization effective. Policies provide an administrative framework for the CEO; they do not lead her by the hand. Thus, the final interpretation and application of policy is often at the discretion of the CEO, regardless of how closely the board guards its policy-making responsibility.

OVERLAPPING FUNCTIONS

The argument is not that the CEO should wrest control from the board, but that she should be given a substantive voice on policy. A close and *cooperative* relationship should exist between a CEO and the board. A joint policy-making relationship does not imply impingement by either party on the other's prime responsibility; rather, it implies a sharing of knowledge and experience and a logical overlap of functions for organizational enhancement.

The combination of interested and concerned board members and a professionally astute staff can be a catalyst for achieving quality programs. If properly used, a partnership or a "common market" of skills, expertise, and responsibilities between the board and CEO can be successful.

Whatever theory of policy making one espouses, it is important to avoid circumscribing the responsibilities and duties of the board and CEO too strictly. When it is of value to have the CEO recommend policy, he should be allowed to do so. When the board has greater insight into administrative development, its views should be considered. However, this overlap in function should include a clear perspective of the major concerns and responsibilities of each party. It is not suggested that the CEO spend all of his time creating policy, nor that the board spend all its time on administrative issues. It is recommended, however, that the arbitrary separation between these functions be loosened.

Although the line drawn between the CEO's policy and administrative functions should not be too firm, it should *not* be the concern of the board to administer or implement policies for the organization. Reliance by the board on the CEO and his staff for these tasks is a sign of confidence. Such confidence is, of course, greatly enhanced when the CEO has formal input into the policies in the first place.

The overlap for the CEO between his primary administrative function, his administrative function, and his actual policy-making involvement, if agreed to by the board, can be exceedingly rewarding for the entire organization. In many organizations, the director can, and does, do more than administer. The board should encourage the CEO's expansion of responsibility and involvement.

Another way of viewing CEO/board relationships, often cited in public administration, is that the board generally makes *value* decisions, and the director makes *factual* decisions formulated on these values. That is, the policies of an organization reflect primarily the values that the board views as essential to its public charge. The factual decisions or policies suggested by the CEO generally draw on management, with the organizational values of the board serving as a base.

A board and CEO must, therefore, resolve any differences in values and objectives prior to participation in policy formation. Constituents bring to the board service values that may not be consonant with the values of a professional CEO. Frequently, these divergent values make it difficult to separate value and factual decisions clearly because the values themselves are in question. However, awareness that such differences may exist is a first step toward resolving them.

Critics of this position feel there is always the danger of a powerful CEO relegating the board to a rubber-stamp capacity. The CEO should engage a free flow of ideas and use the expertise of her board members to the utmost. On the other hand, she should also strive to obtain a voice in policy making when the board may have relegated her to a rubber-stamp position by underestimating or misconstruing her true value.

LEADERSHIP

Strong leadership by the CEO does not mean dictatorship. It means that mutual confidence must exist between the board and CEO regarding their abilities to carry out their prime tasks, and that both parties must recognize the strategic role of the CEO's policy.

A major task of the CEO, and one that justifies her leadership, is to keep the board fully informed on all matters that concern the organization. If the board is not properly advised or is ignorant of some facet of an issue under consideration, the CEO is generally at fault. It is her duty to be certain that the board receives as much background information as is necessary for a considered judgment.

The CEO is expected to be more informed of the issues of his profession than the board, because this is his career field and the prime reason for his presence in the organization. However, a board is not a group of amateurs and should not be viewed as such. Information should be available for board members to whatever degree they desire. There is no point in the director cultivating a "dumb" board, because he is the one who will ultimately lose. Remember, the board has the final say in policy matters.

The board, on the other hand, should expect and demand valid and substantive information for use in reaching decisions or making policies. The board should never determine or evaluate policy recommendations without a thorough comprehension of the issue. An effective CEO will ensure that information is available well in advance of board meetings. She should further assure board members that they will always receive accurate information. A board cannot function effectively unless it receives reliable information.

How does a CEO cultivate a well-informed and helpful board? The simplest way is to channel all pertinent information on a given subject to the board verbally by written reports and statements, with film presentations, or through the use of professional speakers. In addition, board members should be encouraged to attend professional meetings and conferences where they can gather what current information is related. When this is not feasible, board members should receive copies of materials distributed at such conferences or summations of these activities from the CEO (or staff).

Board participation in such activities keeps board members informed about the interests and concerns as well as binding them closer to the CEO in a mutual pursuit. This common interest is more than a desire for *esprit de corps*: it is essential for a successful board/CEO relationship.

Often we hear of the need for more informal interaction and communication within organizations. Here is an example of where such interaction can and does occur in organizations throughout the country. To ignore an overlap in organizational functions is to focus too heavily on clearly drawn lines of responsibility, ignoring the potential benefits and achievements of functional overlap. Organizations should not bind themselves to any tight organizational structure because of theoretical neatness. The workings of an organization should be judged on the basis of the organization's effectiveness, not on the basis of its structure.

SUGGESTIONS FOR SUCCESSFUL CEO AND BOARD RELATIONS

- Once each year, the CEO and the board should meet to discuss their existing and future relationship.
- The CEO should be present at all meetings of the board except when discussing her own employment, salary, or successor.
- The best way to get an apathetic board member to take responsibility is to persuade him to accept an assignment that he understands needs to be done, can be achieved, and to provide recognition.
- The CEO should be the intermediary figure between the staff and the board. The CEO should be concerned with the work of the staff so that the mission of the organization can best be achieved.
- The CEO may designate topics to be discussed between board members and staff, but the CEO authorizes such contacts and knows of their results.
- The board should extend to the executive a great deal of latitude in administrating staff responsibilities. The internal communications should be clear to the board and they should be responsible for any structural changes in the organization.

- It should be understood that the CEO's accountability is to the board, not the officers of the board, nor to board committees. The CEO should be instructed by the board as a whole.
- At no time shall the CEO have full authority or control over the financial situation of the organization.
- The CEO should keep the board informed by providing internal reports, the use of consultants such as auditors, real estate experts, attorneys, architects and engineers, and on-site monitoring and inspection.
- The CEO must be able to rely on the board to confront and resolve issues of government while staying out of management and operations.

SUGGESTED READINGS

Arelord, N. R. (1991, October). Your significant others. *Association Management*, 22-26.

Asman, D., & Meyerson, A. (1985). *The Wall Street Journal on management*. Homewood, IL: Dow Jones-Irwin.

Bannon, J. (1973). Who really makes policy? *Parks and Recreation, 8*(7), 30-31.

Levinson, H. (1981). *Executive.* Cambridge, MA: Harvard University Press.

Schupp, R. W. (1991). When is a union not a union? Good faith doubt by an employer. *Labor Law Journal, 42*.

Chapter 23

DEVELOPING PERSONNEL POLICIES

Employees and their contribution to the organization are affected by the conditions in their work environment. Personnel policies reflect management philosophy and concern for its employees. They should be designed to influence directly and indirectly the match between the employees and their jobs.

Personnel policies act as a guide to actions for organizations, large or small, to achieve their goals and objectives. Objectives establish the tasks that should be done, while policies describe how they are to be done. Therefore, personnel policies refer to standing plans that furnish broad guidelines that direct the thinking of managers about personnel issues. Some of the most common issues discussed in personnel policies are problems such as absenteeism, tardiness, and insubordination. Personnel policies serve three major purposes: (1) to reassure employees that they will be treated fairly and objectively, (2) to assist managers to make rapid and consistent decisions, and (3) to give managers the confidence to resolve problems and to defend their decisions.

The smaller the organization, the more the belief that policies are "understood" by all employees. This may hold true only if the enterprise is a sole proprietorship and the person in charge has an excellent computer-like memory. Management in these settings has been by instinct rather than a clearly written set of personnel policies.

Some companies have a great deal of negativism for personnel policies. This may be true because of some people in the company who understand what a policy is, or who know existing policies are so poorly conceived and written that they are difficult to understand or administer. Policies are often confused with plans, objectives, rules, and procedures. In order to clarify the distinction, the following definitions are provided:

- A *procedure* is a series of steps and/or techniques for accomplishing a task.
- An *objective* is a specific goal the organization sets for itself. It should have a precise result, a deadline for accomplishment, and cost related to achieving it.
- A *rule* is a declaration about the conduct that is acceptable or not acceptable in certain situations.
- A *plan* is a set of tasks established to assist in the accomplishment of a goal or objective.
- A *policy* is a guide to decision making under a given set of circumstances. If administered properly, it should assume consistency and fairness within the philosophy of the organization.

It is important to put personnel policies in writing. Doing so provides a structure for decision making and a reason why particular decisions are made. The information contained in personnel policies is important when explaining an unwelcome decision to employees.

To achieve these purposes, personnel policies should be written and made available to the employees and their families. Written policies tend to be more authoritative than verbal ones and can serve as valuable aids in orienting and training new personnel, in administrating disciplinary actions, and in resolving grievance issues. It should be remembered, however, since first-level supervisors are often involved in administering these policies, input obtained from them can be useful in formulating new policies.

Written personnel policies have a number of advantages to management, supervisors, and employees. The following are often cited by management consultants:

1. All members of the organizations have an understanding as to the potential decisions in circumstances covered by the policies.

Developing Personnel Policies 185

2. Policies produce consistency in decision making. They reinforce employee perceptions of fairness and equitable treatment.
3. Established personnel policies provide advanced information and predictable decisions for situations that are repetitive or that occur frequently. They avoid the need for dealing with questions one at a time.
4. Employee relations become stable and, as a result, organization members know that the company has considered and anticipated problems and has a planned reaction.
5. Personnel policies bring consistency between different types of executives decisions, even where executives move from one position to another.
6. Policies produce an environment in which growth and expansion can develop on an orderly basis.
7. Personnel policy development makes management aware of problems requiring coordination and control. As a result, it requires management to consider the consequence of actions and decisions.
8. Approval by top management of personnel policies makes a useful means to exert its influence upon actions of employees.
9. Personnel policies provide the basis for the organization's commitment to fair employment practices and equal employment opportunity. They may not stop grievances or court actions, but it is certain proof of your attempt to be in compliance with local, state, and federal rules and regulations.

When an organization puts into writing its philosophy and employee expectations, managers and employees are encouraged to act and perform their responsibilities under conditions of orderly coordination and control. Inconsistent treatment of employees is thus at a minimum, and equitable relationships can be maintained.

Organizations vary considerably in whether they consider their policies unchallengeable. Some adhere strongly to policies following a literal interpretation. Others allow a certain amount of flexibility, letting each situation determine the application of each policy. However, in most organizations, policies are firmly enforced and exceptions are made only in rare situations.

Who is the primary benefactor of the organization's personnel policies? The primary users will include the new employee, who will find it useful during his orientation period. It describes to the new employee what rules and regulations he is expected to follow and explains that these rules and regulations are meant to protect the rights of all. Another benefactor is the husband or wife of the employee. By becoming aware of the company personnel policies, the spouse gains the feeling of involvement through knowledge of what the employee benefits are and the environment in which the employee works. The current employee also finds the written personnel policies useful in describing all the benefits as well as the rules and regulations of the company. The general public will also be interested in personnel polices. Many organizations use them as a recruiting tool. Some companies send copies of personnel policies to employment agencies and to schools from whose students they hope to receive applications.

Personnel policies often represent the organization's general character and the nature of the work environment. Rigid policies or capricious handling of policies tends to create resentment and hostility among employees. Systematic and rational execution of personnel policies is essential if they are to be accepted by both management and employees.

Personnel policies should keep all employees informed about company regulations and give supervisors support in enforcing the regulations and policies. Personnel policies should not be thought of as just a list of benefits and do's and don'ts. They should describe the expectations management has for its employees. Written personnel policies can reduce the feeling of insecurity many employees feel. Policies can ensure employees that the organization is attempting to operate efficiently. Equally important, they save management's time for other organizational issues.

Today, many companies have half-written policy handbooks in their desk drawers because they did not have the time to complete them. They felt unsure of their ability to write well or they were not sure of what should be included. Consultants in this area recommend that any company with as few as six employees should have a handbook that includes policies affecting all personnel. They do not have to be the glossy productions common in large organizations. For the small firm, seven or eight pages stapled together may be sufficient. The main point is that a policy instituted by management assures employees they are being treated fairly. If employees are perceived as being treated unfairly, the organization could have the beginning of a silent grievance in the works.

Dissatisfaction with the organization can be minimized by having policies so defined that individual employees' differences are recognized and provided for. Management of organizations would be much simpler if all employees had identical values, desires, and needs and could be motivated in the same manner. However, such is not the case, and any policies that deny individual differences are doomed to at least partial failure. Inflexible, compulsory, and arbitrary policies can result in only frustrated, indifferent, and alienated employees.

The checklist below should remind you of the issues that need to be addressed in developing personnel policies. It is not intended that this list be all inclusive. Local trends and state regulations will add to the items suggested on this list. If you decide to proceed in the development of a set of personnel policies for your organization, it is suggested that you make a copy of this checklist to assure that you have addressed all items that are appropriate for your organization.

PERSONNEL POLICY CHECKLIST

- Absenteeism
- Protective Equipment
- Work Rules
- Grievance Procedure
- Hours of work
- Polygraph Test
- Affirmative Action
- Military Service
- Educational Assistance
- Employment of Family
- Orientation
- Security Regulations
- Terminations
- Vacations
- Bonding
- Medical Benefits
- Access to Files
- Sick Leave
- Garnishments
- Jury Duty
- Holidays
- Drug Testing
- Performance Appraisal
- Promotion
- Medical Physical
- Probationary Period
- Family Death
- Safety Regulations
- Leaves of Absence
- Parental Leave
- Employee Classification
- Retirement
- Layoffs & Recall
- Pay Days
- Lunch Periods
- Wages & Salary

- Conflict of Interest
- Overtime Pay
- Group Insurance
- Ethical Standards
- Coffee Breaks
- Education Subsistence
- IRA Savings
- Compensation
- Unemployment
- Payroll Deduction
- Accident Reports
- Accepting Gifts
- Disability Time
- Parking Regulations
- Funerals
- Profit Sharing
- Sexual Harassment

SUGGESTIONS FOR DEVELOPING PERSONNEL POLICIES

1. Make your personnel policies realistic for the type and size of your organization. Keep policies competitive by checking local trends.
2. Encourage optimum participation in the design and modification of your personnel policies. This includes employee involvement surveys, committees, internal communication, and feedback.
3. The personnel policy handbook should be prepared in a looseleaf binder in order to permit material to be revised without reprinting the entire book. The handbook should also be divided into special sections so as to limit the number of pages printed if a minor or major revision becomes necessary.
4. Be sure to give your supervisors an advanced copy of the policy handbook before you issue it to all your employees. They often can come up with good ideas. They should be given a chance to understand and review each policy so they can be in a better position to sell it to the subordinates.
5. Make your policy statements clear and easy to read. Employees will not read them if you use too much technical jargon.
6. Before beginning to write your personnel policies, check with state and federal regulations so that you will be in compliance. A number of regulations have been enacted, particularly at the federal level, which will affect your statements.

7. Personnel policies must be explained, interpreted, and taught. What people do not understand they cannot use correctly. Effective managers never assume that the issuance of a written statement is enough.
8. While many policies are, in effect, permanent, it should never be assumed that they represent laws engraved in stone. If the goals of the organization's major plans change, personnel should be reconsidered to meet the new situation.
9. Always put your personnel policies in writing. The fact that policies are in writing has a way of eliminating fuzziness and inconsistency. A policy that cannot be put in writing is at best an unclear one.
10. Most importantly, have the organization's attorney review all personnel policies before they are submitted to employees. The attorney should check to make sure that they are in compliance with state and federal laws.

SUGGESTED READINGS

Cheorington, D. J. (1982). *Personnel management.* Dubuque, IA: William C. Brown.

Glueck, W. F. (1982). *Personnel: A diagnostic approach.* Plano, TX: Business Publications, Inc.

King, A. S. (1982, September). A programmatic procedure for evaluating personnel policies. *Personnel Administrator*.

Matherly, T. A., & Slepina, L. P. (1985, December). Public sector personnel policies may threaten D.P. operations. *Journal of System Management*, 36.

Soukup, W. R., & Rothman, D. R. (1987, August). Outplacement services: A vital component of personnel policy. *Advanced Management Journal*, 52.

Chapter 24

EMPLOYEE APPRAISALS

Managers who say they do not have time for setting goals, employee appraisals, and evaluations do not understand the manager's role in the organization.

Practically every major business organization makes provisions for some form of regular personnel performance appraisal and review. How do you rate your employees? By their personalities, past performances, or what they accomplish? Unfortunately, no matter which method you use, it could be worthless. For decades organizations have said, "People are the most important part of our organization." Then, in an effort to learn more about their people, they set up an appraisal program.

Usually this means an annual review and filling out a form to evaluate an employee's work. In theory, it is an all-important management tool. Its purpose is to tell you whether an employee should be promoted, trained, given a raise, or fired. It is also supposed to help the employees develop their skills and abilities to the fullest.

In the past, it was thought that if people are the most important part of an organization, they should be evaluated to determine just what kind of people they have been during the past performance period. Characteristics such as initiative, common sense, ambition, tact, sincerity, and drive are very difficult to measure. Numerical values cannot be assigned to these traits.

There are many other pitfalls to this approach:

- Who decides which traits are to be rated?
- Are some traits an asset in one job and a liability in another?

- Should all traits have the same basic value?
- Can prejudice by the rater be overcome?

So, the designers of the appraisal forms said, "Let's change the questions. Let's be concerned about performance itself, not personality traits." Thus, instead of an arbitrary list of personality traits, the appraiser tried to answer questions such as:

- How well does the employee overcome problems?
- Does the employee display any special qualities when working with peers?
- How does the employee perform under pressure?

But typical answers read thus: This employee shows common sense in overcoming problems. She is very cooperative with her peers. The employee works calmly under pressure.

What happens is obvious. The questions do not get any better and the answers are still trait oriented.

Many organizations have turned to a different type of appraisal system. They feel that the focus should be on the measurable accomplishments. At any given moment an employee should be working at something that has a quantifiable goal that contributes to the organization's major objectives. The advantages of such an approach are many. Goals describe why a job exists in the first place. We do not hire a maintenance person to operate a lawn mower; we hire him to mow so many acres of land or keep vehicles in repair at a certain cost and within a certain length of time. Goals are thus both definable and measurable.

Are there any drawbacks to this approach? Unfortunately, yes. There is at least one serious, built-in pitfall that cannot be overcome: it has to do with timing. Performance appraisal usually calls for an evaluation of an employee's work on an annual and semiannual basis. Unfortunately, the goals that any employee works toward seldom have exactly a 12-month deadline. Some goals will take one month to achieve, others three months, a year, or more. As good management technique points out, praise for accomplishment of a task should take place at the time of completion of the accomplishment, not some months later.

At best, the formal evaluation of one human being by another is bound to be contaminated by some degree of subjective bias. When it comes to peer ratings, and especially ratings by subordinates, these prejudices often run wild.

Apart from human bias, employee appraisal systems suffer from other common flaws. Among them are the following:

- *Lack of top-management support.* Most supervisors do not like to evaluate other people. When they know that top management gives the program lip service, they find that the solution is easier to ignore it.
- *No reward.* Unless a supervisor has recognized the long-range satisfaction to be gained in helping an employee develop through frank appraisal of her performance, he sees no personal advantage to himself for the time and effort required to conduct one.
- *Advantages not made clear.* Everyone should understand the benefits of performance, employee development, and better utilization of human-resource power.
- *Inadequate training.* Becoming skilled at sitting opposite a subordinate and fairly evaluating his past performance requires considerable training and practice.

Often supervisors look forward to appraisal interviews with trepidation, conduct them reluctantly, and become immensely relieved when they are over. With two strikes against it from the start, it is hardly surprising that the interview is often not a success. Aside from this general mutual apprehension that so often permeates the interview, there are some specific reasons why these face-to-face discussions about the employee's past performance do not come off well:

- Too often the interview has no specific agenda. Instead of using the time to summarize what the employee has accomplished, the interview often is used to discuss whatever comes to mind.
- Too often the interview is personality centered. Personality traits are important, but they have been mostly debunked as yardsticks to measure past performance.
- Any discussion between boss and employee tends to be one-sided. If the employee suddenly is invited to speak up on this one occasion, the result is likely to be bewilderment rather than a free exchange of ideas.
- Forms are nearly as much at fault as the interview itself for failure of employee-rating systems. Too many managers'

performance appraisals mean only one thing—filling out a form.

If performance appraisal has not proven the answer to a healthy employee development program, what will? To answer the question, we must first make certain what we want performance appraisal to achieve. In general the objectives include the following:

- better use of existing personnel,
- improved performance on the job, and
- periodic feedback to let the employee know how he or she is doing.

A manager should hold frank talks with an employee about his shortcomings—but only when the shortcomings occur. If supervisors postpone discussions because organization policy calls for a get-together only once a year, then the discussion becomes a mere formality conducted on an untimely basis, covering things that are difficult to recall, no longer of immediate concern, and properly discussed only when they happen. To establish a climate of understanding, rapport, friendliness, and team spirit with employees, a manager must do it on a day-to-day basis. Mutual discussions between the manager and subordinate; working together; and timely, informal, progress appraisals should relegate the unsuccessful, outdated, performance appraisal systems to management's junk heap.

It is important for the manager to determine accurately the employee's level of performance in order to establish suitable objectives for the appraisal process. When the manager meets with the employee, she should mutually agree upon the ways that the employee's performance can be improved, maintained, or changed in direction.

When an employee has been determined to be outstanding, he should be congratulated for his effort. Reinforce how the employee's performance affects the success of the whole organization. Discuss the forms of recognition the employee will receive in terms of increased responsibility, promotion, or additional training. Review with the employee any potential problems and assist him with a plan to solve them. Most importantly, establish a specific time for a follow-up meeting.

When dealing with an employee who has received an unsatisfactory performance review, it is important to examine the individual's duties and discuss specific examples where less than satisfactory behavior was displayed. Discuss each item and corresponding improvement activity in depth. This review is critical in improving performance of below-average employees. The appraiser should look for potential problems. Be alert to *excuses* the employee may give as justification for unsatisfactory performance. Above all, listen to the employee's comments. He may have some legitimate reasons why his performance is below par. If appropriate, express confidence in the employee's ability to improve. You must be sincere in expressing this feeling.

SUGGESTIONS FOR IMPROVING EMPLOYEE APPRAISALS

- Have each employee prepare a written statement of the important factors, responsibility, and goals of her job as she perceives them. Discuss them until you can reach some substantial agreement as to what they should be.
- When you have new employees, frequent performance reviews are advised. This will allow each individual to know where he stands and where improvement is needed. New employees usually do not know their jobs very well and need coaching, mentoring, and some training in how to do their jobs correctly.
- Do not limit performance reviews just to making judgments about past results without adjusting goals and objectives of the future. This is also an opportunity for management coaching and mentoring employees.
- Do not discuss pay raises during performance reviews. Raises are usually based on a number of factors, with performance being only one of them. Discussing salary diverts attention from the real issues of improved performance.
- At the end of the performance review, it is a good idea to let the employee summarize what has been said during the session. This gives the individual and the manager a chance to check if there were any misunderstanding about the review discussion.

- Use active listening techniques during the review session. Encourage the employee to "tell her side of the story." It will reduce defensiveness, clarify the situation, and provide both parties the opportunity to think through the issues of the performance review.
- Do not wait for the performance review to raise problems. The employees should not be hit with surprises. This supports the practice of frequent reviews not done just on an annual basis.
- Be prepared for the review session. Think out the points you want to discuss. This will take time and planning. However, it will make a much more productive performance review for both the employee and the manager.
- Praise people as soon as possible after you discover praiseworthy behavior or work. Tell people in specific terms what they did right, outstanding, or of value. If they know exactly what pleased you, they will want to repeat that behavior.
- Do not threaten people. Threats either immobilize them with fear or cause resentment. Point out the error in incorrect behavior, but tell the individual that he can be successful with practice, training, and possibly by modifying his behavior.

SUGGESTED READINGS

Evans, E. M. (1991, March/April). Designing an effective performance management system. *Journal of Compensation and Benefits, 6*, 25-29.

Garfield, C. A. (1987, April). Peak performance in business. *Training and Development Journal,* 54-59.

Heneman, H. G., & Schwab, D. P. (1986). *Perspectives on personnel/human resource management* (3rd ed.). Homewood, IL: Irwin.

Landy, F. J., & Farr, J. C. (1980). *Performance rating psychological bulletin, 87,* 72-107.

McAffee, R. B. (1982). Using performance appraisal to enhance training programs. *Personnel Administration, 27*(11), 31-34.

Chapter 25

EMPLOYEE FITNESS

Companies with cardiovascular fitness and cardiac rehabilitation programs are helping to save lives while improving employee relations and reducing health care costs.

Many physicians and exercise physiologists believe that proper, well-planned physical activity and exercise may not only increase length of life, but also will improve quality of life. Today, exercise experts have entered the boardrooms of corporate America where they espouse the values of fitness, low-fat diets, and the elimination of health-threatening habits such as smoking, excessive alcohol consumption, overeating, and the dangers of maintaining an unhealthy lifestyle. Studies indicate that healthy employees are more productive, have less absenteeism, and have improved morale.

According to statistics released by the U.S. Center for Disease Control, 54% of all deaths of people under the age of 65 are directly attributed to unhealthy lifestyles. Health costs in the U.S. have increased tremendously and are continuing to increase. Health insurance premiums faced by corporations doubled from 1977 to 1984, when they reached over $80 billion. The Health Research Institute performed a cost-control survey of 1,500 of the largest corporations. Companies without employee fitness programs paid $1,456 per employee in total annual health-care costs, while those with fitness programs paid $1,061—a savings of $395 per employee. Another study of 15,000 employees of Control Data Corporation showed that people with high-risk behavioral profiles generate much higher health costs than those with low-risk profiles. Studies also indicate that high-risk lifestyles generate 68% more annual claims of more than $5,000 than do low-risk lifestyles. Employees who smoke an average of one pack of cigarettes a day generate medical claims 118% higher than nonsmokers. People who fail to wear seat belts use 54% more hospital days per 1,000 than did those who regularly buckle up.

Many corporations see the advantages of keeping their executives fit. For example, the New York office of White and Case, the 37th largest law firm in America, opened a private health and fitness center for its 127 associates and 56 partners. The center operates from 7 a.m. to 7 p.m. on weekdays and 10 a.m. to 5 p.m. on Saturdays. The firm provides shoes and fitness gear. The staff at the center includes two full-time exercise physiologists, a part-time laundry person, and maintenance personnel. The staff operates a supervised, medically based program. Each participant receives an individualized exercise program and a daily review of exercise goals.

One of the most well-known corporate fitness programs is Campbell Soup Company's Turnaround Program. This is not just an exercise program. Participants are encouraged to reassess their entire lifestyles. Following the evaluation, participants select programs in stress management, lower-back care, relaxation techniques, self-defense, nutrition, and dance. One of the most active of Campbell's programs is at its headquarters office in Camden, New Jersey. Here, a 10,000-square-foot facility features motorized treadmills, an indoor jogging track, stationary exercise bicycles, and extensive weight-training equipment. The center is open from 6 a.m. to 8 p.m. and is open to employees' families. While participation is voluntary, Campbell claims 70% of its employees at Campbell's headquarters are enrolled.

L. L. Bean, one of America's largest outdoor fitters, operates two employee fitness centers: one in Freeport, Maine, and another in Northport, Maine. These centers hold a variety of classes. Smoking cessation and stress management classes are open to employees and their families. Other offerings for employees only include a 15-week "heart club" program, a walking program, jogging, cross-country skiing, kayaking, biking, aerobic exercise, ballroom dancing, and yoga. Twenty-five to 30% of L. L. Bean's employees participate in fitness center activities, and close to 80% have taken part in cholesterol and high blood pressure screening programs.

The question often arises, "How much exercise and what kind is needed to maintain a reasonable fitness level?" The American College of Sports Medicine has recommended 90 to 120 minutes of aerobic activity each week for developing and maintaining cardiovascular fitness and body composition. Two to four sessions, each lasting from 20 to 30 minutes, are recommended for proper cardiovascular functioning. Physical activity must be done at a threshold level so that the heart and vascular system are challenged to develop and maintain their proper functions.

Duration is also a key element in an effective exercise program. The duration of aerobic exercise should be 15 to 30 continuous minutes. Start gradually and work up to the 30-minute level. For example, in an initial walking exercise session, a beginner would perhaps walk briskly from five to 15 minutes. Each session thereafter would involve a slight increase until one could perform the exercise at an elevated heart rate for the necessary 20 to 30 continuous minutes.

The mode of exercise for aerobic activity should be activities involving large body muscles. Activities such as walking, jogging, running, cycling, and swimming are best. These exercises all involve large muscle activities and allow the individual to get his or her heart rate elevated. It is important that you consult your physician regarding your heart-rate level during exercise.

Many physicians point to the value of home and office exercise equipment such as stationary bicycles, rowing machines, cross-country skiing machines, and exercise treadmills as an idea to put exercise in your life.

Exercise alone cannot guarantee good health. Regular exercise is essential but your diet, because of its effect on your blood cholesterol levels, can be equally important. Obesity affects about 14% of men and 24% of women in the United States, and is associated with a variety of health risks. When losing weight, a gradual loss is best. A good guideline is one to two pounds or one percent of the total body weight per week.

You can reduce your caloric intake in several ways. Eat smaller portions, more salads, and chicken without skin. Bake, broil, or steam foods instead of frying. Avoid too much saturated fat and cholesterol. A small amount of fat is needed in everyone's diet, but many Americans go overboard. Fats are a very concentrated source of calories. In addition, too many fats in the diet can result in high blood cholesterol levels that contribute to heart disease risk. Health and government agencies recommend that Americans reduce their fat intake to 30 to 35% of their total calories, with saturated fats making up no more than 10% of the total.

Another suggestion is to eat foods with adequate starch and fiber. The best way to ensure the proper amount of starch and fiber in your diet is to eat four servings of fruit and vegetables and four servings of whole grains each day. Avoid too much sugar. Sweets contribute calories to the diet without contributing important nutrients. Sugar also causes tooth decay. To cut down on sugar, reduce the amount added to food and beverages and decrease sugar in food preparation.

Avoiding too much sodium is also important. Evidence shows that high sodium intake may contribute to hypertension. The body needs only about 200mg of sodium each day, but the average daily intake of sodium in the American diet is between 2,300 and 6,900mg. Drink alcohol in moderation. Alcohol contains seven calories per gram and offers virtually no nutrients. Alcohol can also increase the body's need for vitamins and can impair the absorption of other nutrients in the body.

For many executives, travel schedules interrupt their exercise programs. Uncertainty about exercise facilities in hotels, uneasiness about jogging in strange places—especially before dawn or after dark—and the lack of control over food choices and preparation lead many executives to abandon their programs of exercise and healthy eating temporarily. If you are faced with this dilemma, check the fitness facilities at your destination point. There are a number of guidebooks that describe the layout of the facilities. If you fly frequently, order low-fat meals through your travel agent. Remember to bring your exercise clothes. This will motivate you to get up and get going. If you have extensive layovers at the airport, take a brisk walk. Avoid alcohol, which has a dehydrating effect. When at restaurants, order a child's portion or half portions to limit your caloric intake.

When people enhance their health and fitness, they earn a whole set of hidden benefits and values that go beyond the purely physical. Studies have shown physically fit people have an improved self-image. Physical fitness is no longer a trend; it is a way of life. Organizations now see its advantages. Research is overwhelming in describing its benefits, but only the individual can provide the motivation. By maintaining a healthy lifestyle and fitness program, you probably will live longer, enjoy your work more, have more energy for leisure pursuits, and watch your grandchildren grow up.

SUGGESTIONS FOR IMPROVING YOUR FITNESS AND LIFESTYLE

- Set goals for the level of fitness you want to achieve. When you have attained these, establish new and more ambitious ones.
- Select a type of exercise that you enjoy and is fun. If you like competitive sports, give racquetball a try. Running and cycling are excellent aerobic activities and give you a chance to view the scenery.
- If you like to listen to music when you work out, develop your own customized audiocassette with your favorite music. Listening to these can be inspiring and motivating.
- Establish a set time to do your daily exercise. Make it an important part of your day. Let other people know you will not be available at this time for meetings or other commitments. Most people will respect your request.
- If you are just beginning an exercise program, you should receive a medical exam from your physician. Your physician knows your medical history and can properly screen you and prescribe an exercise program that meets your needs.
- Limit your consumption of high-fat foods such as fatty luncheon meats, nuts, granola cereals, avocados, quiche, and croissants. These are high in cholesterol. If you need to reorganize your eating habits, consult with a registered dietician for guidelines.
- Do not give up all the foods you enjoy. Just eat less of them. Share some fries with another person, eat only one slice of pizza, and eat less meat and more fish, turkey, and chicken.
- Keep a balance between work and recreation. Every individual requires a different balance of fun and work; overdoing either will impair your effectiveness and efficiency.
- Take your lunch away from the workplace. Get your mind off the business of the day. Center your conversations on matters that do not concern work.
- Keep a daily journal of your stressful situations. Write down your life feelings and reactions to these experiences. A journal is one way to get out on paper feelings you might normally keep bottled up inside.

SUGGESTED READINGS

Dun's mouth. (1986, July). *Executive Health Update*, 128.

Frew, D. R., & Bouning, N. S. (1987, June/July). The fitness/work connection. *Corporate Fitness*, 16-18.

Lydecker, T. W. (1987, March). Winning the battle of the bulge. *Association Management*, 64-75.

Pollock, M. L. , Foster, C., Salisburg, R., et al. (1982). Effects of a YMCA starter fitness program. *The Physician and Sport Medicine, 10*, 89-99.

Udeleff, M. (1985, October/November). Extending a helping hand. *Corporate Fitness and Recreation*, 29-32.

Chapter 26

EMPLOYMENT INTERVIEWING

The interview is probably the most important step in the whole selection procedure, for all the relevant information is brought into focus. Often the final decision to hire an individual is made during the interview.

One of the most important functions of a manager is to employ personnel to carry out tasks that will ultimately achieve organizational goals. A variety of methods are used to recruit, select, and employ personnel. The employment interview is one technique used by most organizations. Surveys have shown that employment interviews are the most important of the selection process. In a recent survey, over 90% reported that they had more confidence in the interview than any other method. Yet reviewers of over 150 research studies conducted over 20 years conclude it is rarely a valid predictor of job success. However, recent work suggests that interviews can demonstrate validity if conducted properly. Personal interviews are usually the final step in the selection process.

There are usually two stages of the interview process. The first is the screening interview. If the applicant passes this phase of the selection process, he is interviewed further by upper management and coworkers. In the screening interview, someone usually spends a short time with applicants in what is called the "initial screen test." Here, the organization develops some preliminary guidelines to be applied in order to reduce the time and expense of actual selection. These guidelines could specify, for example, the minimum amount of education required, or in the case of secretarial personnel, the number of words typed per minute. Only those who meet these minimum criteria are considered potential employees and are given more in-depth interviews.

Many managers mistake the results of the personal interview because they are not well conceived. Many interviewers are inclined to interrogate, rather than converse, so they learn what they want to hear. Many job candidates will give the answers they think will get them in the position. The interviewer is cautioned not to sell the job or company until she is sure the candidate is the right one for the job.

Interviewers should do only 10 to 20 percent of the talking. Their first remark should be intended to alleviate tension by exuding calmness and a nonthreatening attitude. The interviewer should interpret the candidate's behavior. It is important to pay attention to the way the interviewee responds not so much to the context of what is said, but the way it is said. Always make the first interpretation tentative and verify it three or four times during the interview.

Most experts in the field of personnel administration identify three types of interviews: (1) structured, (2) semi-structured, and (3) unstructured.

STRUCTURED INTERVIEW

In the structured interview, the interviewer prepares a set of questions in advance of the session. Usually a standard form is used to record the interviewee's response. The structured interview is very limited, and there is little opportunity to adapt to individual applicants. This technique is restrictive to the interviewee also. It gives him little chance to elaborate on his response to questions.

SEMI-STRUCTURED INTERVIEW

In the semi-structured interview, only the major questions to be asked are prepared in advance. This approach calls for greater preparation; however, it allows for more flexibility than the structured technique. The interviewer is free to probe areas that seem to merit additional examination. Since these interviews have less structure, it is more difficult to replicate them. This approach combines enough structure to facilitate the exchange of information with adequate freedom to develop insights.

UNSTRUCTURED INTERVIEW

Little preparation is required for the unstructured interview. A list of possible topics to be addressed is prepared. The main advantage of using this technique is it allows the interviewer and the interviewee to adapt to the situation and the dialogue that takes place. Freedom and openness describe the approach of this technique. However, when utilized by an inexperienced interviewer, the session may result in chaos, frustration, and confusion. When used by a highly skilled interviewer, the unstructured interview can lead to significant insight that might enable the interviewer to make a distinction among applicants.

Preparing for the interview will vary depending on whether the structured, semi-structured, or unstructured technique is used. It is also important to know how much time is available before and after the interview. The following are steps that should be taken prior to the interview:

1. Pertinent information should be collected about the candidate. This preparation saves time and effort prior to the interview and enables the interviewer to determine in advance a general picture of the interviewee.
2. The prospective interviewee should be given advance notice of the time and place of the interview. This should be done in a way that is reassuring rather than alarming.
3. Privacy and comfort are important. The surroundings should allow the interviewee to relax and be assured that the discussion will not be overheard. When possible, interviews should be held in a private office with comfortable chairs. If a private office is not available, adequate privacy can be found in a corner of a room, in a restaurant, or anywhere no one else can hear the discussion. The relationship between physical and psychological relaxation should not be overlooked.
4. Regardless of the type of interview, it is suggested that it begin with a few general remarks that may put the interviewee at ease and establish rapport at the very beginning.
5. It is all-important to assure the interviewee of the confidentiality of the interview process. If the interviewee is open and honest, confidentiality of what is said is essential. Many candidates have withdrawn from consideration because confidentiality has not been maintained.

6. Interruptions during the interview should not be allowed. Such occurrences will make the interviewee feel unimportant and not the central focus of the moment. Interruptions will also cause the interviewer and interviewee to lose their trains of thought and the emphasis on which the discussion is focused.
7. Whether the interview is structured or unstructured, it is important that the interviewer practice good listening skills. Many managers feel that they need to assume the responsibilities for all the dialogue and give little chance to the interviewee to express himself. Responsive listening can often help a candidate to overcome interview "jitters," anxieties, and nervousness.
8. It is impossible for an interviewer to remember all the facts and discussion points that take place during the interview. It may be necessary to take notes. Many managers hesitate to write anything down in fear it will be incompatible with the informal atmosphere of the interview. Some interviewers believe it is impossible to write and observe at the same time. However, it should be remembered that notes are useful for follow-up and evaluation, especially when it is necessary to conduct a number of interviews during a short period of time.
9. When an interview is over, an interviewer's task is still unfinished. To learn all one can, notes should be reviewed to see what went right, what might have gone better, and where interview skills could be improved. Every interview offers an opportunity to get new ideas about company policies and practice—what is needed to make discussions between two people a worthwhile experience for both.

Numerous research studies have been conducted on the interview process during the past 20 years. Neil Schmitt, a researcher in personnel administration, identifies a number of sources of error in the interview process. These include the following:

1. *Overemphasis on negative information.* This occurs when there is a search for negative information. Often the finding of a small amount of negative information can lead to the rejection of a candidate.
2. *Interview stereotypes.* Often interviewers develop a stereotype of an ideal job candidate; successful interviewees are

then not the ones best qualified, but the ones who conform to the stereotype.
3. *Job information.* Lack of relevant job information can increase the use of irrelevant attributes of interviewees in decision making.
4. *Different use of cues by interviewers.* Some interviewers may place more weight on certain attributes than others or they may combine attributes differently as they make their overall decisions.
5. *Visual cues.* Interviewees' appearance and nonverbal behavior can influence their evaluation in an interview, yet be unrelated to job success.
6. *Similarity to interviewer.* Sex, race, and/or attitude similarity to interviewers may lead to favorable evaluations.
7. *Contrast effects.* The order of interviews influences ratings. For example, strong candidates who succeed weak ones look even stronger by comparison.

What Questions Do You Ask?

The interviewer should prepare a set of questions that will elicit information that reveals the candidate appropriateness for the position. It is not suggested that these questions be asked in any particular sequence; however, the questions listed below should help retrieve important information about the candidate.

1. Why are you interested in this position? Why are you leaving your present job?
2. If you are selected for the position, where do you see this job fitting with your career goals?
3. If selected, when could you begin work?
4. Explain in detail your current job responsibilities. Describe a typical day in your current job.
5. Describe the people (not by name) you have employed in your current job. How long have they been with you?
6. What do you think it takes personally, and professionally, for a person to be successful as (name job)?
7. What specific strength would you bring to the position?
8. What do you think is the single most important idea or accomplishment you have provided in your current position?
9. Are there any projects or ideas you were unable to implement in your current job?

10. Explain the process by which you go about solving a problem or making important decisions.
11. What do you do for recreation?
12. How do you ensure that you and your staff stay up-to-date on professional developments in your area of expertise?
13. What type of quality control measure do you implement to ensure the quality of your staff's work?
14. What can your present employer do to be more successful?
15. In your opinion, what leadership style do you use in your present situation?
16. Where do you see yourself in three years? Five years?
17. What type of job did you work at while you were attending school?
18. What extracurricular activities were you involved in while in school?
19. What is the most difficult aspect of your present job?
20. Why do you feel you are the best candidate for the job?

What Questions You Should Not Ask

Certain questions that might have been asked routinely a few years ago are now questionable legally. Questions about race, sex, national origin, or age are illegal. The following is a list of questionable areas of investigation. The interviewer should investigate state and federal laws related to this issue.

1. Inquiry into any title that indicates race, color, religion, sex, national origin or ancestry.
2. Specific inquiry into foreign addresses that would indicate national origin.
3. Any inquiry that would indicate sex.
4. Any inquiry to indicate or identify denomination or customs; may not be told this is a Protestant, Catholic, or Jewish organization; request of a recommendation or reference from someone in clergy.
5. Any inquiry into place of birth; any inquiry into place of birth of parents, grandparents, or spouse; any other inquiry into national origin.
6. Any inquiry that would indicate race or color.
7. If native-born or naturalized; proof of citizenship before hiring; whether parents or spouse are native-born or naturalized.

8. Require birth certificate or baptismal record before hiring.
9. Require photograph *before* hiring.
10. Any inquiry asking specifically the nationality, race, religious or school affiliation; inquiry as to what is his mother tongue or how foreign-language ability was acquired, unless necessary for job.
11. Any inquiry about a relative that is unlawful (e.g., race or religion inquiries).
12. Inquiry into all clubs and organizations where membership is held.
13. Inquiry into military service in armed services of any country but U.S.; request military service records.
14. Any inquiry into willingness to work any particular religious holiday.
15. Any non-job-related inquiry that may present information permitting unlawful discrimination.
16. Request references specifically from clergymen or any other persons who might reflect race, color, religion, sex, national origin, or ancestry of applicant.

SUGGESTIONS TO IMPROVE YOUR INTERVIEWING SKILLS

- *Interviewing is not something that can be done casually.* Each interview should have a specific objective. Prior preparation is all important. Never try to "wing it."
- *Let the candidate do most of the talking.* Give the candidate a chance to answer your questions fully. Do not interrupt if there is a pause. The candidate may be thinking of what to say and how to say it.
- *No interview is complete unless there is a follow-up.* Inform the candidate you will keep her posted as to developments in the job search. Nothing can be more frustrating for an interviewee than not to hear anything about her candidacy from the prospective employer.
- *Ask questions in an open-ended way.* Answers should not be just a simple yes or no. Try to begin each question with a key word like why, what, where, or when. This will give the candidate the widest possible choices of answers.

- *Be sure to allow enough time for the interview.* It should last at least a half hour. If the interview is rushed, the candidate will not relax long enough to show you what he is really like.
- *Never ask trick questions; the candidate will know when you do.* If there is a conflict in the candidate's statements, explore the discrepancy discreetly.
- *Do not be misled by your prejudices.* Keep an open mind. Never allow your biases to cloud your judgment.
- *Avoid leading questions.* They give the candidate a cue to what the interviewer expects to hear. This can often provide a ready-made answer from the interviewee.
- *When using a structured interview, change the order of questioning.* If, for example, the candidate is discussing work history and she refers to education, that information should be recorded. The flow of conversation should never be halted nor should the interviewee be told to hold some item until later at a more appropriate point of the interview.
- *The first few minutes of an interview can be difficult.* You may want to introduce the discussion with talk about family or personal recreation. This will break the ice and put the interview in a comfortable situation for both the interviewer and interviewee.

SUGGESTED READINGS

Diffie-Couch, P. (1984). Building a feeling of trust in the company. *Supervisory trust (29)*4, 31-36.

Ewing, D. (1977). *Freedom within the organization.* New York: McGraw-Hill.

Frierson, J. G. (1988, December). New polygraph test limits. *Personnel Journal 67*(12), 84-88.

Glueck, W. F. (1982). *Personnel—A diagnostic approach,* (3rd ed.). Plano, TX: Business Publications, Inc.

Heneman, H. G., Schwab, D.P., Fossum, J. A., & Dyer, L. D. (1986). *Personnel/human resources management* (3rd ed.). Homewood, IL: Irwin.

Chapter 27

HIRING THE DISABLED

Many companies have found that what they once viewed as an obligation has become a sound, mutually beneficial hiring practice.

Employment is critical in the lives of most Americans, regardless of whether their work is in upper management or labor. People work not just because they need money to support themselves and their families, but also because they need and enjoy the intangible benefits of their jobs—the opportunity to interact with others, form relationships, develop self-esteem, and give something back to society.

The disabled are no exception. They, too, desire the independence and satisfaction work provides, and the majority of them want to find jobs. In the past, however, a lack of educational opportunities, support services, and physical accommodations—as well as employer attitudes—severely limited the number of jobs available to the disabled. As a result, their unemployment rate has been one of the highest of any groups in the country. Among those who are employed, only a tiny percentage work full-time, the rest have part-time, low-income jobs.

In recent years, federal legislation and increasing pressure from the public have induced employers to hire more disabled workers. In the process, many companies have found that what they once viewed as an obligation has become a sound, mutually beneficial hiring practice. With the tightening of the labor market, managers are relying more heavily on the large pool of disabled people to meet their staffing needs. Employers have a source of eager, motivated workers, and people with disabilities have the opportunity to become tax producers instead of tax consumers.

LEGAL ISSUES

Once a company makes a commitment to hiring the disabled, management needs to develop a thorough understanding of the numerous federal and state laws that govern hiring practices and physical accommodations. Failure to comply with these laws, which have become much more complex in recent years, could lead to costly litigation.

The most sweeping federal legislation affecting today's employers is the 1990 Americans with Disabilities Act (ADA), which prohibits discrimination against any qualified person with a disability. It also requires employers to provide "reasonable accommodations" to help physically or mentally disabled workers perform their jobs.

The first step in complying with this law is identifying which employees and job applicants are "disabled" and which impairments and diseases are considered disabilities. The ADA defines a disabled individual as "any person who (1) has a physical or mental impairment which substantially limits one or more of such person's major life activities, (2) has a record of such an impairment, or (3) is regarded as having such an impairment."

The breadth of the government's definition suggests that large numbers of employees and job applicants are potentially "disabled." It includes people with visual, speech, hearing, and orthopedic impairments and developmental disabilities such as learning disorders, mental retardation, epilepsy, cerebral palsy and autism. It can also apply to people with diseases that affect their daily lives, including multiple sclerosis, cancer, heart disease, diabetes, muscular dystrophy, mental and emotional illness, and AIDS.

The third part of the ADA's definition—"is regarded as having such an impairment"—covers conditions that do not affect a person's current ability to work, but may in the future. This definition includes workers who have been diagnosed with a potentially disabling disease (cancer or HIV-positive status, for example) and those with a history of disablement (including mental illness and drug and alcohol addiction).

MYTHS ABOUT THE DISABLED

The ADA's purpose was to open up new job opportunities for the disabled. In doing so, however, it heightened employers' worries about compliance and possible lawsuits, and made some even more wary about hiring the disabled.

Much of this fear stems from the negative stereotypes that abound about the work habits of the disabled. Many employers believe that disabled workers will have higher rates of absenteeism and job turnover, that they will suffer more on-the-job injuries, and that they will be less productive than other workers. Previous research, however, shows that disabled workers' records in these areas are as good as or better than those of unimpaired personnel in labor-intensive jobs.

The laudable record of the disabled in safety, attendance, and production is attributed to a number of factors. First, disabled workers seek no special privileges. Generally, they want to be treated the same as other employees. Even though special considerations (such as parking privileges or modified hours) are necessary, these are not resented by other workers. Second, handicapped workers tend to be more cautious than their fellow employees because they are more aware of their physical and/or mental limitations. Thus, they have better safety records.

SPECIAL ACCOMMODATIONS

One of employers' biggest worries about the ADA is its requirement that companies make "reasonable accommodations" for the disabled. Specifically, the ADA states that employers must make existing facilities accessible, provide special equipment and training, arrange part-time or modified work schedules, and provide readers for the blind. These accommodations can be costly, especially if the company needs to remodel its physical facilities. A company may need to widen its doorways, build wheelchair ramps, expand and install new fixtures in bathrooms, and even install elevators.

Every employer has heard the horror stories about small businesses that had to close because they could not afford the major remodeling required by the ADA. In most cases, however, the cost of modifications is minor, and the law does provide relief on a case-by-case basis. A firm can avoid this requirement if it demonstrates that such an accommodation would impose an "undue hardship." Among the factors considered are the size of the employer's operations, the composition and structure of the workforce, and the nature and cost of the needed accommodation.

In practice, the most cost-effective method of solving many barrier problems is to handle them after the disabled worker begins his job. Allow the person to enter the work environment and interact with all the work stimuli. If a problem arises, let the person with the disability suggest a solution first, then work with him to implement it.

TRAINING

Every company incurs costs in training new employees, and these can rise for disabled workers. People who are developmentally disabled may have difficulty learning the job the first time and require repeated training efforts.

Special types of training may also be required. In many cases, people without disabilities receive the majority of their training from their supervisor. However, when training individuals with mental or developmental disabilities, the most effective way for them to learn and remember the job is by actually walking through their duties with them. As the time of training increases, so do the costs.

There are, however, tax incentives for hiring and training disabled employees. State agencies provide funds to compensate the employer at least in part for the cost of supplies and the time spent training disabled employees.

Once a disabled person is hired, the employer may also find it beneficial to provide training for other employees to help them understand how such workers can contribute to the organization and how to interact with them at work. The employees' increased awareness of the disabled person will aid in cooperation and effectiveness of company operations.

SOCIAL ACCEPTANCE

Legal requirements, workplace accommodations, training costs, and other direct expenses are important considerations for employers who hire the disabled. These are much easier to measure and control than the larger—and in many cases, more critical—issue of how the disabled employee will interact with coworkers and how this interaction will affect everyone's work performance.

One of the most common reasons for the failure of disabled employees—especially those with developmental disabilities—is their inability to fit in socially on the job. Research has shown that people's attitudes toward the developmentally disabled are more negative than towards other persons with different disabilities. The more severe a worker's mental disability, the more likely her coworkers are to judge her as unable to perform her job. As a result, non-disabled workers may feel, rightly or wrongly, that they are bearing an unfair share of the group's workload.

Every group of employees has its own social structure, and it is natural for people to prefer to associate with those who have similar intelligence and interests. This preference can cause the disabled worker to feel left out, and her work can suffer. Many developmentally disabled workers rely heavily on their jobs to provide them with opportunities to interact with other people. If they are not experiencing any social benefits on the job, they become unhappy, and the employer can usually expect decreased production.

If a disabled person does not fit in socially, it should not be assumed that the blame lies with the employer. Employers have no obligation to provide social outlets in the workplace, but they can help disabled workers find outside sources of social activity. Local advocate agencies and sheltered workshops often provide activities for these individuals. Employers may find it beneficial to encourage and assist their disabled workers in participating in these social programs.

A PRACTICAL APPROACH TO HIRING

Employers should consider applicants with disabilities as they would other applicants from a business perspective. Employment should never be a charitable or benevolent offer; it should be based on qualifications.

The ADA requires employers to reevaluate job requirements so that judgments about qualifications are based on the individual skills that are necessary to perform specific jobs. A key factor in complying with ADA provisions is the development of clear, realistic job descriptions that identify the essential functions of each position.

Many organizations have vague job descriptions that are unrelated to specific results. In such cases, new descriptions need to be developed that focus on job-specific activities. Sometimes a person with a disability will be unable to perform the job, even with accommodations, because of his physical or mental limitations. For example, a job description that says a mail-room employee's job is to "handle mail" provides no basis for determining whether a person in a wheelchair could perform the work. However, if the job actually requires hand delivery of heavy parcels to remote locations, this specifically stated activity is a defensible reason for not hiring a person in a wheelchair.

At the other extreme are detailed job descriptions that include long lists of duties, some of which are rarely performed or are not "essential" to job performance. Requirements that are merely marginal for the job in question cannot be a basis for refusing to hire an otherwise qualified disabled worker. For instance, if the mail-room clerk's job description covers dozens of specific tasks, one of which is picking up a monthly shipment of heavy packages, there will be no reason why a person in a wheelchair could not perform this job. The ADA would consider it a "reasonable accommodation" to restructure the job by assigning the monthly trip to another employee.

One approach to developing these new job descriptions is to ask incumbent workers to describe what their work entails. These descriptions may overemphasize what the writers do well rather than the objectives that are truly important to the job, so it is important to have managers review and supplement the descriptions.

Standardized questionnaires can be used to help bring consistency to the process of writing job descriptions. Instructions with the questionnaire should define and explain several key categories of information that are needed. These could include:

- A brief description of the purpose of the job in the organization.
- A section on "essential functions" that includes a list of the major activities of the job, the hours spent each week on each activity, and the key end results of each activity. This section should focus only on activities that are critical to overall job performance.

- For each major activity, a description of how results are measured and success is determined.
- A list of the specific skills, knowledge, physical abilities, and other requirements that are necessary to obtain the results associated with the position.

These job descriptions can improve the process of evaluating all job applicants, but they are especially helpful when interviewing disabled individuals. They provide an objective basis for discussing the specific activities a job requires and the related skills the applicant brings to the position. Most importantly, they help keep the focus where it should be—on what individuals can do to help the organization, rather than on what they cannot do, which may not matter.

If an interviewer believes a disability may be related to job performance, specific questions can be asked. It is important that the interviewer allow applicants to describe how they can perform the job, rather than assuming it is impossible. The person with the disability is often the best source of information about what accommodations might be necessary.

SUPPORTED EMPLOYMENT

One option that many employers may want to consider is supported employment, which involves working with local shelters or workshops to hire severely disabled individuals.

Supported employment is a departure from traditional vocational rehabilitation programs that have served individuals with severe disabilities in sheltered settings. In the past, these people have had little or no opportunity to find work outside of the workshop. Supported employment allows the severely disabled person to get out into the community and interact with non-disabled workers. Because these individuals need intensive, ongoing services to perform such work, the shelter provides job coaches and other support services.

CHANGING ATTITUDES

One of the greatest barriers to employing the disabled are the attitudes of employers, who often see the laws as a burden. As thousands of organizations have discovered, though, compliance is simpler and less costly than they feared, and the benefits are great. Companies have a source of eager, dedicated workers, and millions of Americans now have the opportunity to contribute, perhaps for the first time, to the nation's productivity.

Attitudes of the disabled themselves can also present barriers to employment. Because they have been denied job opportunities in the past, they have not been able to explore career options and they are often unaware of their full capabilities. These attitudes are changing, though, as more and more of the disabled enter the workforce. With training and experience, many of these individuals can quickly develop the skills and the self-confidence they need to become productive workers.

Fully integrating the disabled into the workplace can be a challenge, and it can create some special problems. The first step in a successful relationship is for the employer to realize the positive attributes the disabled can bring to the job. If companies are well prepared—and if managers do their homework on the laws—they can expect a profitable working relationship.

SUGGESTIONS FOR HIRING DISABLED EMPLOYEES

- Do not create new positions just to fulfill your commitment or obligation to hire the disabled. Instead, try to accommodate people in current jobs and offer advancement opportunities through normal promotional procedures.
- Investigate all the resources available to you. There are numerous agencies and organizations at the national, state, and local levels that can help you hire the disabled and make accommodations for them.
- Check out tax incentives for hiring and training disabled employees. Many states compensate the employer at least in part for the cost of training and supplies.
- When hiring a disabled employee from a workshop or other agency, find out if the agency will help meet the person's social and leisure needs. Try to keep the agency involved in the employee's life away from the job.
- In writing your job descriptions, include both the intellectual and physical requirements of every task. Be specific, but keep the focus on *what* is to be accomplished, not on *how*.

- Train your interviewers. Everyone who is involved in recruiting and hiring should be familiar with the laws and the types of questions they can and cannot ask.
- Review your hiring policies for all employees and change any that may cause ADA compliance problems. For example, all pre-offer medical exams (except for drug screening) should be eliminated.
- When considering reasonable accommodations for a disabled employee, always keep a written record of the individual's own suggestions, the costs, the sources you consulted, and the options you investigated.
- Consider offering home-based employment to accommodate physically disabled workers. Make it an option, not a requirement, and try to set it up so the employee has a limited amount of in-office presence on a daily, weekly, or monthly basis.

SUGGESTED READINGS

Boller, H., & Massengill, D. (1992, Fall). Public employers' obligation to reasonably accommodate the disabled. *Public Personnel Management*, 273-300.

Hagner, D. (1990). Vocational evaluation in supported employment. *Journal of Rehabilitation 123*, 45-50.

Meisinger, S. (1992, July). Putting the ADA to work. *Association Management*, 81-84.

Meng, G. J. (1991, Autumn). Definitive job descriptions are key to ADA compliance. *Employment Relations Today*, 285-289.

Noel, R. (1990). Employing the disabled: A how and why approach. *Training & Development Journal 44*, 26-32.

Postol, L., & Kadue, D. (1991). An employer's guide to the Americans with Disabilities Act. *Labor Law Journal*, 323-342.

Chapter 28

The economic importance of learning on the job is increasing. The acquired skills and abilities of the population have become the pivotal resource.

Human resource development, otherwise known as employee training, has become an important part of many companies in the 1990s. The employee training industry in 1991 was a $30 billion-a-year industry. Dollars budgeted by U.S. organizations for training programs in 1992 totaled $45 billion. Training is not only being offered to newly hired employees but also to top-line management. The format varies for training within different companies, even different departments within companies, but the main objective is the same: to improve the employee in some way.

The commitment to training and development displayed by management can have a strong influence on the amount of support from other managers, both before and after the training. If managers see strong commitment, they will usually regard support of training programs as a responsibility they must meet.

There are four factors that can contribute significantly to the in-service program's success. To reach maximum effectiveness there must be strong presence of each of the four factors. They are as follows:

1. There must be strong support from middle managers who supervise those being trained.

2. The in-service training program must show measurable results, and these results must be communicated throughout the organization.
3. There must be top management commitment for in-service training.
4. Management must be involved in all phases of the training and development process from the design to the implication of the program.

There cannot be an absence of any one of these factors and still have a result-oriented program. Each of these factors tends to reinforce one another. If you remove any one of these you will have a missing link.

Training is a broad term that encompasses many different areas and activities. The American Society for Training and Development's *Reference Guide to Professional Training Roles and Competencies* defines human resource development as "organized learning experiences sponsored by an employer and designed and/or conducted for the purpose of improving work performance." Employees at all skill levels can learn how to do their jobs better or learn other jobs in the organization. Training allows an organization to cultivate its employees' talents while oftentimes boosting employee morale. When organizing a training session, management needs to ensure results since these sessions may be very costly and time-consuming. Prior to establishing a training program, management must determine that the economic return is recognizable and that it will improve the performance of employees.

Showing results of in-service training programs can dramatically increase management support. Management is usually eager to support programs that yield a return on investment. Although specific ways of measuring results have been covered in articles and books, some basic concepts suggested by Dugam Laird, in his book *Approaches to Training and Development,* provide an excellent reference on measuring results. Ideally, the results of the training program should be measured by the following table.

DECREASE OR REDUCTION IN INCREASE AND IMPROVEMENT IN:

absenteeism	unit hours
accidents	productivity
unit costs	communication
overtime	on-time shipments
tardiness	dollar savings
employee errors	employee morale
turnover	grievances
work backlog	complaints

The best and immediate evaluation of the in-service training program may come from the supervisors and trainees. Using this approach, the program can be evaluated in the following ways:

1. **Examinations:** Pre-course and post-course examinations are administered to measure knowledge gains and attitude changes during the training program.
2. **Trainee Feedback:** Supervisors complete an evaluation form at the end of each training program. This evaluation determines the supervisor's feelings about what she has learned and how it may be used.
3. **Feedback from Management:** Several months after the employee completes the in-service program, management should complete a written report on the progress made by the employees who participated in the program.
4. **Trainee Follow-up:** The employee is asked to complete a similar questionnaire six months after completing the program. The survey determines to what extent the employee has applied what she has learned during the in-service program.
5. **Performance Contract:** The supervisor and his superior agree on changes or improvements that will be completed after the training program is conducted. The contract calls for utilization of skills learned during the training program.

Training may be formal or informal, one-on-one, or with a large group of people. It may consist of a simple read-through of relevant books or articles, enrollment in a university or special instruction class, special coaching with a superior or colleague knowledgeable in the area, or attendance at an instruction session organized by the company. Training may consist of any combination of these methods.

Many companies are expanding the conventional definition of training. Motorola Company, for example, after realizing that its employees were lacking certain reading and arithmetic skills, established its own university. By joining forces with local community colleges and high schools, Motorola offered classes in remedial elementary education and more advanced technical and business classes. The teaching institutions, which often used outdated technology to teach the lessons, benefited from the use of the current technology of Motorola, while Motorola benefited by training its employees to better face the academic obstacles of the work world. Rather than fire a large percentage of its workforce, Motorola pledged to improve its employee skills, thus improving the company. Similar programs are being offered through different organizations, including Levi Strauss. McDonald's established Hamburg University. All managers of a McDonald's franchise are required to attend sessions and are given certificates when they complete their program. The curriculum provides in-depth training of the operation of a franchise.

Training programs can be organized in-house or by an outside firm. By organizing the program from within the organization, management can assure that the trainer has adequate knowledge of the company and product or service offered, a commitment to the goals of the company, and is seen by the employees as a credible and integral part of the organization. With an outside firm, management can verify the trainer's previous performance records and select the best firm for its needs.

Before following through with any type of training program, it is necessary to establish the goals and needs of the organization. Below are some of the approaches that may be used to determine these goals.

1. Analyze the employee's job.
2. Complete an employee's need assessment.
3. Ask supervision about the employee's needs.
4. Pretest knowledge and/or skill level.
5. Observe employee work performance.

6. Conduct exit interviews.
7. Establish an advisory committee.
8. Review training programs of other organizations within the industry.
9. Conduct exit interviews.

When needs have been determined and objectives set, the next decision concerns the best approach for meeting the need of the organization. Some possible approaches are:

1. On-the-job coaching by the boss. This can be done informally or in connection with a formal program of performance appraisal and review.
2. Reading selected books, articles, manuals, and other relevant material.
3. Completing a programmed instruction course.
4. Completing a correspondence study course.
5. Attending evening classes at a local university or educational institution.
6. Attending daytime institutes, seminars, and conferences.
7. Developing and implementing an in-house training course.

All of these methods can contribute to the training and development of the supervisors. It is helpful to gather information from every possible source about the best way to approach the training program to insure the success of the venture. After the goals and needs of the company have been determined, development of a strategy to reach these goals is organized.

PLANNING YOUR PROGRAM

Many factors must be considered if the in-service program is going to be successful:

1. **Schedule:** How long should the program be? How often should they be held? When should they be scheduled?
2. **Meeting Facilities:** Where should the meetings be held? What kind of room should be used? How should the room be arranged?

3. **Participants:** Who should be included in the program? Should office supervisors be segregated from shop supervisors? Should all participants in a meeting be at the same organizational level? How many participants should be in each group?
4. **Instructors:** Should they be from inside or outside the organization? What criteria should be used to select them? Should a variety of leaders be used or the same instructor for the entire program?
5. **Content:** What material should be covered? How practical or theoretical should it be? How closely should the content be to the organization's situations and problems?
6. **Methods:** What should be the ratio of lecture to discussion? What other methods should be used?
7. **Materials and Aids:** What materials should be given to participants? What audiovisual aids should be used?

Because people may react differently to similar training sessions, the following are some helpful hints to consider when organizing or participating in a training program. Adults generally learn differently than children. Adult learning tends to be more self-directed, is problem oriented, and is influenced by individual experiences and the timing of those experiences. Good trainers will be aware of these differences in learning and will react with effective presentations. Because of different biases and backgrounds, people may have different reactions to similar programs. By identifying and understanding these different biases, the trainer can be in a better position to relay the main goals of the presentation. Good presentation skills are essential for all trainers.

In-house Training Programs

Developing an in-house program is one of the most common and effective ways to organize a training session. There are many things that must be considered when planning a program of this nature. Because the program is organized in-house, it may be easier to control all of the variables that are involved, which include the schedule, meeting facilities, participants, instructors, content, and materials.

- When determining a schedule, it is necessary to find a time that is convenient for all parties of employee management involved. Although training is a very important feature in an

organization, the loss of valuable, productive work time should be minimized.
- Sessions should be organized to eliminate boredom and maintain an alert audience. Many training sessions are becoming more interactive with the use of role-playing, video cameras, and mutual critiquing.
- Meetings should not be too long in duration and should be spaced at intervals that allow the participants to absorb the information.
- Meeting facilities must be comfortable, convenient, and conducive to the thrust of the program.
- Participants and instructors are important, since the dynamics between these two groups can determine the effectiveness of the session. Because many factors may influence the dynamics of the group, it is important to choose willing participants and effective instructors to achieve the best results possible.
- The material at the training sessions should be geared to the level of the participants and should always be pertinent to their jobs. All materials or aids used in the presentation should be concise and applicable. By handing out too many materials, participants may lose the main points and objectives of the session.

As mentioned previously, training may already be apparent in a company without there being a formal training program in place. One of the most common forms of employee training occurs informally between managers or supervisors and their employees. This method is very effective because managers are knowledgeable in the subject matter and are willing to convey the information because of their personal stake in the individual and the company. In order for this form of training to be as effective as possible, open lines of communication between the two parties are essential. Managers must have interpersonal skills to accomplish this task. They must be able to explain what is expected of the employee during the coaching session and allow the employee to express any ideas or concerns he may have. Careful, almost undetectable, observation of the employee performing his duties is also important. After the manager has reached her conclusions, providing proper feedback to the employee is necessary. This information should be conveyed in a non-threatening way with an opportunity for the employee to question the results. Training of this type is useful, as it is available on an ongoing basis, takes little effort to organize, and is not a significant expense to the company.

Use of Outside Firms

Outside sources, an alternative to in-house programs, may include packaged in-house training kits, consultants, or study aids. There are pros and cons to using outside sources; each is dependent on the individual situation and needs of the organization. Many of the outside resources are general in nature and can be applied to various organizations. The most commonly used outside resource is the outside consultant. Outside consultants offer prior experience with the presentation of training sessions, experience with matching programs to organizations, and personal input and feedback about the needs of the organization and the impact of the training session. Once again, the use of this resource is dependent on various factors, including the individual nature of each organization, as well as the quality of the outside program. When hiring an outside firm, it is extremely important to verify the quality and availability of the program or individual to be used. This can be done by requesting references from past clients or contacting the Better Business Bureau to investigate any complaints or problems that have been lodged with the office. Evaluations of all programs are essential, but are even more so with an outside firm to assure quality and satisfaction.

Reactions and suggestions from the trainees, as well as the trainer, must then be taken to evaluate the program for future improvements. This step is the most critical stage in the process because it determines the effectiveness of the presentation. Four areas must be considered when evaluating a presentation: reaction from participants, learning level of participants, behavior changes as related to the work environment, and results in relation to productivity or sales levels. The initial reaction to the program should be recorded on or close to the end of the program. Management must be cautious when reading the evaluations, though, and must remember that it is very hard to please everyone in these situations. Measurements of increased knowledge or skill level as a result of the training session must be taken. Changes in on-the-job behavior are also critical to record how much of what the trainees learned was applied to the job environment. General end results of the presentation should be summarized and analyzed to help plan for similar future endeavors.

With the use of more and more advanced technology in the workplace, trainers must be aware of the current presentation techniques and how they can be related to current or future presentations. The most common medium used to deliver training is by video. By using the most advanced technology, trainers run the risk of confusing trainees due to the lack of understanding and/or knowledge of the equipment. It may be very beneficial, though, to help illustrate or emphasize points. Trainers generally utilize a mix of stand-up presentation and another adjunct instructional aid, such as videos or computers. Many of these instructional aids are available from outside firms, while many others are generated from within companies. The availability of these resources from internal departments depends on the facilities of the company to produce such aids.

SUMMARY

Training has reached the level of a true profession. Many colleges and universities offer programs in training/human resource development. The majority of professional trainers, though, do *not* have specific backgrounds or training in the field. Many are people who have experience in a certain field and enter a training position in that field, using their prior experience to assist them in their job. According to Robert Fenn, "the notion of many coming to training as a second career will be less true than in the past."

In today's recessionary environment, training programs are being cut, but not at the high rate expected. Generally, training programs are being cut in the same proportion to other programs in the organization. Organizations are finding the recession to be the optimal time for employee training. The output is lessened and more time and resources can de devoted to training activities. Training has become an integral part of the needs, goals, and strategies of so many organizations that to cut these programs entirely from the organization would be a mistake.

The importance of training in the future of the business world can be seen clearly with the increased emphasis that has been placed on these types of programs in recent years. The use of training programs helps both the employee and the employer; the continuation of employees' learning will ensure future employability. The emphasis by employers on organizational learning recognizes training as one of the few untapped sources to maintain a competitive advantage. Almost 41 million individuals receive some sort of formal training from their employers each year. Dollars budgeted for outside expenditures such as seminars, computers, and packaged training programs totaled almost $9 billion in 1992. Without these programs, companies may be losing out on valuable resources that are already part of the organization.

SUGGESTIONS FOR DEVELOPING SUCCESSFUL TRAINING PROGRAMS

- Thoroughly evaluate the needs and goals of the company before entering into any training program.
- Question different levels of the organization to include as many perspectives in the evaluation as seem necessary and relevant.
- Establish a tentative training budget before committing to definite plans. It is helpful for a company to try and carefully monitor costs.
- Make sure that all available resources inside the organization are being utilized. This includes one-on-one coaching between managers and employees.
- Establish the source of the training program. This may be in-house or through an outside firm.
- Remember to consider the patterns of adult learning, such as the wide range of reactions to the same presentation and the assumptions that are often made by trainers about the listening quality and capacity of their trainees when choosing a training program.
- Use the latest presentation and training technology that is applicable and available.
- Properly evaluate the training program, looking for all positive and negative outcomes.
- Continue training in a department or area already begun, or find other departments or areas of the organization that may need training assistance.
- Emphasize the importance of training to the organization as a whole and its commitment to ensure its success. By doing this, all employees will realize the necessity and importance of these programs.

SUGGESTED READINGS

Finn, W. T. (1991, July). One-on-one coaching. *Successful Meetings*, 102-104.

Industry Report. (1992, October). *Training*, 25-28.

McIntyre, D. (1992, Summer). Training budgets weather the recession. *Canadian Business Review*, 33, 34, 37.

Stanton, M. (1989, Winter). Workers who train workers. *Occupational Outlook Quarterly*, 3-11.

Watson, K., & Barker, L. (1992, November). Training 101. *Training and Development*, 15-17.

Chapter 29

JOB ENRICHMENT

Many employers have found that happy employees are not always productive workers, and that just meeting an employee's personal needs does not lead to improved performance.

Since the 1960s, hundreds of employers have bought into the promise of improved morale and work performance offered by job-enrichment techniques. Their experiments with different approaches have met with mixed results, and today many questions remain about the relationship between the "enriched" work environment, job satisfaction, and employee productivity.

Does a worker need to perceive challenge or feel a sense of accomplishment before he can be satisfied with his job? Or is a worker's general satisfaction more dependent on how well a job fulfills his personal needs? And which, if either, of these conditions has a positive effect on a worker's performance?

UNDERSTANDING JOB ENRICHMENT

These questions emphasize the need to distinguish between job *satisfaction* and job *enrichment*. The hoped-for results are often similar, but the two approaches involve modifying different aspects of the workplace and the job itself.

Job-satisfaction techniques focus on how company policies and the work environment meet the personal needs of employees. In Maslow's Hierarchy, these are the lower-level needs—physiological, safety, and belonging—that deal with salary, job security, company policy, the social environment, and physical working conditions. Herzberg's two-factor theory categorizes these as "hygiene factors."

Job enrichment, on the other hand, is a management technique that focuses on the task itself. It involves restructuring work responsibilities and activities to make a job inherently more rewarding. This matches up with Maslow's higher-level needs of esteem and self-actualization. Herzberg called them motivators, which included achievement, recognition, and the employee's interest in the work itself.

The relationship between the two approaches has been long debated. Maslow believed that employees must be comfortable with salary, job security, and working conditions before they can gain any satisfaction from the work itself. In contrast, the Hawthorne studies showed that an individual may be satisfied with work even if her needs for pay and good working conditions are not met. Herzberg theorized that these "hygiene factors" contribute little to the most important element of job satisfaction: the ability of a job to give the individual an opportunity to grow, feel needed, and be recognized for her work.

In practice, the causes and effects of job satisfaction—and dissatisfaction—vary depending on each organization's unique situation. Many employers, however, have found that happy workers are not always productive workers, and that simply meeting an employee's personal needs does not lead to improved performance. This has led to the popularity of job-enrichment techniques, which seek to fulfill individuals' needs for challenge, variety, and achievement.

WORK-DESIGN TECHNIQUES

Employers have tried a number of different techniques to create an enriched work environment and overcome the traditional problems of job design. Four of the most popular work-design techniques are:

1. **Job rotation:** moving employees from one task to another. The goal is to reduce monotony by increasing the variety of tasks.
2. **Job enlargement:** combining previously fragmented tasks into one job. This increases variety and adds meaning to repetitive work.

3. **Job enrichment:** giving employees tasks previously performed by inspection and supervisory personnel. In addition to adding variety, its aim is to make work more meaningful by giving employees more responsibility for job outcomes.
4. **Autonomous group working:** allocating an overall task to a work group, then giving it discretion over how the job is accomplished. Workers operate with little or no supervision.

The first three techniques focus on each job or task and redesign it to meet the individual worker's need for challenge, variety, and motivation. With autonomous group working, on the other hand, the work group is the unit of analysis, not the individual job.

BROADER APPROACHES TO JOB ENRICHMENT

These techniques have been used by employers of all sizes. Many, however, have found that the most effective approach to job enrichment is to begin with a general philosophy, then select specific techniques to fit. One of the most well-known approaches, simply termed *job enrichment,* is based on Herzberg's two-factor theory, which assumes that the causes of satisfaction and dissatisfaction are unrelated. It operates on the principle that what is needed to motivate employees is entirely different from what is needed to reduce their levels of dissatisfaction. This two-pronged approach seeks to reduce dissatisfaction by improving "hygiene factors"—physical conditions, company policy, work schedules, supervision—and to increase motivation (and general satisfaction) by changing job content. Changes in job content are achieved mainly by reducing direct supervision. Many supervisory functions become workers' responsibilities, and employees are called upon to perform a greater variety of higher-level tasks. Implementing this approach involves interviewing employees to determine their interests and needs. The results are then grouped into hygiene and motivation categories, and different strategies are used to make improvements in each category. Because it involves reassigning specific job activities, it is probably best suited for operational or production jobs in which tasks can be defined and broken down.

Another approach, called the *job-characteristics model,* emphasizes matching people with jobs. It is based on the theory that individuals will respond differently to the same job, and that a job's character can be altered to increase a worker's motivation, satisfaction, and performance. This strategy assumes that an individual's motivation to perform is based on knowledge of the results and his positive feelings about his responsibility and the meaningfulness of his work. These are achieved by fine-tuning five job characteristics:

1. **Variety:** being able to use different skills and talents to accomplish a wide range of activities.
2. **Task Identity:** doing a job from beginning to end; being responsible for the whole job rather than just a portion of it.
3. **Task Significance:** the degree of meaningful impact the job has; its relative importance to the company or organization.
4. **Autonomy:** freedom to make decisions about how and when the work will be done; discretion in scheduling.
5. **Feedback:** clear and direct information about job outcomes and performance.

Usually, this method involves enlarging the job and giving the employee a greater variety of higher-level tasks.

Although this approach can be used with work groups, its main focus is on structuring the job to meet the needs of the individual performing it. As such, it requires careful assessment of a worker's talents and her desire for more responsibility. It also requires personalized training and much closer communication before and after implementation.

In the last few decades, *Japanese-style management* approaches have received a lot of attention because of their association with high productivity, low turnover, and low absenteeism. This strategy, which often involves major changes in company policy, treats employees like family and emphasizes harmony, teamwork, and group consciousness. Competition among team members is discouraged, and individual performance is not a dominant factor.

Many elements of this approach revolve around the assumption that a worker will be with the organization for life. New employees go through a long socialization process and training program, and promotions are slow. To enhance the variety of work—and immerse employees in the organization's philosophy—companies encourage lateral job rotation, which gives employees the opportunity to learn skills needed for future formal promotions.

Japanese-style management is very heavily culture-bound and often requires difficult changes in values by management and employees. Because so many of the strategies are based on the conditions of lifetime employment, employers in the U.S. have had limited success with these programs.

A fourth job-enrichment strategy is the *quality-of-worklife* approach, which concentrates on improving an organization's overall design. It defines the elements that affect an individual's quality of worklife as the task, the physical and social environments, the administrative system, and the relationship between life on and off the job. The techniques used in this approach depend on the needs of the individual employees and the organization's existing social and technical systems. It relies on individuals to understand the company's philosophy and goals, to evaluate their own capabilities, and to make choices about how their jobs should be redesigned. Typically, changes in job classifications are made to promote teamwork, which may require technical system changes.

GOAL SETTING

The cornerstone of many successful job-enrichment programs is the use of goals. Goals provide structure and direction and create a challenge that is needed to enhance the importance and meaningfulness of a job. They facilitate feedback and enable a manager to give an employee more freedom. The job-characteristics model is particularly well suited to the use of goals. Once management has evaluated the enrichment characteristics of a job—variety, identity, significance, autonomy, and feedback—goals can be set to match. Goals can also be tailored to the abilities of the person performing the job.

When designing goals for a specific job or person, it is important to consider two main characteristics:

- **Goal specificity:** knowing and understanding what specific objectives apply to the job and what their relative priorities should be.
- **Goal difficulty:** the amount of challenge; the degree of uncertainty concerning how and when the goal will be accomplished.

Integrating goals into a job-enrichment program can help management solve the dilemma of how to increase both satisfaction and performance. Experiments with these approaches have shown that when a job is changed to make it more challenging and interesting, it can have a substantial impact on satisfaction but little effect on productivity. When specific goals were set, productivity increased, but there was no change in general job satisfaction.

Although goals can build both challenge and freedom into many jobs, some workers may see them as constraints. An employee who already perceives her job as "enriching" may feel her autonomy is reduced by the addition of goals. This dilemma can be offset by the enhanced feedback created by the goal-setting process, and by using participative goal setting rather than supervisory assigned goals. Employees' sense of autonomy can be preserved if management involves them in the decisions about what their work objectives should be.

TRADEOFFS

Depending on the unique needs of the organization and the individuals in it, any of these job-enrichment strategies may help improve employee motivation and performance. Job expansion can make work easier by making it less dull, but it can also make it more difficult, strenuous, and stressful. Adding responsibilities can push people beyond their levels of competence, and some employees may see the extra responsibilities as a burden. They may perceive the extra work as a "promotion" with no title or pay benefits.

Companies that use goal setting may find that no matter how much worker consultation was involved, the goals are not always accepted, especially by blue-collar workers. Employees often get bogged down in paperwork and reporting, and if jobs are interdependent, performance can be hard to measure. Management is also constantly challenged to reinforce objectives and sustain momentum by setting new goals.

Most job-enrichment programs require a long-term commitment and may involve expensive, company-wide changes in job classifications. Existing jobs and tasks must be carefully analyzed and grouped, retraining programs must be implemented, and management must be ready to make changes. As a result, top executives may be reluctant to invest the resources needed to develop a successful program.

Once management decides on a strategy and begins involving the employees, unions can become a major obstacle. Changes in job classifications may require changes in collective-bargaining contracts. Of the four approaches discussed here, the job-enrichment strategy is the only one where the union probably will not have a role in implementation. The quality-of-worklife approach requires the greatest union involvement.

PAY AND TRAINING ISSUES

With new job classifications and responsibilities, training becomes an important issue. Even if a worker already has the skills needed to assume new responsibilities, there will be an adjustment period during which some extra supervision or training will be needed.

Some managers work personally with employees to teach the new skills. Others elect to have coworkers demonstrate new job duties, sometimes during a transition period where employees temporarily share new tasks. In some cases, such as when a worker's new job duties require proficiency with a new software program or piece of equipment, the company must be ready to invest in outside technical training.

The pay implications of job redesign can also be difficult to handle. A worker's morale may suffer, and he may demand higher pay, if the new job is simply more of the same—if it amounts to a speedup of the old work, or a piling on of routine tasks. To avoid these problems, it is important to do more than just add tasks to an existing job. If management can also build responsibility, autonomy, and meaning into the work, an employee is more likely to feel "enriched" and get true satisfaction out of his new role in the organization.

MAKING THE DECISION

For long-term benefits, job enrichment should not be a hit-or-miss series of experiments. The success of these efforts often depends on the organization's willingness to make significant, company-wide changes in job descriptions and technical processes. Management must make the program a priority and commit the necessary resources to research and retraining. Perhaps most important, managers and top executives need to maintain constant communication with all the individuals who are affected by job redesign.

Many employers' experiences have suggested that a well-designed job-enrichment program can help solve the problem of how to increase both job satisfaction and productivity. Although more research is needed to prove a conclusive relationship, it appears that a combination of job-satisfaction techniques, job redesign, and goal setting can be the right mix for many organizations.

SUGGESTIONS FOR IMPLEMENTING A JOB-ENRICHMENT PLAN

- Treat job enrichment and job satisfaction as two separate challenges to be met by the organization. To improve job satisfaction, focus on how employees' personal needs are met by the work environment: company policies, pay, job security, physical conditions, hours and scheduling, supervision, the social environment. To enrich a job, focus on the characteristics of the task itself: challenge and variety, identity, meaningfulness, autonomy, and feedback.
- Evaluate each task individually, then determine how they can be effectively clustered. Find a meaningful way to group tasks and define them as a job.
- In grouping tasks, address the technical requirements of coordinating people, tools and methods.
- When redesigning job duties, provide clearly defined objectives for each task. Give employees a challenge and a way to measure their success.
- If your job-enrichment program involves changes in job classifications, consult union representatives early in the planning process.
- Involve employees in the process. Conduct open-ended personal interviews to identify a worker's desire for challenge and her talents for meeting it.
- Avoid redesigning a job solely to respond to an individual's personal preferences, idiosyncrasies, or habits. What counts is a balance: effective job design meets both the requirements of the tasks and the people.

SUGGESTED READINGS

Cunningham, J. B., & Eberle, T. (1990, February). A guide to job enrichment and redesign. *Personnel*, 56-61.

Herzberg, F., Mauserer, B., & Snyderman, B. (1959). *The motivation to work*. New York: John Wiley & Sons.

Holt, D. (1990). *Management: Principles and practices* (2nd Ed.). Englewood Cliffs, NJ: Prentice Hall.

Job enrichment is dead: Long live high-performance work design. (1987, May). *Personnel Management*, 40-43.

Maslow, A. H. (1965). Eupsychain Management. Homewood, IL: Dorsey Press.

Yorks, L. (1979). Job Enrichment Revisited. *AMA Management Briefing*.

Chapter 30

JOB SATISFACTION

People find the greatest rewards in jobs where they have freedom within the "law"—when they know the spirit of instructions, not just the letter.

The relationship between job satisfaction and work performance has been long debated, and even after thousands of research studies and published articles, it still leaves us with many interesting questions. One of the most perplexing is "Which comes first?" Does job satisfaction result in improved performance and productivity, or does good performance lead to job satisfaction? And what can managers do ultimately to achieve an optimum mix of the two?

DEFINING JOB SATISFACTION

To be able to make effective management decisions, it is important to have a broad understanding of job satisfaction—what it is, how it is created, and how to benefit from it. One of the biggest problems in developing this understanding is in defining exactly what job satisfaction is.

Job satisfaction has often been used as a substitute for the word *morale,* but morale describes the well-being of the group, not just the individual. It has also been explained as the successful fulfillment of the needs of an individual.

Yet another definition is that job satisfaction is an employee's positive, emotional reaction toward his job. Some equate satisfaction with feelings of happiness, but these are two very different emotions. One of the contributors to job satisfaction is the presence of challenge, which involves risk, and risk itself does not promote happiness. Instead, happiness comes *after* a job is successfully completed. In that light, we can think of job satisfaction as something that is earned; it comes from the feelings of accomplishment an individual experiences after he has successfully met the challenges connected with a job.

WHAT CREATES JOB SATISFACTION?

There are a number of theories and approaches to determining the individual components of job satisfaction. Many are based on the definition of job satisfaction as the successful fulfillment of an individual's basic needs.

Maslow's Hierarchy of Needs, for example, posits that lower-level needs—physiological, safety, and belonging—must be met before the individual can fulfill the higher-level needs of esteem and self-actualization. Translated into workplace components, this would mean that employees must be comfortable with salary, job security, and working conditions before they will gain any true satisfaction from the work itself. Maslow's theories, however, were later disputed by the Hawthorne studies, which showed that an individual may be satisfied with work even if her needs for pay and good working conditions are not met.

Herzberg's two-factor theory is related to Maslow's, but it divides factors into two groups instead of Maslow's five. The first group, motivators, are achievement, recognition, and the interest of the work itself. The second group, hygiene factors, includes pay, security, supervision, and physical working conditions.

Herzberg theorizes that the causes of satisfaction and dissatisfaction are unrelated and separate. If motivators are present, they will lead to job satisfaction, but they will not cause dissatisfaction if they are missing. Similarly, hygiene factors do not deliver job satisfaction, but if they are inadequate, they will lead to dissatisfaction. Herzberg feels that hygiene factors contribute little to the most important element of job satisfaction, which is the ability of a job to give the individual an opportunity to grow, feel needed, and be recognized for his work.

Another way to describe an employee's needs is to classify them as physiological, social, and egoistic. The physiological needs are essentials to live, which are met through money and job security. Social needs include friendships, identification with groups, teamwork, helping others, and being helped. Egoist needs are held within the individual; they are the needs for success, status, power, achievement, and recognition.

Employees should be given the opportunity to reach any of these needs. The job itself can allow an employee to grow. If the job promotes responsibility, freedom, and recognition, more than likely an employee will grow to her own expectations. A limiting job, or one that does not offer opportunity, will not build satisfaction.

Many of these factors are connected with job "empowerment"—the amount of control an employee perceives she has over her work responsibilities and environment. Empowerment factors include individual achievement, perception of importance, variety of responsibilities, advancement potential, and the opportunity for personal and professional growth.

Closely related to this idea is a process theory built on the premise that job satisfaction is not achieved by the job and its workings alone, but by what the individual expects from the job. These expectations are what gives the employee a frame of reference for judging the outcome. The frame of reference will usually be a comparison to someone else, and in the individual's eyes, if the expectation is not met, then he is dissatisfied.

A frequent complaint of employees is "We don't know where we stand." This state of uncertainty leads to anxiety. Employees need clear definitions of not just their roles in the organization and job duties, but the amount of freedom they have in carrying out those duties. If employees do not know what is expected and how far they can go to deliver a service or product, they become fearful of making creative decisions. People find the greatest rewards in jobs where they have freedom within the "law": when they know the spirit of instructions, not just the letter, and have discretion to make decisions on the best way to meet goals.

CONSIDERING PEOPLE BEFORE POLICIES

These theories are helpful when management is deciding how to build satisfaction into a work environment, but the success of any effort will ultimately depend on the unique personal needs of individual employees. Each person will have his own goals and needs for achievement, which may be influenced by education, economic situation, experiences on previous jobs, even his personal life. An employee who is satisfied at home needs less satisfaction from work, and vice versa.

Another variable is gender. Studies have shown that women attach more importance to flexible hours, good physical surroundings, and convenient travel to and from work than men do.

Individuals' personal perceptions of their work-related activities can also affect their feelings of satisfaction. Much may depend on whether an employee thinks of her activities as *work, a job* or a *career*. Some people use the words interchangeably, but in general, *work* and *job* carry more negative connotations: work is a task one has to do, but a career is a field one has chosen. It would seem that an employee who has chosen her work—her career—would be more likely to feel job satisfaction, but it is difficult to identify a direct relationship.

CAN WE MEASURE SATISFACTION?

As we struggle to agree upon definitions and components of job satisfaction, we have also found it difficult to develop an accurate way to measure it. Past research suggests that there is no completely accurate method, but that measurement is easier if job satisfaction can be related to specific variables.

The Cornell Job Descriptive Index (JDI) has been recognized as one of the leading measurement tools for examining job satisfaction. The instrument is a scale with five categories: pay, promotion, people, supervision, and work.

Other tools used to gather information on employees' attitudes about their jobs are questionnaires, interviews, and tailor-made scales. The accuracy of their results depends on the research methodology and the degree of confidentiality. Employees often feel pressure to provide socially acceptable answers rather than the truth. Results can also be skewed by people's natural tendency to complain about their work. People who admit to job satisfaction often do it apologetically or defensively, as if it were disgraceful to be content.

HOW IMPORTANT IS JOB PERFORMANCE?

Performance is defined by *Webster's Dictionary* as functional effectiveness and the act of performing (to do a task or fulfill a promise). An employee's performance can be evaluated or measured qualitatively or quantitatively, and the evaluation will depend on the type of work. An employee's performance in a widget factory will be evaluated by how many widgets he produces in a given period of time. On the other hand, a hairdresser would be evaluated by the quality of his work and the satisfaction level of his customers.

Industry and companies strive to attain high performance by all employees. Many of the factors that make up performance relate directly or indirectly back to job satisfaction. Performance may be related to the difficulty of the task, the conditions of the workplace, the employee's coworkers, job satisfaction, pay scale, flexibility of hours, responsibility, and benefits. Each of these may or may not lead directly to the others, but they all have an impact on each other.

Management's efforts to improve job performance and productivity can, if handled carefully, deliver many benefits to the organization and the employees themselves. An emphasis on productivity can become contagious among employees, thus increasing the production of others.

The best strategy for increasing productivity will depend on the employee population of the company. Many of the strategies are directly related to job-satisfaction factors; these include empowerment, flexible hours, incentives, rewards, and production competitions. Each of these strategies can build performance, but if not carefully implemented, may actually have a negative effect.

Of all of the strategies, *empowerment,* defined as sharing management power with employees, may be the touchiest one of all. Empowerment must be used with caution by someone who understands all the ramifications. Empowerment is used to improve employee and organizational performance and assist employees in attaining certain personal goals.

For an empowerment strategy to find a home within a company, a few conditions must already exist. The most important one is that employees must have the skill, knowledge, and ideas to take part in the scheme. Another key to success is that employees must want power. Some employees may just want to do their jobs and go home; giving this group more power and control will do nothing for production except reduce it. On the other hand, many employees do strive to achieve higher levels of fulfillment, growth, achievement, and recognition. For this group, empowerment can provide satisfaction and challenge.

At the heart of many failed strategies is the common misbelief that job satisfaction will automatically lead to higher production. Many employees are satisfied by just putting in their time, doing work at a slow pace, and going home. Conversely, some employees may dislike their jobs, but perform at high levels. This is the case where an employee wants job security, but has to settle for a dissatisfying job.

WHICH COMES FIRST?

Performance and job satisfaction have long been thought to be related, and there is plenty of real-world evidence to suggest that this is true. Corporations such as McDonald's and Wal-Mart have built their reputations on job satisfaction, and their success speaks for itself.

The exact correlation between satisfaction and performance, however, is far from clear. It was once widely believed that with job satisfaction came motivation and productivity, and some research has supported this finding. Other studies, however, suggest that there is no direct cause-and-effect relationship. One analysis reviewed over 50 studies and concluded that the two are only slightly related.

Other researchers found that in some settings, the inverse correlation—that satisfaction results from greater productivity—can actually be the case. Hawthorne's studies showed that under "polite" supervision, employees' levels of satisfaction increased, but that as their productivity increased, so did polite supervision.

So even with all this in-depth research, we still have not found the basic answer, and we are left with a chicken-or-the-egg conundrum. But does it really matter which comes first? There are so many differences within organizations, individuals, and missions that it is utterly impossible to find comparative ground.

If we can find one point upon which the studies agree, it is that job satisfaction is one of the most important aspects of the workplace. Identifying the correlation between satisfaction and productivity should not take precedence over efforts to actually increase satisfaction. The process of trying to measure the relationship can actually decrease both satisfaction and productivity, and that is something no company can afford.

At best, we can only conclude that high performance *can* lead to satisfaction, and vice versa. It all comes down to dealing with people and their unique personalities, needs, and personal situations. Every organization will have its own special conditions that dictate how much emphasis can be placed on productivity and what elements are necessary to enhance job satisfaction. Designing the right mix of the two—and customizing it to the people involved—is the true challenge for management.

SUGGESTIONS FOR INTEGRATING JOB SATISFACTION INTO YOUR MANAGEMENT PLAN

- Understand that attracting new employees is only the first step in designing a satisfying package of salary, benefits, and working conditions. Once on the job, employees must be given motivation to become committed to the organization. For every project or job activity, provide clearly defined objectives and goals. Build in a way for employees to measure their own productivity and success.
- When designing procedures and giving instructions, define employees' discretionary power. Emphasize the freedoms they have to carry out the instructions and meet goals.
- Understand that most employees need to find ways to go beyond just doing their jobs dependably. Offer challenges and build in opportunities for employees to use creative, innovative behavior.
- Give as much attention as possible to individuals and small employee groups. Design job-satisfaction programs to meet the unique needs of the people in the organization.
- Keep employees constantly informed of management's job-improvement decisions. Make new policies highly visible, and solicit employees' input and feedback for any decision that affects them.
- Avoid trial-and-error approaches. If new strategies are frequently tried out and then dropped, employees will become confused and suspicious of management's commitment to improving their job satisfaction.
- Make the performance-evaluation process a two-way communication. Give employees the opportunity to evaluate the organization, their supervisors, and their work duties.

Chapter 31

MOTIVATION

Motivation is not the simple result of anything that anyone does to other people. One cannot 'motivate' another. Motivation must come from within the individual.

The topic of motivation has received attention from people throughout history. Motivation is defined as the drive to work toward certain goals and to expend considerable energy in reaching them. Motivation must be initiated within the individual, although much attention has been paid to the subject of how to motivate others. Many people, academic and professional, have developed their own definition and theory on the subject. Some of the most famous have been developed by philosophers such as Abraham Maslow, Frederick Herzberg, and Douglas McGregor.

In the 1940s, Maslow identified five sets of needs and arranged them in a hierarchy. The needs on the bottom of the hierarchy must be met before moving to the upper levels. The illustration on page 252 describes these needs. The five levels in the hierarchy represent basic need categories that can be described as follows:

1. **Physiological needs.** These are the most basic needs, often referred to as unlearned, primary needs. Typical examples include food, clothing, and shelter. Once these needs are basically satisfied, the next higher level of needs comes into play.
2. **Safety needs.** This level represents the individual's need for security or protection. It extends not only to physical safety but to emotional safety as well.

3. **Social needs.** This level represents the individual's need for affiliation, social interaction, and a sense of belonging. The person needs to give and receive friendship.
4. **Esteem needs.** This level represents the individual's need to "feel good" about himself. Often referred to as *ego needs*, the needs at this level are satisfied through the acquisition of power and status. The individual needs to feel important or worthwhile, and power and status provide a basis for these feelings.
5. **Self-Actualization needs.** This level represents the apex of human needs. While it is the most difficult to describe, it is that level at which a person tries to become all she is capable of becoming. Many people pursue this need when they strive for competence or achievement in some area. They are trying to realize all of their potential.

**Maslow's Need Hierarchy
(and the Way These Needs Are Satisfied on the Job)**

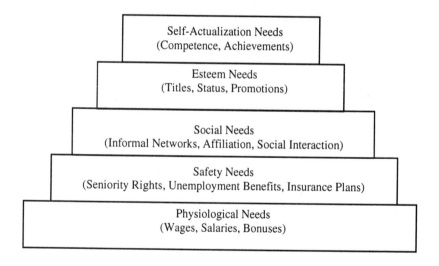

Maslow's theory states that unsatisfied needs are more highly motivating than satisfied ones and they act as the driving force for the individual. Herzberg's theory, from the 1960s, developed a two-factor concept that expanded on Maslow's theory. The two essential components to motivation, according to Herzberg, are *dissatisfiers* and *satisfiers*. *Dissatisfiers* are defined as hygiene or maintenance factors and *satisfiers* as motivating factors. In any given situation, if all of the factors leading to dissatisfaction were removed, no feelings of dissatisfaction would be experienced, but true satisfaction would not be either.

Herzberg's Theory
Typical Organizational Profile

HYGIENES Job Dissatisfaction		MOTIVATORS Job Satisfaction
Company Policy and Administration Supervision Interpersonal Relations Working Conditions Status Security	S A L A R Y	Achievement Recognition for Achievement Work Itself Responsibility Advancement Growth

McGregor classified all individuals into two categories: X and Y. Members of the X group are lazy by nature and avoid all types of work if they can. These people must be threatened and coerced into working. Individuals of the Y group, on the other hand, are not naturally lazy and will work if they find what they are doing is satisfying and suitable to their needs. Punishment and threats only help to thwart their effort. These people may even take up their own initiative under certain conditions on some projects.

McGregor's Theory X and Theory Y

Theory X

1. Naturally lazy	1. Motivate them
2. Dependent	2. Direct them
3. Irresponsible	3. Closely supervise them
4. Hostile	4. Mistrust them
5. Lack imagination	5. Outline their work in detail
6. Lack vision	6. Plan their work for them

Theory Y

1. Dynamic—self-motivated	1. Guide or provide guidelines
2. Independent	2. Provide opportunities for self-direction
3. Responsible	3. Trust them
4. Allies	4. Cooperate—collaborate with them
5. Creative	5. Establish environment
6. Imagination with vision	6. Plan with them

Robert A. Stringer, Jr., believes that there is a network of basic motives within every individual that must be fulfilled. The four main categories are the need for achievement, the need for power, the need for affiliation, and the fear of failure. These motives are always present in the individual. The motivation of the individual results from the interplay between these internal motives and the environment. Stringer believes that individuals will be highly motivated to achieve under certain working conditions and situations. Achievement motivation will be aroused if the goals of management are clearly defined, if the goals involve a moderate degree of risk for the individual, if managers are evaluated in terms of their goal-setting behaviors, if feedback is regularly given to employees, if the working climate emphasizes responsibility, if rewards and punishments are consistent with the goals, and if the employee feels supported and encouraged.

The editors of *Success* magazine conclude that there are three ways of motivating another individual: by force, either physical or psychic, by manipulation, and by persuasion. These situations are often found between parents and children or husbands and wives. These methods may be effective in some situations, but they are hardly beneficial. Motivation by manipulation applies to many situations in everyday life. It may be used in sales contests or special incentive programs. Motivation in these instances is ephemeral. As soon as the benefits do not exist for the individual, the motivation is gone. Motivation by persuasion is the most ethical and effective of the methods. Motivation of an individual is achieved by the sharing of thoughts and opinions to change an individual's perspective, thereby possibly changing his behavior.

Hensleigh Wedgwood, in *Executive Motivation: How It Is Changing*, offers another set of approaches to motivate subordinates. Wedgwood identifies seven distinct approaches:

1. The *authoritarian approach* uses strict orders and very little communication between the manager and the subordinate.
2. The *carrot and stick approach* is a system of rewards and punishments that reinforces correct behavior.
3. The *manipulation approach* places the manager in a position where she sees herself as smarter than the subordinates and able to trick them into doing what she wants.
4. The *money approach* places money as the reward that many employees find most appealing.
5. The *"nice guy" approach* is when the manager sees himself as being so liked by his subordinates that they will be willing to do anything for him.
6. The *rational approach* establishes rules and procedures that cover all scenarios; these rules in themselves reach the objectives.
7. The *paternalistic approach* regards the organization as one big, happy family, therefore making the employees as dedicated to the goals of the company as management.

All of these approaches are relevant, but many of the effects of these approaches are only short term in nature. In an ideal setting, motivation should come from the individual.

Some of the accepted, conventional ways of stimulating employee motivation are through raises, promotions, and other tangible rewards such as contests. Money, power, and prestige are powerful tools, but the effectiveness of these tools is highly dependent on the individual to whom these rewards are being targeted.

Material belongings and wealth are regarded very highly and may be very effective in motivating the individual to elicit the desired responses. In order to determine if these rewards will be effective in motivating their employees, management must diagnose the motives and needs of the worker. Employees who are highly motivated by material rewards will flourish under the material-rewards system. Employees who are motivated by the achievement level of their job, on the other hand, will not.

Contests are different in their approach to employee motivation. The employees who are motivated under these programs have a high need for achievement and recognition. These employees may be motivated by the material rewards that are present at the end of the contest, but most are looking to beat out the other office competition and be recognized for reaching the set goals. Contests can be whatever the management of the organization desires. Many are organized with large budgets, others with very small ones; some with grand prizes, others with small trinkets. Contest themes may surround many different ideas; holidays, anniversaries, sports, office culture, or special events within the organization. The key to the success of such contests is adequate publicity and adequate appraisal of the needs of the employees. One major drawback with the use of contests is post-contest letdown. After periods of intense productivity to reach contest goals, the productivity levels often decrease dramatically. To avoid this problem, it may be important to de-emphasize productivity levels immediately after a contest.

Employee ownership is another powerful tool in motivating employees in an organization. When eight recent major management buy outs were studied in England, employees with ownership were found to be more hard-working and more innovative than before the buy out. The fact that the new owners were now dealing with their own money and personal investment made a significant difference in their motivation levels. The destiny of the organization now rested in their hands and they felt the importance of achieving all that they could with it. These new owners also participated in more teamwork rather than being a small individual competing for a higher-level position in the organization.

Specific motivational programs are also put in effect in various organizations. The Zero Defects program emphasizes the idea the employees should aim to "do it right the first time." This program challenges the individual to complete the task of her job without error. If the employee does make an error, the reaction of management is not to punish, but rather to explain what went wrong and how the employee can avoid the situation in the future. The result of this program is a consistently higher-quality end product. It eliminates the need for massive error correction at the end of production. The elements of this program can be varied, but three consistent components are error cause removal, awards, and communications. The *error cause removal* is an elaboration of the suggestion box. This element continues the quest to identify and correct errors. The *awards system* rewards employees for their consistently high production rate. All employees are judged by the same standards and have equal opportunities. The *communications* element strives to maintain the importance of the program to the employees. The Zero Defects Program is a motivational approach that must be sustained after implementation to be effective.

The Motivation Work Design Program is another program that aims to motivate the employee but from a structural approach. The first area that is improved is the job design of the employee. Certain characteristics are investigated and may be changed: the variety of skills used, the type and timing of performance feedback, and the level of autonomy. The physical layout, work flow, and the organizational relationships are then changed. As a result of these replacements, changes in the management process come about, which then leads to a clearer sense of the mission of the organization. The goal of this program is to see how changing the structure of a job may increase the level of satisfaction.

When setting up a motivational program, it is important to take into consideration a few basic assumptions. While an individual is largely responsible for his own behavior, external factors in the work environment influence behavior as well. Employees make decisions about their membership role in the work setting; whether they will stay, how they will become a member. They also decide how much of an effort they will put into their jobs. Employees may approach their jobs with different expectations, needs, desires, and goals; therefore, managers must be aware of these differences. Employees will engage in behaviors that they feel will reward them in desirable ways and will not engage in behaviors that they feel will lead to punishment.

Many external factors may influence employees' motivation. The "excitement factor" of a job may be important to employees. Management must be able to generate excitement by the determined pursuit of a challenging goal. The importance of a unified sense of mission must be emphasized by management. The environment, both psychological and physical, of an organization is important as well. Employees and others involved may get a certain feeling about an organization without knowing exactly why. This feeling can have a definite impact on their interaction with that organization.

Special motivational programs and contests may serve to engage an employee, but the most important element in employee motivation is a good leader. Management must be able to motivate their employees effectively. A good manager helps her employees feel responsible and strong, rewards them for good performance, and is able to communicate her needs to them coherently. She should also be able to organize her employees into a strong team unit and be willing to listen to suggestions and complaints from her employees. Cooperation at all levels of an organization are imperative.

The following are ways for good leaders to inspire motivation from their employees:

1. Communicate standards and be consistent.
2. Be aware of your own biases and prejudices.
3. Let people know where they stand.
4. Give praise when it is appropriate.
5. Keep employees informed of changes that are relevant to them.
6. Care about your employees.
7. Perceive people as ends, not means.
8. Go out of your way to help subordinates.
9. Take responsibility for your employees.
10. Build independence.
11. Exhibit personal diligence.
12. Be tactful with your employees.
13. Be willing to learn from others.
14. Demonstrate confidence.
15. Allow freedom of expression.
16. DELEGATE!
17. Encourage ingenuity.

The following list gives a few examples of how management should not act. These efforts act to demotivate employees.

1. Never belittle an employee.
2. Never criticize an employee in front of others.
3. Never fail to give employees your undivided attention.
4. Never seem preoccupied with your own interests.
5. Never play favorites.
6. Never fail to help your employees grow.
7. Never be insensitive to small things.
8. Never embarrass weak employees.
9. Never vacillate in making a decision.

Motivational programs are being established not only to increase productivity but also to improve employees' well-being. To increase the exercise level of employees, many companies have started incentive programs. At one company, employees are encouraged to use the stairs instead of the elevator and for doing so receive monetary benefits. Many companies have either built health clubs or gotten membership status at nearby health clubs for their employees. Other companies offer monetary benefits for every pound that employees lose. These programs have had high success rates and provide long-term benefits for not only the companies but the employees, too.

In these times, with many companies feeling the pressure of the faltering economy, downsizing is becoming all too common. Morale and motivation levels at these companies are difficult to maintain. It is important for the executives to revitalize the employees who are still left after the shake-up. The first step is to determine whether the problems plaguing the company are new or are debris from the old management. The second step is to find solutions to these problems. And the third step in maintaining management motivation is to increase the skills of the employees who are left through training programs.

Motivational programs are extremely successful in attaining management goals, but management must be wary of some recent problems. Allegations have been made against certain outside services charging that motivational sessions thought to be helpful by management were "cult-like" in nature. Many employees were emotionally scarred by these experiences and many refused to attend the sessions. Management must be careful to determine that the services they are using are trustworthy and effective. Without such caution, good employees could be lost. Motivation is an important aspect of all organizations and must be accessed to the fullest extent possible.

SUGGESTIONS FOR MOTIVATING EMPLOYEES

- When possible, let people know that there are upward-mobility opportunities within the organization.
- Never belittle a subordinate. No person likes to think that others regard him as stupid. He may have doubts of his own, but does not like others agreeing with him.
- Never criticize a subordinate in front of others.
- Never fail to give subordinates your undivided attention. You do not have to devote every waking moment to your employees. From time to time, however, it is important to give your undivided attention to every person under your control.
- Never seem preoccupied with your own interests. Your own future may very well be your primary concern, but try not to communicate this to others.
- Never play favorites. This is another cardinal rule of good motivation. When you start to make exceptions because of personal preference, the rest of the staff resents it.
- Never fail to help your subordinates grow. Do not try to hold back a good person, even if it means transfer out of your department into another.
- Never be insensitive to small things. Never make loose or rash statements. You will regret it. No problem is too small.
- Never embarrass weak employees. It is embarrassing to an employee to have her supervisor show off at her expense. It is important to a person's dignity to be able to do something well on her own; when you take her work away you also remove self-respect. You are, in short, demotivating her.

SUGGESTED READINGS

Imberman, W. (1989, September/October). Managers and downsizing. *Business Horizons*.

Johnson, A. (1992, January). Mind cults invade the boardroom. *Canadian Business*.

Leposky, R. E. (1987, April/May). The art of motivation. *Corporate Fitness*.

Meet, C. M. (1991, December). Everyone needs attention. *Small Business Reports*.

Nisbet, M. A. (1992, May/June). People aren't wheelbarrows. *Canadian Banker*.

Chapter 32

Participative management involves all levels of workers in the planning and control of their own work activities.

Participation management is a style of management that has gained popularity in recent years. It has become a necessity in the face of increased technology and expanded markets. Employees and management must work together to ensure success for the organization. Rather than using the standard autocratic style of management, organizations need to involve employees in the decision-making process. Many companies are finding that employees are open and willing to the change in management styles. Others have to sell the idea and coax the employees into trusting the positive attributes. Many organizations, eager to maintain a competitive edge, have instituted employee participation programs, such as labor-management cooperative groups and employee teams. In general, these programs aim to involve nonmanagerial employees in the process of making decisions about the work that they perform, the condition under which they perform, and the benefits and other rewards that they receive for such performance. The goals of these programs include the improvement of the workplace, efficiency, productivity, and morale.

In one year, IBM paid out $107 million in awards to employees who made suggestions on how the company could improve its operations. That may seem like an enormous amount of money, but IBM received a substantial return. The company benefited by a saving of $98 million by acting on employee suggestions. The Lockheed Corporation spent over $700,000 to establish quality circles. As a result, over a four-year period employees generated ideas that saved the aerospace giant over $5 million. Blue Cross/Blue Shield estimates it saved approximately $1.65 million in four years by designing a wellness program for employees. As a result, the number of insurance claims filed by employees dropped while absenteeism was reduced.

These companies quickly realized that an employee-involvement program is not a luxury—it is a necessity. Organizations are confronted with change in technology and markets that require joint efforts between employees and management. With assistance from supervisors, employees can solve significant problems and develop new work methods. The delegation of responsibility to participative groups can work to the advantages of both labor and management. It should be emphasized that the participative process exposes workers to the expertise, knowledge, and work styles of many different people, allowing them to adopt successful ones for themselves. This system increases the sophistication and effectiveness of the group working together.

WAYS TO INVOLVE EMPLOYEES

Many managers have trouble adjusting to the participatory style. They are used to delegating and ordering rather than discussing and cooperating. Participatory management involves changes throughout the whole organization, and these changes must first be evaluated to ensure that they are right for the organization. Involving employees in management can help companies cut expenses, increase revenues, and develop new lines of business. Consultants who have had prior experience with participatory management programs should be questioned and consulted. After consultation, a thorough plan should be put into writing. This plan should deal with possible pitfalls and should accurately describe each step of implementation.

An organization can do several things to foster greater employee involvement. These options are not the only ones; there are others. Every employee involvement program must be adapted to fit the unique characteristics of the organization. Here are some ways to involve employees in the decision-making process:

- **Suggestion systems.** Individual ideas can save companies millions of dollars a year. Most savings come through a series of modest suggestions that, taken collectively, add up to impressive amounts. Encourage employees to come up with ideas that they can implement to make their work more efficient or to serve customers better.
- **Quality circles.** These are some of the most popular employee involvement techniques. They are simple, effective, and easy to implement because you do not have to change your organizational structure. Employees typically meet once a week to examine quality issues and recommend improvements. A committee can coordinate work among the various circles and spearhead quality control ideas that cut across organization lines.

Self-managed teams are more sophisticated than the quality circle groups. These groups of employees set their own agenda, decide the tasks to work on, develop job descriptions, and may even hire new members. This is the most hands-off management style. Therefore, management must be sure that employees will rise to the challenge of managing themselves. They must be able to spot and correct their own errors and be truthful to their superiors. Often with these groups, the members have a variety of different backgrounds and must understand the other points of view in order to arrive at consensual decisions. Decisions are often better, solutions are reached faster, team members feel more ownership in the organization, and the work environment is more positive. Employees are able to see the effect of their individual efforts on the company.

Gainsharing offers employees the opportunity to become involved in management while giving them a tangible reward for becoming involved. Employees are asked to give management advice on how to utilize company resources better. Gains that the company experiences because of the improvements that are above a certain base level are split between the company and the employees.

Improvement teams are set up to solve specific problems in the organization. These teams may be composed of employees from one unit or they may come from a cross-section of various departments. They may be set up to solve only one problem or may be in place for long periods of time to do continual problem solving. For true success, every member of the team must play an active role; therefore group membership should stay fairly small.

Organizational surveys are used to gather information about the company from employees. The information is gathered from written questionnaires or from group interviews and is then analyzed. The results are used to solve problems or make improvements.

Participative goal setting involves employees in setting the goals of the organization. This method allows them to have an input on the goal level. The employees and management regularly meet to discuss the progress of the goals. Employees feel more motivated and have a stronger sense of control of their work levels.

The *concept of empowerment* is another important management technique that is similar to participatory management. The idea behind empowerment is to give employees the freedom to make decisions without any supervision. The degree of freedom must be carefully analyzed and decided upon. An increase in the vested interest of the employees in the day-to-day operations of an organization will increase their stake in the company and their effort to improve results. Empowerment can be carried out by giving employees more say in the organization and by getting managers to delegate more responsibility. Often with the use of empowerment, managers feel that in order to be effective, they must become submissive, overlook poor performance, relax standards, and become overall "softer" managers. All of these ideas help to create ineffective and inefficient empowerment programs. To be effective, empowerment programs must set expectations. These standards must be communicated to employees so they understand their role.

As soon as an organization decides to establish an employee involvement program, it should bring together all those involved into the planning process. Employees, as well as management, must give 100% effort if the process is to succeed. It is suggested that a steering committee be created to help devise and implement the necessary changes. Allow this committee to make decisions and give it a significant voice in the process while defining the boundaries of its responsibilities. It is also important to encourage people who have strong viewpoints and who may disagree with the process. It is better to hear their arguments now than to have them undermine the process later.

Matthew Juechter, in his article "The Pros and Cons of Participative Management," discusses Velma Lashbrook's series of questions that access the degree to which an organization is managed participatorily. Below is a list of these questions (see pages 270-271 for scoring and results).

Is Your Work Unit Participative?

Take this quiz to assess the degree to which your work unit managed participatorily. Indicate as *true* (T) or *false* (F) the applicability of each of the 15 statements to your situation. You may answer by (1) using your peers and immediate managers as the work unit or (2) if you are the manager, looking at the work unit consisting of your subordinates and you.

T F 1. Members of my work unit receive regular feedback about their performance in relation to goals.
T F 2. There is more independent effort than teamwork in my work unit.
T F 3. Members of my work unit are encouraged to think for themselves and suggest ways to improve the work environment.
T F 4. Major decisions affecting our work unit require group consensus before they are adopted.
T F 5. Management often makes decisions without explaining their rationale.
T F 6. Members of my work unit are committed to achieving our group's goals.
T F 7. When a problem needs solving, members of my group usually work together.
T F 8. Members of my work unit are seldom asked for their opinions before decisions are made.
T F 9. My coworkers and I help make the decisions about the important issues affecting us.
T F 10. The rumor mill is often a better source for information than our managers.
T F 11. Members of my work unit meet regularly to review how we are doing on our goals.
T F 12. There is little sense of togetherness in my work unit.
T F 13. Management appreciates and uses our suggestions for improving the work culture.
T F 14. The authority and power for decision making rests clearly with management in this organization.
T F 15. I feel free to go to management to find out about decisions that are being made.

Instituting a participatory style in an organization does not mean that success comes immediately. In some instances, there has been a decline in productivity. Management should be cognizant of the problems it may face and the reasons why participative management may fail.

PITFALLS OF PARTICIPATORY MANAGEMENT

One of the common mistakes management makes when instituting an employee-involvement program is not informing and preparing employees for it. Historically, case studies have revealed that it is not smart to disagree with management. Subordinates know that managers often take confrontation personally, even managers who profess to welcome disagreement.

A participative management style can also be a problem for managers who have big egos. Sometimes employees feel that when group members are interacting with each other's ideas, they will not receive recognition for the ideas they contribute. Managers in participation management systems must be constantly aware of innovative ideas and recognize the employees who submit them.

Another pitfall of participative management is when senior management is accustomed to using the authoritarian style and now must give up some power to their subordinates. Some managers feel threatened by this change.

Also, some management experts fear that participatory groups lack leadership. No one individual wants to accept the responsibility to make final decisions; as a result, it is difficult to get anything accomplished.

When you decide to use the participatory style, you should articulate the purpose of the organization and identify the role employees will play in the decision-making process. For example, ask yourself whether the purpose is to share information, to brainstorm, or to seek input. Is the role of the employee to help make decisions, or will he be making the decisions himself? Writing down the answers to these questions is an excellent way to help clarify the purpose and roles.

As mentioned previously, effectively operating a participatory system can sometimes be difficult. However, it is worthwhile to expand the effort given the growing demand that management tap its human resources, make employees more self-reliant, and promote teamwork. In most cases, being autocratic is less efficient. It is well accepted that you get better results when you get people involved. Participatory style lets employees influence results. Using the participatory approach does not cause management to lose control of the organization. In fact, it is a way of gaining control. Without a good two-way relationship between management and employees, the manager usually ends up out of touch with the organization. The participatory style is most effective when applied to situations where change is taking place, when consensus and commitment is needed, and when the manager does not know the decision to be made.

Often with participative management strategies, the "leaders" of an organization must question and reevaluate their role. The truth is that in any organization, there are not individuals who can achieve their goals without the input and assistance of other people. The best leaders use other people to solicit ideas and assist with problems. No one in a position of authority can ever do too much listening. Individual situations may require people other than the designated leader to work in positions of leadership. In any given group, there may be dozens of leaders, not just one. Leaders have vision, either for the organization or for their own department, the drive to share the vision with others, the courage to stimulate change, the ability to inspire change and action, the foresight to motivate people to new levels of achievement, the wisdom to learn from others, the integrity to serve as a good example, and the willingness to acknowledge the efforts of others in the organization or team.

Where do you fall within the participatory management spectrum?

Tell	Sell	Invite Comments	Seek Input	Join the Group	Authorize Group
You make a decision and announce it, assuming it will be accepted.	You announce your decision but also use persuasion to ensure its acceptance.	You announce your decision but remain open to questions and comments that may help you fine-tune or change it.	You present an issue before any decision has been made to ask for ideas and suggestions. You are still the one who makes the decision.	You empower your group to make the decision, with you as one of the participants. You define the limits.	You define limits but leave the decision to the group. You do not participate in making the decision.

SUGGESTIONS FOR ESTABLISHING A PARTICIPATORY STYLE OF MANAGEMENT IN YOUR ORGANIZATION

- As soon as you decide to go ahead with an employee involvement program, bring subordinates into the planning process. Your employees will have to give you 100% effort if the project is to succeed.
- Consider creating a steering committee to help you devise and implement the necessary changes. Allow this committee to make decisions and give it a significant voice in the process while defining the boundaries of its responsibility.
- If you are attempting to increase the level of participation but historically had an autocratic style, participants are likely to withhold their ideas until they feel confident about your intentions. Be patient; it will take a period of time for the group to work with you.
- Participation management must be perceived as being important to the attainment of organized goals by both management and employees.
- All involved in the participatory process must accept the problems inherent in the implementation of participative schemes and must set realistic goals.
- Accurate information about the organizations should be made available to all involved in the participative process. This requires effective communication between all levels in the organization.
- In order for participation management to be successful, the organization must give employees more say, get managers to delegate more, and provide overall direction, priorities, and ground rules of the organization.
- Participation groups need to set expectations relative to the desired outputs and results. Management should convey to the group what it hopes it will accomplish and check for understanding and agreement.
- Managers of participatory style of management must communicate the organization's business goals and accomplishments on a regular basis and share the success stories with work team members.

> - It should be remembered that no one in authority can ever do too much listening. The best leaders know that people become successful not based on what information they impart, but on what they elicit.

SUGGESTED READINGS

Schmidt, W. H., & Finnigan, J. P. (1993). *TQM manager.* San Francisco: Jossey Bass.

Naisbitt, J., & Aburdene, P. (1985). *Reinventing the corporation.* New York: Warner.

Bouening, J. C. (1990, March). Employee participation strengthens incentive programs. *Occupational Hazards.*

Lawler, E. E. (1986). *High involvement management: Participative strategies for improving organizational performance.* San Francisco: Jossey Bass.

Towe, L. A. (1989/90, Winter). Survey finds employee involvement a priority for necessary innovation. *National Productivity Review, 9*(1).

ANSWERS TO "IS YOUR WORK UNIT PARTICIPATIVE?" QUIZ

Five major management criteria determine the degree of participation in this quiz, and three questions were used to tap each. Score yourself on each area, using the following key:

Area	Item	Points
Goal directed	1	T=1, F=0
	6	T=1, F=0
	11	T=1, F=0
Input seeking	3	T=1, F=0
	8	T=0, F=1
Information openness	5	T=0, F=1
	10	T=0, F=1
	15	T=1, F=0
Group oriented	2	T=0, F=1
	7	T=1, F=0
	12	T=0, F=1
Shared decision making	4	T=1, F=0
	9	T=1, F=0
	14	T=0, F=1

These five criteria do not include all of the characteristics associated with participative management. However, they are representative of some key values believed to influence managerial effectiveness.

When a group is *goal directed*, its members have a context in which to channel their participation. Goals help distinguish relevant from irrelevant issues.

A *group orientation* improves the quality of decision making and the ease of implementation. It also helps improve morale and develop a commitment to the work unit and organization.

When management is *input seeking*, it both gathers and uses the ideas of relevant parties to aid decision making. This improves trust in management and also aids quality of decision making.

Shared decision making involves reaching a consensus prior to decision making. The most obvious results of this practice are commitment to goals and decisions, increased acceptance of change, and improved implementation.

Finally, *information openness* is a key to effective participative management. Well-informed employees can give more relevant and useful feedback. They are also more likely to trust management when it's clear that management places some faith in them.

Adding across the five dimensions will give you an overall participative management score:

13-15 *High participation.* Scores in this range indicate that participative management has been clearly established in your work unit.

10-12 *Moderate participation.* Scores in this range indicate that participative management is definitely making headway in your work unit. Look at the five subscores to identify problem areas.

7-9 *Some participation.* Scores in this range indicate emergence of some participative values. Examine subscores to see which areas are developing and which ones need work.

4-6 *Low participation.* Scores in this range indicate a lack of appreciation for the values of participation. Major organizational or personnel pressures may be required before a shift will occur.

0-3 *No participation.* Scores in this range are unbelievable in contemporary American business. If the work unit is to survive, these values need to be reexamined.

Chapter 33

PROBLEM EMPLOYEES

Employees will experience problems on and off the job. The astute manager will recognize the problem employee ... know when to confront and comfort, so that problems will not become permanent and result in destructive behavior.

The problem employee is annoying, antagonizing, obstructive, frustrating, and often an impediment to organizational goals. There are many different kinds of problem employees: those with no ambition, those who are lazy, those who are confrontational, and those who create conflict among employees. The common denominator among all problem employees is they are difficult to manage. Managers may delay taking action with these individuals for a period of time, but eventually they must be dealt with for the good of the organization. If they are not, they will reflect badly on management and the organization. One of the primary causes of the existence of problem employees is the incompatibility between their job assignment and their personal goals. In some cases, an individual may be overqualified for her job or her work style may not fit the organizational culture. In some cases the problem employee may be dissatisfied due to job reassignment because the organization may be in a downsizing mode, or he is in disagreement with company personnel policies. It is also possible that the job situation creates a high stress environment and causes the employee to behave abnormally.

The manager should be constantly alert the characteristics of the problem employee. By doing so, steps can be taken before a crisis occurs. Managers might ask the following questions to detect whether problem employees exist within the organization: (1) Does any particular individual spread gossip within the organization? (2) Is there continued complaining by employees? (3) Does an employee find it difficult to accept constructive criticism? (4) Does an employee's action affect others adversely? (5) Has an employee's productivity level decreased? (6) Are employees seeking jobs elsewhere? (7) Are employees absent or habitually late for work?

Dealing with employees who are in some way unsatisfactory or ineffective can be a time-consuming experience. For the manager, problems among subordinates are generally limited to a small percentage of the total group. However, a small number of persons can account for a large percentage of the problems and concerns and take a large share of the manager's time. It is imperative to deal effectively with this group, no matter how small, since the impact of its dissatisfaction can become multiplied throughout the group and organization. Let us consider examples of problem employees in four distinct categories: (1) the agitator, (2) the goldbricker, (3) the energizer, and (4) the career coaster.

The Agitator

The agitator is present in every organization and represents a substantial problem. This individual is often treated severely by the manager. In many cases the agitator does not recognize the trouble he causes. He is usually an insecure person and in need of constant attention. Substantial efforts have often been made by the manager to meet his needs and to provide counsel, correction, and even severe discipline.

The agitator is usually subtle in the trouble he causes. He is likely to cause trouble in the form of dissension in the work group or to have interpersonal conflicts with individuals than to create trouble for their immediate superior. He in many instances may "brown nose" the boss or put on a good appearance to cover his tracks. The best course of action for the manager in this situation is to let everyone know that she is not unaware of the source of trouble. This results in the winning of support of the agitator's coworkers, since they are often held responsible for the disruption of their work. The manager may lose respect from other employees if he does not deal with the agitator. It is important to identify the agitator, not to brand or fire him but to find the causes of his disruptions and to correct them. No doubt the manager must spend a great deal of time with the agitator. She will be required to instruct, correct, listen, and introduce recommendations to correct the situation. She also must be firm and demand improved behavior.

Being objective and fair will be difficult but is essential if the agitator is to be made a productive employee.

The Goldbricker

The problem of the goldbricker not contributing enough to the work effort has been a concern since the initial studies of human resource management began in the early 1900s. Even now limited progress has been made. Many management scientists believe the problem lies primarily in the area of employee motivation.

Absenteeism represents a behavior adopted by goldbrickers. The chronic absentee, though she may have seemingly good reasons for her behavior, is often showing hostility to the organization and/or her colleagues. As indicated, this represents more of a case of motivation than a situation requiring discipline action. However, many organizations still approach this problem from the latter point of view rather than the former.

In former years there was a tendency to view the goldbricker as a disruptive employee or a person with poor personality traits. The revitalization of this individual was not frequently considered, since many psychologists felt it was not possible to change an individual's personality. We now know that with assistance and guidance from the manager, the goldbricker can often become a motivated individual and a productive employee. In order to accomplish this task, the manager must conduct accurate and periodic performance appraisals, interviews, and discuss the findings with the problem employee.

The Energizer

This individual is often the most creative person in the organization. The worst scenario is when a moderately competent individual is also the energizer. The energizer who is highly intelligent can often be disruptive, but in many cases can do the organization much good. However, the manager must guide this individual's energies in a direction that will result in a productive effort. The energizer can often accomplish difficult goals if the manager provides firm direction and strong leadership.

The energizer is a person who has high physical and mental energies. As a result, he will place excessive demands on other individuals and organizational resources. As mentioned previously, these individuals are often creative and appear to be self-centered. They want to be involved in all organizational activity. Many of their colleagues may view them as opportunists or manipulators. When an individual has a high level of intrinsic motivation, the problem becomes one of direction and guidance rather than to push and motivate, as in the case of the goldbricker. When the demands for resources by the energizer are not met, this is when he becomes a potential problem for the organization. The manager should try to be encouraging to the energizer but not at the expense of all others in the organization. The approach should be supportive and reassuring. Frequently the energizer is an insecure person, wanting constant attention and praise, nonthreatening situations, and approval from colleagues and management. The energizer is often thought of as the "square peg in the round hole"; however, if properly managed he can become the most productive individual in the organization.

The Career Coaster

Individuals faced with career standstill are interesting in view of the pressure our society places on many of us to climb the career ladder until we have reached the top of the organization. The individual who does not progress to the next level of "success" may present a unique problem for the manager. Whenever our expectations of not moving to the next level of the hierarchy are not realized, our behavior may be disruptive and our productivity level can decrease. "Organizational political" problems and strains of interpersonal problems may also increase. The person in a career standstill encounters fears, stresses, tensions, and dangers that may influence her to accept a plateau of job dissatisfaction. This can happen particularly to older employees. However, this can happen regardless of chronological age, even to a person in her early thirties. Even after only a few years, the career coaster may question the time she has spent with the organization, as well as the purpose and importance of her job. She believes that the passing of time threatens to create new discomforts, disappointments, and defeats. If the employee is chronologically older, she may become increasingly concerned about her downhill slide in the organization. The older employee may feel stagnated and can feel a sense of frustration, believing that her career aspirations will not be attained. This is especially true when younger individuals gain promotions to which she aspired. The career coaster will feel angry and guilty about not achieving her career goals. The anger these employees feel can be demonstrated in the form of irritability and oversensitivity, resulting in a disruptive employee completely unmotivated in her job. This employee refuses to extend herself in any way, either pacing her job, giving extra time to her work or taking time to work through problems. Finally, the disenchanted older employee can become a thorn in the manager's side. Her negative attitude and continual interference can affect the morale of other employees. She is not one you want other employees to emulate.

At some point in time all employees face problems on the job. Usually through the help of their superior or fellow employees, they resolve their problems. However, some situations cannot be resolved, resulting in a problem employee. This is not a desirable situation for the organization, the manager, or the employee. Eventually problem employees will demand action.

In-service training becomes an important aid in minimizing problem employees. These programs can impact greatly problem employees who (1) have low interest in their jobs, (2) have negative work attitudes, (3) lack job enthusiasm, (4) continually bring complaints, or (5) need to increase their productivity. For the most part, the employee who has received adequate training does not contribute to these problem areas.

As soon as an individual has been employed by the organization, he should participate in training and orientation programs. Rules, regulations, and policies should be discussed, specific job tasks and responsibilities should be reviewed, organizational goals need to be understood, and staff reporting responsibilities should be clear. An understanding of the organization's purpose and function should reduce situations that lead to problem employees.

SUGGESTIONS TO IMPROVE YOUR MANAGEMENT OF PROBLEM EMPLOYEES

- Do not over-criticize, embarrass, intimidate, dominate, or over-control your employees. In many instances, a manager can create problem employees without realizing it. This will affect their morale and their productivity.
- Listen to complaints; do not become defensive about them. If an employee expresses to you what is troubling him or her, it will give you the opportunity to learn something about your employees. Remember, an employee who expresses a com-

plaint is giving you a chance to help when your help is most wanted and needed.
- When dealing with the older employee, encourage her to capitalize on her seniority rather than becoming a problem employee. She can assume the role of an elder statesman and a mentor to younger employees.
- A manager should ask the following question when analyzing the problem employee: Am I communicating with him? Remember, communication is a process culminating in a change of the receiver. If no change has occurred, communication has not taken place.
- Once you have decided a problem employee needs help, do not waste time. Plan for an informal interview. Bad habits should be stopped immediately, not delayed until a scheduled formal appraisal. It takes courage, sense of responsibility, and maturity to take forceful action to straighten out a problem employee.
- Try to see the situation from the problem employee's viewpoint. If the solution to the problem is to be successful, it must recognize the difficulties faced by the problem employee. However, recognize that differences in experience, knowledge, and attitude exist and that complete understanding is not fully achieved.
- Exert authority only from reason. Dependence upon power invites rebellion from the problem employee. A manager's authority should make sense in terms of its humaneness to employees.
- Assist in removing obstacles that confront the problem employee. It has been shown that many problem employees are unmotivated because of the obstacles presented by the organization's bureaucratic structure. By eliminating these barriers, these employees can perform more effectively and increase their productivity.
- Be fair with the problem employee: conduct effective performance appraisals. It has been my experience that many managers do not take performance appraisals seriously. If the manager does not, we cannot expect the employee to do so. Remember, the purpose of performance appraisal is to improve performance and productivity as well as reduce problems faced by employees.

SUGGESTED READINGS

Alliore, P., & Richard, N. E. (1989, December). Quality and participation at Xerox. *Journal for Quality and Participation.*

Milkovich, G. T., & Glueck, W. F. (1985). *Personnel: Human resource management. A diagnostic approach* (4th ed.). Plano, TX: Business Publications.

Parker, M., & Slaughter, J. (1988, Fall). Managing by stress: The dark side of team concept. *IIR Report.*

Schuler, R. S., & Huse, E. F. (1979). *Case problems in management.* New York: West Publishing.

Weiss, W. H. (1982). *The supervising problem solver.* Chicago: AMACOM.

Chapter 34

SEXUAL HARASSMENT

In an era when men and women work closely together, claims of sexual harassment are more prevalent. A strong anti-harassment policy supported by effective and well-planned preventative and/or corrective action will go a long way to insulate your organization from liability.

No event riveted the attention of the American public as did the 1991 Clarence Thomas/Anita Hill hearings. The televised U.S. Senate Judiciary Committee hearing on the appointment of Clarence Thomas to the Supreme Court was reported to have drawn the largest audience of any program in television's history. While some of the allure of the hearings may have been simple titillation or racial concern, by far the most riveted members of the audience were women familiar with the burdens of sexism. What they were hearing was, unfortunately, too familiar. As the shock of the public hearing faded, however, the topic of sexual harassment did not, as the covert or hidden experiences and thoughts of women began to surface. Although sexual harassment charges were already rising before the Thomas/Hill controversy, dramatically more women began to speak, write, and complain of sexual harassment. And the flow of revelations and charges reported in the national media continue—against U.S. senators, the FBI, the military, and the Veterans Administration. In less than a year after the Anita Hill/Clarence Thomas hearings, sexual harassment charges filed with the EEOC increased by more than 50 percent.

Most females of all ages have experienced some form of sexual harassment. Much of this harassment, unfortunately, occurs at home or work. It is not the stranger stalking females who is the threat, but someone known, trusted, or more critically, someone with power over the female. At a time when women and men work more closely together and women begin to defend themselves, claims of sexual harassment are more likely and prevalent. Responsibility for minimizing such claims lies with the manager of any organization where women and men work or interact together. To meet this responsibility, a manager must specifically know the legal and social dimensions of sexual harassment.

DEFINITION OF SEXUAL HARASSMENT

Sexual harassment can be defined as any unwelcome advances or requests for sexual favors or any conduct of a sexual nature when:

1. Submission to such conduct is made, either explicitly or implicitly, a term or condition of an individual's employment.
2. Submission to or rejection of such conduct by an individual is used as the basis for employment decisions affecting such individual.
3. Such conduct has the purpose or effect of substantially interfering with an individual's work performance or creating an intimidating, hostile, or offensive working environment.

The courts have determined that sexual harassment is a form of discrimination under Title VII of the U.S. Civil Rights Act of 1964, as amended in 1991.

What Is Potentially Harassing?

Sexual harassment is not simply a personal problem between men and women but is a form of discrimination based on different treatment of both sexes, or sex-role stereotyping. In particular, sexual harassment can simply involve:

- Obscene pictures in the workplace or at a work-sponsored activity.
- Leering at a coworker's body.
- Sexually explicit or derogatory remarks.
- Unnecessary touching, patting, or pinching of another's body.
- Subtle pressure for sexual activities.

- Demanding sexual favors in return for a good work assignment, performance rating, and promotion.

More specifically, sexual harassment includes:

Deliberate or repeated unsolicited verbal, visual, or written comments, gestures or physical contacts of a sexual nature, as well as sexual advances or requests for sexual favors; that is, any sexually imbued messages or actions that are unwelcome or offensive. Although men can also experience sexual harassment, it is predominantly women who are targeted.

A 1993 Supreme Court decision relieved women of having to *prove* they suffered psychologically from sexually abusive language or actions. Rather, the justices listed specific factors to guide the lower courts: Is the conduct frequent, severe, physically threatening, or humiliating? Does it interfere unreasonably in a worker's job performance? Would the conduct seem hostile or abusive to a reasonable person? Or does the individual worker who is the target view the workplace environment as abusive?

Some of the claims of sexual harassment that federal and state judges have considered valid are: unwelcome sexual advances, coercion, paramour preference (organizations where there are intimate relations between supervisors and employees), indirect sexual harassment (having to witness someone else being harassed), improper conduct, and visual harassment. In addition to its own employees, an organization can also be liable for the inappropriate behavior of customers, clients, service people, and visitors!

Sexual harassment also includes a manager who knowingly permits such behavior on the part of employees, ignores complaints about sexual harassment, or is remiss in promulgating a sexual harassment code.

Because of the impact of such events as the controversial Thomas/Hill hearing—and the sharp increase in sexual harassment complaints following it—men are more noticeably ill at ease about what constitutes proper behavior in the workplace. Managers have even greater cause for concern because of the legal and financial implications of sexual harassment complaints and charges brought against their agencies or businesses. Sexual harassment charges filed against a manager or other employee can quickly escalate into a class-action suit as others experiencing harassment come forward.

Managers must be fully aware of the range of sexual harassment, from the subtle to the grossly overt or assaultive. It is surprising to hear managers respond to sexual harassment charges by exhorting: "It can't happen here!" Or managers and supervisors who are not familiar enough with obscene language to realize a sexual insult has been made then judge the woman who describes the harassment as herself depraved! Managers and supervisors are themselves too often guilty of sexual harassment, as their more powerful position in a business or agency lends itself to such abuse. Thus, it is important that an organization policy impartially cover and protect *all* employees.

PREVENTION IS BETTER THAN A LAWSUIT

The problem of sexual harassment begins and ends with the manager. As stated in the U.S. Equal Employment Opportunity Commission's *Interpretive Guidelines on Sexual Harassment*, "the employer has an affirmative duty to maintain a workplace free from sexual harassment and intimidation." A manager who is alert to the potential discrimination charges, and the problems and cost of sexual harassment, will focus primarily on training employees how to behave and treat others in the workplace. There are various ways in which a manager of any organization can generally discourage sexual harassment. By far, the most effective step is to adopt a strong and enforced *Sexual Harassment Policy*. Such a policy must be more than a notice on a bulletin board, or merely lip and pen service, but one strictly and fairly enforced in a timely manner, regardless of the parties involved.

The policy should succinctly and clearly state the organization's policy, perhaps tying it in with the organization's overall values; include a promise that all grievances will be promptly and thoroughly investigated; protect the complainant from reprisal or retaliation of any sort or further harassment; and explain the disciplinary measures to be taken against those who harass and those who bring false charges. Finally, incorporate this policy into all orientations for new employees.

A suggested policy is:

> - Sexual harassment of employees or applicants for employment will not be tolerated.
> - Any employee who feels that he or she is a victim of sexual harassment by a supervisor, coworker, or customer should bring the matter to the immediate attention of (the person's supervisor or a designated person in the personnel department). An employee who is uncomfortable for any reason about bringing the matter to the attention of his or her supervisor should report the matter to (the designated person in the personnel department).
> - Complaints of sexual harassment will receive prompt attention and be handled in a confidential manner to the extent possible. Prompt disciplinary action will be taken against persons who engage in sexual harassment.

In addition, provide employees with meaningful recourses for sexual harassment. As often with charges of rape, the victim is put on trial, where mere association or familiarity with obscene language or behavior tars the complainant with the same brush as the offender. For this reason, an organization should provide two neutral but powerful persons to whom employees can bring grievances. The head of the organization or one of its managers is a logical choice; the second person could be an outside advisor. Or the policy statement on sexual harassment can clearly state how one can contact the Equal Employment Opportunity Commission (EEOC) and what the function of the commission is.

Investigate promptly and take appropriate action. Managers must handle all complaints seriously and impartially. Sexual harassment is a veritable legal mine field, so if at all possible, an organization should have a special EEOC officer or someone assigned to that function. This person should also receive appropriate and ongoing training to ensure the best legal defense and handling of such grievances. In no way give the impression to the complainant of not taking him seriously, or treating him in a perfunctory manner.

In reviewing recent class-action suits of sexual harassment at larger organizations, much effort and expense went into attempting to disprove the charges. In the end, the cases are frequently settled in favor of the complainants, who have by then often left the organization. The money wasted in defensive posturing could have been more effectively used to take appropriate action promptly, as well as perhaps retaining all the lost employees. Shooting the messenger is the worst response to what is a costly and demoralizing organizational problem. Sexual harassment, whether reported or not, causes financial and emotional costs for an organization and the individual victim—legal fees and settlements, loss of contracts, absenteeism, damage to the organization's reputation, and poor morale.

FILING A COMPLAINT

The process for making a complaint about sexual harassment falls into several stages:

- *Direct Communication.* If there is sexually harassing behavior in the workplace, the harassed employee should directly and clearly express her objection that the conduct is unwelcome and request that the offending behavior stop. The initial message may be verbal. If subsequent messages are needed, they should be put in writing in a note or a memo.
- *Contact with Supervisory Personnel.* At the same time direct communication is undertaken, or in the event the employee feels threatened or intimidated by the situation, the problem must be promptly reported to the immediate supervisor; then the problem should be reported to the next level of supervision or with the EEOC officer.
- *Formal Written Complaint.* An employee may also report incidents of sexual harassment directly to the EEOC officer. The EEOC officer will counsel the reporting employee and be available to assist with filing a formal complaint. The department will fully investigate the complaint and advise the complainant and the alleged harasser of the results of the investigation.
- *Resolution outside Department.* It is hoped that most sexual harassment complaints and incidents can be resolved within an agency. However, an employee has the right to contact the State Department of Human Rights or the Equal Employment Opportunity Commission (EEOC) about filing a formal complaint.

The Need for Training

Sexual harassment policies and clear grievance and disciplinary procedures are invaluable in responding to sexual harassment in an organization. But to prevent such harassment and to encourage the necessary changes in employee attitudes and behavior requires training. Small organizations and agencies can join together to hire a consultant on sexual harassment, or on the broader topic of gender equity; also, there are videos and other educational materials available on the topic.

Education and training are essential to reveal not only the wide disparity between the way in which men and women view and judge certain behaviors and body language, but to describe the discrimination that many endure that encourages unwanted sexual attentions. Training employees, supervisors, and managers is critical to inform them of their rights and responsibilities with respect to the prevention of sexual harassment. As organizations are held liable for the behavior of their supervisory or management staff, it is essential they be trained in how to deal with issues of sexual harassment. If a complaint goes unattended and the complainant files a charge with the EEOC, the organization can be held liable. The manager who fails to get to the bottom of a complaint in a timely manner may wind up with the court doing so instead. The court will assess not only the alleged conduct, but whether the employer knew about, or should have known about, the alleged conduct.

It is generally believed that sexual harassment evolves from the sex-role stereotypes or expectations men have about women that are transferred to the workplace. A confusion occurs between what is perceived to be a person's sex role with his work role. These stereotypic roles, as well as their confusion with each other, should be dealt with in training, which puts the problem of sexual harassment in the broader social context from which it springs. The growth and successes of women's liberation have been met with unprecedented levels of violence, abuse, and harassment toward women and children, which ultimately surface as well in the workplace.

As mentioned previously, many people are not clearly aware of what constitutes sexual harassment. In many cases, the perceived harassment is a misunderstanding of expectations between parties, often based on stereotypic role expectations. For this purpose, a professional trainer or facilitator is invaluable. The range of topics such training should encompass are: what causes sexual harassment; its impact on the workplace; definitions of sexual harassment; sexism; sexual politics; types of sexual harassment and their implications for the organization, especially liability; what constitutes a good policy statement; distributing and communicating sexual harassment information to employees; the scope of a grievance-punishment system; how to investigate complaints; and strategies for preventing sexual harassment.

PROPER BEHAVIOR IN THE WORKPLACE

As media attention continues to focus on prevalent examples of alleged sexual harassment in the workplace—and at the highest levels—managers must begin to focus on guidelines for workplace behavior. At the very least, an employer or manager must not lower standards of behavior or condone their slippage:

For instance, monitor behavior rather than simply language: the "body English" of male-female interactions. Workers need to learn the difference between a pleasant attitude and one that implies or assumes sexual availability. All too often the "language" of sexuality is translated by men for women. Therefore, the burden for avoiding even the appearance of sexual interest has rested with women. While women are being encouraged and trained to be more assertive in stopping unwelcome sexual advances, men also require training in how they incorrectly interpret women's body-English "messages" and clothing.

Employers and managers who neglect to promulgate explicit and detailed anti-harassment policies and guidelines may find themselves at particular risk. While non-federal employers are not required to have such guidelines, without them the presumption is that the employer condones sexual harassment in the workplace. There should be no room for doubt as to a proactive policy by management on this volatile issue. Although sexual harassment is not a new workplace issue, organizations have more recently been made aware of its effect on the bottom line as a costly and demoralizing organizational problem. Even if an organization is confident it is not at fault in a particular charge of sexual harassment, it frequently settles out of court if it is apparent that its defense is going to be very expensive. As organizations continue to downsize (or rightsize) their staffs to reduce expenses, issues of discrimination and workers' behavior will be a deciding factor in who keeps a job. Organizations can no longer afford to ignore the high cost of discrimination nor pay the penalty for employees who perpetrate such behavior.

SUGGESTED GUIDELINES FOR SEXUAL HARASSMENT PLANNING

- All workers need to be informed of their rights and responsibilities with regard to preventing sexual harassment.
- Be alert to any discomfort on the part of coworkers about being complimented on their wardrobes or personal appearances.
- Try to keep business and personal life separate. If a worker or manager asks someone out and he is refused, he should not ask again.
- Discourage gender-based hostility when women enter a previously all-male work environment.
- Conduct yourself in a professional manner, treating others with respect and courtesy.
- Pay attention to how others respond to what you say and do.
- Do not assume your coworkers or employees enjoy comments about their appearance, hearing sexually oriented jokes, being touched, stared at, or propositioned.
- Designate someone in the organization to handle the issue, which includes clear guidelines for grievance procedures and a clear statement of disciplinary actions when anyone is found guilty of such behavior or of false allegations.
- Prevent the occurrence or repetition of sexually harassing behavior by avoiding persons who exhibit such behavior, and especially avoid being alone with them after business hours. Instead, confront the issue head-on by meeting with the person, keep a written record (a paper trail) of each incident (where it occurred, time, date, what happened), and take the problem to personnel or other designated officer within the company or agency. If that fails to curtail the offensive behavior, a worker can then formally file a complaint with the EEOC.
- Protect workers who file informal or formal complaints. Remember, three-quarters of those filing such complaints have experienced retaliation.

SUGGESTED READINGS

American Management Association, courses on Managers and Sexual Harassment, 135 West 50th Street, New York, NY 10020 (212-586-8100/Fax 212-903-8168).

Bravo, E., & Cassedy, E. (1995). *The 9 to 5 guide to combating sexual harassment: Candid advice from 9 to 5. The National Association of Working Women.* New York: John Wiley & Sons.

Eskenazi M., & Gallen, D. (1996). *Sexual harassment: Know your rights!* New York: Carroll & Graf Publishers.

Gomez-Preston, C., & Reisfeld, R. (1995). *When no means no: How to protect yourself from sexual harassment by a woman who won a million-dollar verdict.*

Petrocelli, W., & Repa, B. K. (1994). *Sexual harassment on the job: What it is and how to stop it.* Berkeley, CA: Nolo Press.

Chapter 35

STAFF DEVELOPMENT

The function of training is to ensure that each employee makes a required contribution to the achievement of the goals of the organization.

Those in charge of staff development training must find ways to capture the processes by which people at work increase their skills and expand their insights. Management must develop monitoring and measuring systems that motivate people to learn more and better. Trainers must create training courses, conferences, manuals, videos, and one-to-one learning situations that are geared to efficient and purposeful learning results. Especially in growing organizations such as park and recreation organizations, trainers have an opportunity to influence the direction of growth through training policy. Robert Reich, one of the foremost economists in the U.S., states that the success of American enterprise no longer depends on the private investments of highly motivated entrepreneurs, but rather on the unique attributes of our workforce and their accrued skills and insights of American workers. He further points out that what really identifies U.S. economic strength is its educated brain power.

It has been reported that American organizations spend a lot of money each year on training their employees. One conservative estimate puts the figure at $25 billion. Knowing how to develop a diversified workforce by meeting job training needs is a strength of a number of organizations. In the next decade, organizations will need to extend themselves to understand the workforce training needs of business, industry, and public service as they prepare for the 21st century.

Trends now affecting workforce-training needs include the growing complexity of many jobs, a drop in the number of entry-level employees available and the quality of their basic skills, and the need for constant retraining as whole classes of jobs are made obsolete due to changing technology and organizational restructuring. Trends related to the graying of America and increased use of technology to address workforce needs created by these and other trends will require additional efforts by organizations that will encompass renovations in existing programs.

Among the challenges facing those in charge of staff development is designing and delivering high-quality programs based on need and staying abreast of an explosion of information and new technologies. At a recent meeting of the American Society for Training and Development, Donald Peterson, former CEO of the Ford Motor Company, emphasized the needs for trainers and consultants to maintain a strong commitment to lifelong learning in their training programs.

To demonstrate training and development importance, trainers must not only present excellent programs but also must prove the programs get results, improve job performance, and make more efficient use of resources and satisfactory returns on the training dollars invested. Those who plan, design, present, or market programs must start building a strong and lasting relationship with management by asking themselves some questions: "Does the return on our training investment include increased productivity and cost containment?" "Does our training truly benefit the organization and its workers?" "Do our programs result in the acquisition and use of new and marketable skills—the kind the organization will need over the next three to five years?"

MANAGEMENT SUPPORT FOR STAFF TRAINING PROGRAMS

We are constantly reminded that staff development programs must be supported by management to be effective. Management must: (1) Give enthusiastic endorsement and approval when staff and employees are involved in the program; (2) Make a commitment and a statement outlining what changes should take place after the training has been accomplished; (3) Reinforce the behavior change taught in the training program; and (4) insist on follow-up on the results and evaluation of the program for each employee. When management displays this kind of support for the training program, it will be effective. This, of course, is assuming that the training program is properly designed and addresses the needs of the employees. Jack J. Phillips, a training expert, has classified management into four different types according to the degree of its support. These include:

1. *Supportive.* Here management becomes a strong, active supporter of all training and development efforts. Management wants to be involved in the training programs and wants its employees to take advantage of every program that applies to the organization. Management reinforces the material discussed in the training program and voices approval. It gives positive feedback to those in charge of training, frequently calls on the training staff for assistance, advice, and counsel.
2. *Responsive.* Here management is supportive of training programs but not as supportive as the above style. Responsive management allows its staff to participate in staff-development programs and will encourage them to get the most out of the program. If questioned about the program, it will usually voice support but does not go out of its way to show strong interest.
3. *Nonsupportive.* In this situation, management will privately voice displeasure at staff development on a formal basis. It will reluctantly send staff to the program. It will probably think it is spending too much money on training. There is little, if any, reinforcement for the training program. In fact, management's action may destroy any value of the training.
4. *Irresponsive.* In this case, management will actively work against the staff development programs. It will try to discourage staff from attending the program, even suggesting reasons why staff should not attend the programs. It will openly criticize the training program and the training department's efforts, both in public and in private. The management philosophy is that all training should be accomplished on the job. There is negative reinforcement when employees attend programs.

By assessing each of the management philosophies and classifying them according to the suggested categories, you can develop a strategy for working with each. Ideally, you need supportive and responsive management. Realistically, you will have some of all. The most important thing is to eliminate or change irresponsive management.

Showing results of training programs can dramatically increase the management support of those programs. Management is usually eager to participate in a program that yields results and a return on investment. Dugan Laird, in his book *Approaches to Training and Development*, provides an excellent reference on measuring results. The illustration below describes his suggested measures.

Measuring Results of Training Programs	
Decrease or Reduction in	Increase or Improvements in
absenteeism	total output
accidents	productivity levels
overtime required	on-time work
tardiness	cost savings
employee errors	employee morale and attitudes
turnover	
grievances	work backlog
complaints	

In order to gain management support, those in charge of the training programs must be very clear as to the principles they will employ with regard to the developmental processes and training programs that will be initiated. Suggested planning principles should include:

1. Whenever possible, all training programs should be the product of participative decision making between those responsible for conducting the training and those being trained.
2. All training programs will be developed according to the state of readiness of the trainees rather than trying to sell the latest and/or most novel approach.
3. All training programs will be built upon the results of need analysis, which indicates where the problems are and can be dealt with.

4. All programs will attempt to be on the "cutting edge" of the organization's needs.
5. Consistently develop credibility for the training efforts and not just be viewed as the "ivory tower bunch" who do not understand the real world.

With these principles in mind, top management support can be secured, while at the same time become self-sustaining from a cost standpoint.

Determining Training Needs

As mentioned previously, the purpose of staff development is to improve the performance of employees in the organization. The initial step in the planning of the program is to determine training needs—the knowledge, skills, and attitudes that should be changed to improve performance. There are many approaches to determining these needs. Some of these are:

1. Analyzing the job descriptions of all employees. Be sure job descriptions are up to date. This analysis will identify the duties and responsibilities of each employee and the skills necessary to perform their job effectively.
2. Analyze problems in the organization. High turnover, excessive accidents, high costs, and low productivity will be good indicators of possible pitfalls in the organization. The causes of each problem should be determined and steps taken to correct them. One of these steps may be instituting a staff development program in such areas as employee orientation, communications, and motivational techniques.
3. Conduct a survey of employees. Who knows better than the employees themselves what they need? Here we are looking for *perceived needs*, areas in which the employees feel they need training. This information is helpful for two reasons. First, the very process of asking them what they need helps establish a rapport with the employees and those in charge of training. Second, the survey tells employees, "We want to help *you*, but you have to tell us what will help. Figure 35.1 is a suggested form that may be used to survey employees. This information may be obtained by interviewing employees. Interviews take much longer but would provide more in-depth information. A written survey can obtain a great deal of data in a much shorter time.

Figure 35.1: Employee Training Needs Assessment

Place an "X" after each topic in the column of your choice.

	No Need	Some Need	Great Need
Programming principles and concepts			
Setting program goals and objectives			
Development of programs			
Market needs assessment			
Program plan preparation			
Advertising and promotion			
Registration techniques and procedures			
Staffing			
Program pricing techniques			
Determining and tracking program costs			
Equipment purchasing			
Program evaluation techniques			
Sports programming			
Cultural arts programming			
Special events			
Social recreation programming			
Tournaments and competition			
Innovative activities and trends			
Managing and managers			
Social responsibility and ethics			
Problem solving and decision making			
Strategic planning and management			
Implementation techniques			
Authority, delegation, and decentralization			

Figure 35.1 Continued

	No Need	Some Need	Great Need
Motivation, performance, and satisfaction			
Leadership techniques			
Groups and committees			
Interpersonal and organizational communication			
Effective control techniques			
Methods of fiscal control			
Entrepreneurship			
Employee training			
Budget process			
Internal and external networking			
Volunteers: the new employees			
Safety			
Risk management			
Supervision			
Employee performance appraisal			
Interviewing techniques			
Time management			
Management by objectives			
Employee and participant discipline			
Majorities and minorities			
Sexual harassment			
Public relations			
Other:			

4. Ask management their opinions. Many feel that management know more about what training needs there are than do the employees. This information can be obtained by using the survey form or by personal interview.
5. Test for the skill and knowledge level of all employees. Before instituting the training program, those in charge of training should know at what level employees are in terms of skill, knowledge, and attitudes. A quiz or test could be constructed that would determine these levels. It is important that any test used be job related. These paper-and-pencil tests are valuable in determining attitudes and knowledge. Performance tests are needed to measure skills.
6. Observe employees at their work stations. This approach is not as objective as some of the previously described techniques but can provide some indicators to training needs. For example, those in charge of training may periodically observe employees in situations such as presenting ideas, listening, obeying rules and regulations, making out reports, planning the use of time, and attitudes toward other employees. A summary of these observations may suggest some training needs.
7. Analyzing performance-appraisal forms can be a good source of determining training needs. Where there is an effective performance appraisal in place, management has probably spent a great deal of time reviewing employees. This information can be helpful in a determining training needs. Unfortunately, this approach is rarely used for two reasons. First, the performance-appraisal system is not effective; therefore, the information is not useful. Second, those in charge of training do not go to the trouble of studying the information. Also, in many cases, the performance review system is under the direction of someone other than the training personnel.
8. Conducting exit interviews can elicit valuable information. Some of the best information regarding training needs can come from employees who leave their positions. If the interviews are well conducted, the data can become valuable. The interview must be conducted by a neutral person who knows what information to get and how to get it.
9. Using a management- and employee-advisory committee can be helpful. The committee should consist of about six to 10 individuals representing all departments. The purpose is to

give advice, not to make decisions or take action. The data from the surveys and interviews should be presented to the committee for analysis and recommendations. The committee can help sell the program and in many cases take part in it.
10. It is important to look at other organizations' training program. Many organizations exchange ideas with one another. This approach provides suggestions, but care should be taken that training needs are the same.

Establishing Training Objectives

Once specific training needs have been identified, training objectives should be formulated. A definition of training needs is not a statement of training objectives. The need must be restated as a training objective or a series of training objectives. Many training objectives have been loosely labeled as "training objectives" with little thought given to the relative complexities of attaining them. The major classification of objectives that have been training objectives are:

- *Operational objectives.* These are measured in terms of organizational outputs, reduced costs, and increased productivity.
- *Performance objectives.* These have to do with an individual's performance, such as achieving efficiency standards and staying within budget.
- *Instructional objectives.* These can be measured by some type of test or examination.
- *Reaction objectives.* These are subjective in nature and can be obtained by asking participants to report their feelings and reactions to the training program. There is considerable merit in obtaining feedback from participants for use in improving the program content and instructional methods. In many cases, these reactions are used to justify training expenditures or to sell training to top management.
- *Personal growth objectives.* These objectives have to do with feelings of self-confidence, self-competence, self-image, and other aspects of self-realization.

Use of Outside Training Resources

Most organization training programs make use of outside training resources simply because they do not have enough expertise to conduct the courses on their own. These outside resources can be divided into four categories: equipment and aids, package training programs, consultants, and outside training and development programs.

1. Equipment and aids. Trainers need to buy or rent equipment and aids necessary to conduct effective training. The first step is to know what is available. The second step is to determine which ones are desirable and necessary. The final decision is whether to buy or rent. The list keeps changing as new products are developed. Some of the products that could be helpful are: projectors, overheads, videotapes, movies, slides, audiotapes, easels, flip charts, hook-and-loops boards, chalkboards, books, games, tests, and case studies.

The best way to keep up to date on what is available is to be an active member in associations that promote staff development. You will be kept informed through ads in various training and development publications, exhibits at national and state meetings, and mail you receive from manufacturers and distributors.

2. Package training program. A number of organizations have developed package programs that can be purchased for immediate use. Some of these programs come with leaders as well as with materials. Others include only materials and a detailed lesson plan on how to conduct the program. The value of these programs depends on two factors: first, what alternative do you have? If you have the knowledge and ability to plan and conduct the program yourself, the package program will be of little value. If you do not have the internal resources to plan and implement a program, the purchase of a package program may be to your advantage. Second, do the package programs meet your needs? A careful study should be made of the objectives, content, and approaches to see whether it would fit your organization's need. Also, various programs should be considered to compare costs and benefits.

Remember, if you purchase a package program, do not be afraid to adapt it to fit your organization. You may wish to eliminate, change, or add to the content, aids, or approach. This customizing can make the difference between an effective program and one that is not.

3. Use of consultants. Books have been written on the selection and use of outside consultants. Some organizations have been very happy with the contributions made by outside consultants and others have been disappointed. The process of selecting an outside consultant should be essentially the same as hiring an employee. The following six steps can be helpful.

- Determine organizational need. What kind of help are you looking for? Do you want someone to help you evaluate your approach, develop a program, or teach a seminar?
- Know which consultants can meet your needs.
- Evaluate the consultants by means of correspondence, telephone, conversations, interviews, and reference checks.
- Negotiate a fair and clear agreement of exactly what the consultant will do. Know before the work begins what it will cost.
- If you are not positive that you want a long-term arrangement, hire the consultant to do one job. If it works out well, then you can negotiate a long-term agreement.
- Thoroughly orient the consultant to the organization. Be sure that she knows the things about your organization that are related to the work she will do.

4. Use of outside programs. Most organizations make use of outside programs. Organizations that offer training programs fall into the following categories: (1) universities and colleges, (2) professional associations, (3) employee trade associations, and (4) private training organizations.

In order to select the best programs, several steps should be followed: (1) Find out what programs are available that meet the needs of your organization, (2) Evaluate the training programs—careful reading of program bulletins will provide some indication of the content, quality of leaders, and pertinent information on cost and potential benefits. Contact other organizations that have used the program. In checking out these references, care should be taken to get objective and specific evaluations.

Evaluation of Training Programs

To demonstrate training and development's importance, trainers must not only present excellent programs but also must prove the programs get results, improve job performance, make more efficient use of resources, and achieve a satisfactory return on the dollar invested. The proof of the above is revealed through program evaluation. The concept of evaluation has received widespread recognition as beneficial but the practice of evaluation has lagged behind. Few reports of actual program evaluation have been published compared to the number of training programs. According to the report of one study, less than 12% of the 285 organizations studied evaluated the results of training programs. In most cases, one or more of the following excuses is offered by organizations for not evaluating their training programs.

- "We know training works." The contention here is that since training is developed by the organization experts in response to specific needs, no evaluation is needed.
- "Evaluation is too expensive." Prohibitive cost is frequently cited as the prime reason for not evaluating training. This argument can be particularly strong when it is coupled with the "we know it works" philosophy.
- "Why risk self-evaluation?" In many organizations, the evaluation function exists as an appendage to the training department. It is not surprising that training personnel are hesitant to undertake evaluation efforts that might appear to invalidate the training department's own activities.
- "The human element is not treated as an asset in many organizations." In this context, training is often viewed as an expensive, necessary evil. The primary objective becomes avoidance of any nonessential program spending—namely evaluation costs.
- "If it doesn't work, I don't want to know."

Why evaluate? Most organization experts agree that competent evaluation is the cornerstone of meeting organizational and individual educational needs and improving the cost effectiveness of the training function. Although a single evaluation provides answers to all the questions relevant to return on the training program dollars, carefully planned and controlled research enables one to monitor and justify productive training expenditures and to avoid or reduce unnecessary losses.

Questions frequently answered through evaluation include:

- Is the problem amenable to a training solution?
- Which training method is most appropriate for the material and target audiences?
- Was the course material learned? Did the training result in the desired immediate demonstration of behavior change?
- Did the immediate learning or behavior change translate into the desired job performance?
- What changes could be made to improve the training programs?
- Are there particular types of trainees for whom the training is more or less effective?
- What are the implications of the induced behavior change relative to meeting long-range organizational objectives?

SUGGESTIONS TO IMPROVE YOUR STAFF-TRAINING PROGRAMS

- On-the-job training, where possible, is especially effective because it gives the employee direct work experience along with specific guidance.
- Training lectures should be combined with case studies, role-playing games, and other teaching methods for maximum effectiveness.
- Group interaction at a classy resort has a great deal of glamour, but as a training method is seldom worth the cost.
- If you are to build on success, training programs must be evaluated for effectiveness. They may need to be adjusted or redesigned to make them as valid as possible.
- Managers require training somewhat similar to that of their subordinates but with heavier emphasis on dealing with people.
- Employee training should focus on specific needs of the organization and not on current fads.
- Training must be honest and perceptive; no one should suggest that every aspect of the system works perfectly.
- To facilitate the transfer of training from the classroom to worksite, management can use a number of methods such as discussing objectives, providing role models, nonperformance, and praising successful behavior.

- New employees require special attention. They are often receptive to any help received. Orientation training is a valuable tool for sharing important information regarding pay, benefits, and safe work practices with workers and for developing a relationship with them.
- Expect progress to be slow at first as the trainees struggle to set aside old ways of doing things.

SUGGESTED READINGS

American Society of Training and Development (ASTD). (1989). Training America: Learning to Work in the 21st Century. Alexandria, VA: Author.

Hudson Institute. (1987). Work Force 2000. *Work and Workers for the 21st Century.* W. O. Johnson, Project Director and A. H. Parker, Co-Project Director. Indianapolis, IN:Author.

McCarthy. (1989).*Training Program Workbook and Kit.* Englewood Cliffs, NJ: Prentice Hall.

Nelson, C. (1992). *How to start a training program in your growing business.* Chicago: AMACOM.

Wexley, K. H., & Latham, G. P. (1981). *Developing and training human resources in organizations.* Glenview, IL: Scott, Foresman.

Chapter 36

SUBSTANCE ABUSE

No matter who you are or what your company does, your workplace can be affected by drug abuse. And if your image of the typical drug abuser is a minority, inner-city teenager, think again.

Substance abuse is broadly defined as the illegal use of controlled substances or the immoderate use of legal substances that can affect performance. Researchers estimate that the first category is far less damaging to companies than the second: people who drink excessively or take some drugs but are not readily classified as "addicts."

Some managers choose to deal with the issue of substance abuse by simply assuming that their workplace is free of problems or by sidestepping the issue until termination is the only remaining option. A more beneficial approach is to address the problem directly and take corrective steps at an early stage.

Because most substance abusers are employed full time, problems in the workplace occur in roughly the same proportions as in the general population. That means most managers are probably dealing with a problem right now, whether they know it or not.

The pattern of substance abuse is to worsen gradually to the point where the employee may no longer be salvageable. Left undetected or ignored, substance abuse can have negative effects on the workplace through increased absenteeism, abuse-related health problems, and judgment errors; decreased productivity, initiative, and overall morale; and, potentially, the loss of employees in which the company has invested valuable training and experience.

Managers, who have supervisory control over employees' livelihoods, have a great deal of leverage in addressing problems of substance abuse. The best management tool in which to apply that leverage is a current substance-abuse policy that is applied fairly and communicated clearly throughout the company.

DEVELOPING OR REVISING A SUBSTANCE-ABUSE POLICY

A policy on substance abuse will be most effective if it has broad-based support throughout the company. In developing or revising a policy, it is wise to convene a working group with broadly based representatives: top management, company health professionals, employee or union representatives, and human resources personnel.

The first task of the working group is to consider what the company hopes to accomplish with the new policy. Is productivity the main concern? Reducing accident-related costs? Preserving the company's investment in valued employees? By determining the relative importance of each goal, the working group can build a policy that serves both the company and the employees.

In developing the policy and related programs, the working group should draw on the experiences of other companies. There is, for example, a general consensus that the best policies include the following:

- Approach the issue as a job-related health problem affecting job performance.
- Adopt as a priority the protection of employees' privacy and confidentiality.
- Develop strategies for identifying substance abuse.
- Develop strategies for addressing substance-abuse problems.
- Include initial and ongoing education as a prevention method.

Approaching substance abuse as a job-related health problem restricts the company's policy to the area in which it has a legitimate interest: job performance. Very few jobs require that the company police its employees' behavior after hours, and employee resentment of such intrusions can undermine support for the entire program. Questions of morality or criminality, therefore, are best avoided.

Similarly, protecting the confidentiality of employees is an important part of building support for the program. Employees who feel they are placed in jeopardy by acknowledging a problem will naturally hesitate to seek assistance. Employees will also hesitate to report any problems they identify in coworkers, for reasons of both loyalty and fear of damaging a fellow employee's reputation and career.

There are a number of ways to identify substance abuse in the workplace. Supervisors, who should already be able to identify work-related problems, can be given additional training in how various forms of substance abuse can affect work performance. Every employee benefits from a safe and prosperous workplace; educating employees about how substance abuse endangers the work environment gives them a stake in spotting problems at an early stage.

Some companies have adopted various drug-testing programs to identify substance abuse. The testing can be administered as an across-the-board screening device for job applicants. It can also be required of existing employees in periodic random checks. Other companies have rejected drug testing, citing the expense, employee resentment, and the cumbersome monitoring involved in collecting valid urine samples. In addition, while drug testing is able to identify the presence of certain substances, it does not reveal how the substance is affecting work performance. And drug testing has almost no practical use in detecting problems with alcohol, the most commonly abused substance of all.

Some companies have introduced random searches of employees' belongings and brought in dogs trained to sniff for controlled substances. With these practices and with drug testing, the company should move cautiously and consider all possible legal implications.

Identifying substance abuse is only the first step in a well-conceived, coordinated substance-abuse policy. The company must also have strategies in place to address any problems that are discovered. Few companies can afford to simply terminate valuable employees—and few would want to even if they could. One solution that has evolved over the past few decades is known broadly as "employee assistance programs" (EAPs).

EMPLOYEE ASSISTANCE PROGRAMS

Programs to assist employees with substance-abuse problems or a wider range of problems affecting job performance come in as many varieties as there are companies. The common element is recognition that companies have a stake in restoring job performance to an acceptable level and a willingness to invest corporate resources to preserve valuable employees.

Some companies have found that offering assistance for a wide range of problems gives employees greater confidence in making use of the program. An employee might ask for financial assistance, for example, and in the intake and counseling process reveal that the actual issue is cocaine addiction, the source of his financial plight. With the EAP defined broadly, the employee has a ready-made explanation for coworkers (he needs some money). And because employees may access the program for any number of reasons, confidentiality is maintained more easily.

Large companies may decide they have the resources to deal with substance abuse internally through their own EAP and in-house staff. Some smaller companies have formed consortiums for this purpose, and some unions have established their own EAP for members. Some companies have opted to contract substance-abuse services to outside specialists.

The experiences of EAPs with established track records indicate that efforts to treat substance abuse are most effective in the early stages of abuse. Best, of course, is educating employees about the effects of substance abuse and the possible sanctions if abuse is discovered. If prevention does not work, supervisors should try to detect the abuse before it becomes a large-scale problem. Very, very few cases of substance abuse are self-correcting, and the longer the problem persists the more difficult it is to address.

Evidence suggests that EAPs work best when employee participation is voluntary. Ideally, employees will refer themselves when they begin to have problems. If they do not and a performance problem is detected, referral to the EAP should still be voluntary.

EAPs have reported enormous success in dealing with substance abuse and can be an important component of any substance-abuse policy. If the company is not prepared to institute a large-scale EAP, it may still be helpful to learn more about how they work and then incorporate useful elements into the company's procedures.

ESTABLISHING THE NEW POLICY IN THE WORKPLACE

Procedures for detecting and addressing substance-abuse problems (through an EAP or other means) should be fully in place before the new policy is announced. Supervisors must be alerted to the new procedures and trained to detect early symptoms of substance abuse (see Table 36.1).

Unless the supervisor is also trained to deal with substance abuse (and most are not), it is not appropriate for him to address that part of the problem. When the supervisor detects a performance problem, he should address the performance with the employee and attempt to determine the cause. If substance abuse is indicated, the supervisor should refer the employee as described in the policy and follow up administratively to ensure the situation is in hand.

When the company introduces the new policy, it is important that every employee understands the new procedures and knows how they apply to her. It may be appropriate to distribute copies of the policy to all employees and to post notices on company bulletin boards or include information in the company newsletter.

Part of communicating the new policy is helping employees understand the reasons for it. Substance abuse has a negative effect on everyone in the workplace, through higher costs for benefits, greater hazards for abusers and nonabusers alike, inequitable distribution of the workload, and more. Generating support among the employees will go a long way in paving success for the new policy. Recognizing how it affects them personally, for example, will encourage employees to participate in early detection of substance abuse.

Continue communicating with employees after the program is up and running. New employees should be oriented to the policy as part of their introduction to the company. Continuing employees can benefit from occasional workshops about trends in substance abuse or new information about how it affects the workplace and the company. Without compromising anyone's confidentiality, seek opportunities to present the program as a positive benefit for all employees. Over time, if the program succeeds in rehabilitating employees who abused substances, word of mouth should reduce any remaining employee resistance.

Table 36.1
Substances and Their Effects

Alcohol, the substance abused most widely in the workplace, leads to a variety of problems with job performance. Interestingly, on-the-job accidents are *not* highly associated with alcohol problems, although accidents do occur disproportionately outside of work. Easily confused with other causes, warning signs of alcohol abuse include poor attention and difficulty concentrating, tremors, absenteeism, long lunch hours, poor job performance, and rapid changes in work pace. In management positions, alcohol abuse can be manifested as an extreme reluctance to take risks or make mistakes. As with all substance-abuse problems, the pattern of abuse and related problems worsens over time.

Marijuana, hashish, mescaline, and LSD distort perceptions and alter moods, occasionally causing hallucinations. Users can become psychologically dependent on the substances. Unless the actual use takes place at work, users generally can hide their abuse of these substances with some success.

Heroin, codeine, morphine, methadone, and other narcotic drugs dull perceptions, cause faulty judgment, and induce sleep. Users can become psychologically and physically dependent on the substances, which leads to painful withdrawal if the substance is not readily available. As these substances are expensive; abusers can resort to theft to finance their habits.

Barbiturates, tranquilizers, and other depressants (including Valium and Librium) produce effects similar to narcotics, with dulled perceptions, limited attention scope, and induced sleep. Extreme symptoms may resemble drunkenness. Users can become psychologically and physically dependent on the substances, and withdrawal is difficult and painful. A significant number of people abusing these substances began—and may be continuing—with legal prescriptions.

Cocaine, amphetamines, and other stimulants present a particularly difficult problem in the workplace because symptoms so closely resemble those of a highly productive worker. Users are alert, take initiative, and may put in long hours and show above-average achievement. In rebounding from the substances, users experience the opposite effects (only worse, for a net loss). Another symptom is suppressed appetite and sometimes weight loss. Users can become psychologically and physically dependent on the substances. It is particularly difficult to withdraw from cocaine, which is also very expensive.

SUGGESTIONS FOR MANAGING SUBSTANCE ABUSE

- Recognize that all companies, large and small, at some point are likely to experience problems with substance abuse. Remember that alcohol is the substance most widely abused.
- Establish or revise a substance-abuse policy and communicate it clearly to everyone in the company.
- Approach the issue as a job-related health problem affecting job performance.
- Intervene as soon as possible: prevention first, then early intervention, then full-scale treatment.
- Train supervisors and coworkers to detect signs of substance abuse.
- Protect employees' privacy and confidentiality at all times.
- Establish methods of addressing substance-abuse problems through an employee assistance program, outside referrals to qualified professionals, or some combination of the two.
- Educate employees about the hazards of tolerating substance abuse in the workplace.
- Emphasize the benefits of addressing substance abuse for the abuser, the other employees, and the company.
- Adopt ongoing substance-abuse education as a prevention method.

SUGGESTED READINGS

American Management Association. (1987). *Drug abuse: The workplace issues*. New York: AMA Membership Publications Division.

Lawless, G. (1990, March). Cost containment through outpatient substance abuse services. *Employee Benefit Journal, 15*, 6-10.

Moriarty, A., & Field, M. (1990). A new approach to police EAP programs. *Public Personnel Management, 19*(2), 155-161.

Wrich, J. (1982). *Guidelines for developing an employee assistance program*. New York: AMA Membership Publications Division.

Zetlin, M. (1991, August). Combating drugs in the workplace. *Management Review*, 17-24.

Chapter 37

TEMPORARY EMPLOYEES

Prudent managers realize when they use temporary help they pay only for productive time and avoid the expense when there's no work to do.

How do you cope with unexpected personnel shortages? You can prepare to some degree for seasonal peaking of the workload and inventory taking, but what about the time when several employees are on sick leave at the same time, or an unexpected flood of orders pours in?

Many companies are finding a satisfactory solution to such problems by using part-time and temporary employees.

The growth in part-time work that many critics say marks a long-term trend toward less secure and low-paying jobs is neither unprecedented nor necessarily undesirable, according to a study of national wage and employment data. While the number of part-time workers continues to grow, most choose their style of work voluntarily, according to research at the Employee Benefits Research Institute in Washington, D.C., a nonpartisan think tank whose studies are widely respected by organized labor, major corporations, and the government.

There are factors leading to the spontaneous and rapid growth of the temporary help industry, such as:

- The dramatic shift from a production-oriented to a marketing-oriented economy.
- The shift to a white-collar world. The factory worker to the office worker.

- A change in social attitudes—women forced from the drudgeries of housework and entering the workforce.
- Emphasis on using temporary help to reduce costs, taxes, benefits, and balancing resources against fluctuations in the workload.

The issue of part-time and other forms of contingent labor has captured the attention of policymakers struck by the continuing wave of layoffs at American corporations and the reluctance of many firms to hire permanent replacements. In testimony before the House Subcommittee on Human Resources, former Secretary of Labor Robert Reich stated, "a big percentage of the new jobs has been temporary or part-time jobs, offering more of the job security that American workers once took for granted." These trends are likely to continue as large corporations continue to restructure and downsize as technology continues to evolve and the defense-related sector of the economy continues to contract.

Part-time workers—those who work fewer than 35 hours a week—numbered 20 million in 1992 compared with 88 million full-time workers. Their proportion to the total workforce, which reached 18.9% in 1996, has grown only 3.4% points in the past 23 years. Much of that modest rise reflected the normal jump in involuntary, part-time work during and immediately after a recession. Other factors include the decision of many older employees to keep working on a reduced schedule past retirement and the enormous growth in the employment of women, many of whom work shorter hours to reconcile the competing demands of home and job.

Part-time workers are disproportionately employed in retail and service industries, where pay rates tend to be low. Median hourly earnings for part-time workers was $5.40 per hour, compared with $8.67 for full-time workers in 1992.

Organizations struggle over whether or not to hire temporary employees. Does it take more time than it is worth to train and develop these individuals? A recent survey revealed why organizations employ temporary workers. Some of these include:

- Hiring for specific skills that are not needed on an ongoing basis.
- Supplementing professional staff to give management adequate time to prepare and establish new corporate initiatives.

- Launching special projects so permanent staff can be free to carry out everyday operations.
- Supplementing staff during peak periods, such as year or quarterly end closing.
- Designing and integrating new computer systems within each department.

Who are the part-time workers? They come from every area of the workforce and represent all ages and both sexes. Often they are students working after school, moonlighters (people who hold down full-time jobs and supplement these with additional part-time work), retirees who either need the income or are bored with their lack of activity, and unemployed people working part-time while seeking full-time employment.

What kind of work do they perform? Almost every kind of work there is from the professional to the most tedious. How much money can temporary employees earn? As in full-time work, the rates of pay range from minimum wage to very high for professional and managerial work. However, unlike most full-time jobs, part-time work usually offers very limited or no benefits such as medical insurance and hospitalization coverage and vacations.

Temporary employees are often used to provide skills that do not exist within the company, are not needed on a permanent basis, or both. Intimate knowledge of the company is not required to perform some jobs. In addition, temporary workers bring a fresh and creative approach to assignments. Unconcerned with how a job has been done in the past, they look at each assignment as an opportunity to broaden their exposure to a variety of companies.

ADVANTAGES AND DISADVANTAGES OF USING A TEMPORARY HELP SERVICE

When an organization decides to engage temporary workers, it often turns to temporary help services. Workers supplied by a temporary help service are quickly available. Experienced and well qualified, they need little if any "breaking in." They usually walk in and begin to function right away. By using workers from this source, you can adjust to fast-breaking opportunities or problems without interrupting your regular production schedule.

Some companies need temporary help every week for a few hours. Others need temporary help for full days at various times, regularly or occasionally.

The hourly rate you may pay a temporary help firm may be higher than you would pay an employee you hired yourself, but the total cost of getting the work done will probably be less. Using a temporary help service does away with many personnel and record-keeping operations that are costly and time-consuming. Your agency does not have to advertise for help, screen responses, interview and test applicants, and check references, perhaps without producing a single qualified worker. You can save the cost of training, of overtime and idle periods, of paid employee absences, and other fringe benefits. Many services pay the premiums for worker compensation, unemployment, and other insurance. They handle the reporting and deductions for Social Security and income taxes.

In considering whether to use a temporary help service, list the disadvantages as well as the advantages. In some instances, the disadvantages may involve answering the complaint of regular employees who lose overtime pay because of the temporary workers. However, permanent employees' attitudes toward their temporary counterparts has changed significantly in the last several years. Ten years ago there was a general attitude that temporary employees were taking away the job of permanent employees. Today, prudent use of temporary employees tends to instill confidence in permanent employees that they are insulated from layoffs more than they would be if everyone in the company were on permanent status. With everyone trying to accomplish more with fewer people, they appreciate management's understanding that sometimes a few extra hands are appreciated.

It is possible, through circumstances beyond the control of the temporary help service firm with which you are doing business, that a change in personnel may occur during the assignment. If this should occur, it may require that you go through the orientation process again. In still other cases, the work itself may mean that temporary help cannot be used as effectively as regular employees. For example, the job may be complex and require a great deal of supervision for a worker who is unfamiliar with your way of doing it. In such instances, it may be more economical to pay overtime to a regular employee than to use a temporary worker.

If you decide to use a temporary help service firm, there are a number of questions you will need to have answered. These include:

- Is performance guaranteed?
- Does the service test its employees?
- Does it train its employees?
- What are the firm's recruiting and retention methods?
- What are the available job classifications and skills?
- Does the firm follow up to check on the temporary employees assigned to your organization?
- What method does the temporary service use to ensure the availability of temporary employees when needed?
- Is the firm financially sound?
- Does the firm provide insurance, such as worker compensation, for its employees?
- Does the firm have adequate liability insurance, and are its employees bonded?

Successful Use of Temporary Personnel

The key to the successful use of temporary help is in planning what type of help you need, how much, and when. You need answers to such questions as:

- How seasonal is my business?
- Do any of my employees have to work overtime to meet peak workloads? If so, what does overtime cost?
- If an extensive amount of overtime is needed, will there be a performance log and possible morale problems during regular working hours?
- With better planning, could any of the peak workloads be spread out throughout the year?
- Do customers come up with rush jobs? If so, can I get them to plan their needs farther ahead?
- Are employee vacations scheduled so as not to interfere with peak seasons?
- What extra help is needed to cope with these problems and reduce costs?

It is important to discuss the planning of day-to-day operations with key personnel. Study the work needed to be done, not peak periods. Compare this with previous years. A pattern will begin to emerge and you will be able to see where some firms will advise in this regard.

What to Do When the Time Comes

When the time comes that you need temporary help and you want to place an order with a temporary-help service, take these steps:

Estimate Your Needs

Decide what the specific requirements of the job are. Exactly what talents do you need? How long will you need the employee? Do not ask someone with higher qualifications than the work calls for—the cost will be unnecessarily high. On the other hand, do not try to economize by getting underqualified help and then expect the worker to carry out tasks that he is not prepared to handle.

If You Use a Temporary-Help Service

If the temporary-personnel firm is to help you get the best results at the lowest possible cost, you must give detailed information about the work to be done. Tell it the nature of your business, the working hours, when and how long you will need help, the skills required, and the types of equipment to be operated. You may want to send samples of the work to be done, if it is feasible, for example, with various clerical tasks. Be sure to give the exact location of your business, transportation available, parking information, and the name and title of the person to whom the temporary employee is to report.

Arrange for Supervision

Appoint one of your permanent employees to supervise the temporary employee and check on the progress of the work from time to time. Be sure this supervisor understands the job to be done and just what her own responsibility is.

Tell Your Permanent Employees

It is a good idea to let your staff know that you are taking on extra help and that it will be temporary. Explain why the extra help is needed and ask them to cooperate with the new employees in any way possible.

Prepare the Physical Facilities

Have everything ready before the temporary worker arrives. The work to be done should be organized and laid out so that the employee can begin producing with a minimum of time spent in adjusting to the job and the surroundings. See that the materials needed are available and the equipment is in place and in good working order.

Plan the Workload

Do not set up schedules that are impossible to complete within the time you allot. Try to stay within the time limits you gave the temporary-help service, but plan to extend the time period, if necessary, rather than crowd the employee. Rushing and overwork can result in costly mistakes.

Prepare Detailed Instructions

Describe your type of business, the products you manufacture, or the services you offer. Be specific in outlining the procedures your company follows. Most employees of temporary help adjust quickly to the methods of an individual firm.

Help the Employee Settle In

Receive the temporary employee as you would receive one hired on a permanent basis. Make him feel like a member of your team. Explain where to hang coats, the location of the washroom, the lunch hour, coffee breaks, and so on.

Introduce the temporary employee to the permanent employee who will supervise the work.

Introduce the temporary employee also to permanent employees in the same department. Explain that the temporary employee will be here for a few days to help out.

Explain the Job

Go over work assignment and the instructions. Explain company routines. Make your directions as simple as possible and provide samples of the work to be done. If the work is complex, explain it clearly and make certain that your explanation is understood. Assure the temporary employee of your staff's cooperation and willingness to help, and show your own interest and concern.

Do not Expect the Impossible

How much can you expect from a temporary employee? Fully as much as you contracted for with the service firm. Most employees of temporary-personnel services perform well. They are experienced and versatile. Because they have worked for a variety of businesses, they have learned to adapt quickly to a new situation, and they know that future assignments depend on their doing satisfactory work.

But do not expect the impossible. Do not overload temporary employees—make a slight allowance for the fact that she is not familiar with your business and its operations. Check the work occasionally, ask for any questions, never leave the employee feeling stranded or left out. At the same time, do not make her nervous by hovering over her.

SUGGESTIONS FOR IMPROVING YOUR USE OF TEMPORARY HELP

- If you believe that the use of temporary help makes sense for your business, then establish how many employees you need, for how long, and what skills they should have.
- Determine how much work there is to be done and how quickly it must be done. Once you know these factors, you should be able to determine how many temporary employees you need and for how long.
- When a temporary employee has finished his assignment, evaluate the performance level. Compare the original time and cost estimate to the actual expenditures.
- Be sure to inform your full-time staff. Tell them what the temporary help will be doing and how long they will be on the job.
- If you plan on using a temporary-help service firm, check with your chamber of commerce, your attorney, your accountant, and your banker for recommendations for selecting the most appropriate firm.
- In trying to determine whether or not to hire temporary help, conduct a management audit of all positions. A close examination will show those that are less than full-time and can be handled by a temporary employee.

- Direct the temporary employee the same way you do your permanent employees. Be demanding, set standards, but make allowances at the start to give the person a chance to catch on to your way of doing a job.
- Arrange for direct supervision of the temporary employee. You can appoint one of your regular employees to check the progress of the worker and answer any questions the person has.
- A good temporary help service will follow up on its employees. Their representatives should call or visit you in person to find out if you are satisfied or if some correction is required.
- Remember the advantage of using temporary employees is you can get more qualified individuals for special situations than you might have full-time. If it does not work out, you do not have a long-term commitment to the individual.

SUGGESTED READINGS

Brown, D. (1990, December). Temps enter the executive suite. *Management Review, 27,* p. 28.

Cherington, D. J. (1993). *Personnel management.* Dubuque, IA: William C. Brown.

Graham, R. (1979, January). In permanent, part-time work: You can't beat the hours. *Nation's Business, 67,* 65-69.

Pell, A. R. (1984). *The part-time job book.* New York: Monarch Press.

Messmer, M. (1990, October). Strategic staffing for the 1990s. *Personnel Journal, 21,* p. 4.

Chapter 38

TERMINATING EMPLOYEES

Recent studies have shown that the level of emotional stress from termination of employment can be equated with the stress level resulting from an individual being told that he or she is dying of an incurable disease.

The termination of an employee is a managerial responsibility that one would rather not have to face. With the termination of a job comes a great deal of trauma: to the individual losing his or her job, to the employer faced with the unpopular task of firing, and often to the morale of the company. Avoiding this situation altogether, though desirable, is next to impossible. Organizational policy should be clearly mapped out, and great effort should be taken to hire, train, and evaluate effectively in order to ensure employees who perform sufficiently. But if the time does come when an employee must be eliminated for downsizing reasons, or for reasons relating to job performance, a manager must deal with the situation in a skillful manner. Tactful dismissal of employees is of benefit to the manager and the organization in both the short term as well as the long term. By creating stable policies and carefully following a businesslike procedure, employers salvage worthy workers, carefully eliminate a minority of employees, and maintain a good reputation in the business sphere.

Because firing an employee is so unappealing, most managers wait far too long before dealing with the situation. Many managers are reluctant to fire quickly because they see their task as a heartless one, and want to draw out the process to spare feelings for as long as possible. But rather than spend sleepless nights attempting to prolong the inevitable, a manager would be much more wise if he would use a quick firing method.

The decision to fire should be well thought out and discussed in a serious manner with any co-managers or other relevant members of the organization who have a stake in the dismissal. Once the decision to terminate an employee has been made, the process of dismissal should be relatively quick and as painless as possible. According to one management expert, "It should be like taking off a Band-Aid—fast and with as little pain as possible."

Advantages to this quick technique are many. In using a quick method of dismissal, not only would the manager promptly overcome her own unpleasant situation of doing the actual firing, she would also be setting free the employee to begin the process of dealing with feelings and preparing to look for another job. If the employee was dismissed for disciplinary reasons or a shoddy performance, the quick dismissal might alleviate tension and provide a quicker recuperation of the organization, allowing it to get back on track. A fast firing does not, however, mean an insensitive firing.

LEGAL IMPLICATIONS OF EMPLOYEE TERMINATION

In recent years, the legal process has been proactive in the fight for employee rights in termination cases. Historically, the law gave employers the right to discharge an employee "at will" for any reason or for no reason at all, with or without prior notice. This was known as "at-will" employment, but recently many exceptions to this rule have been made at the legislative and judicial levels.

The legislation that affects termination prevents employers from discharging anyone on the grounds of discrimination. Title VII protects against discrimination on the basis of race, creed, color, religion, sex, or national origin. The Immigration Reform Act of 1986 took this a step further by protecting an employee's citizenship as well. And an employee's age and the right to his/her pension is protected under the Employee Retirement Income Security Act. Other areas also considered discrimination in many states, for example, marital status, pregnancy, physical handicap, AIDS, and sexual orientation.

Public Policy, Breach of Contract, and Good Faith and Fair Dealings are the judicial areas that are covered by the law. Under the Public Policy exception, an employee may not be terminated for serving some statutory right, voicing a safety complaint, or for being a "whistle blower." If an employee establishes a contractual arrangement implied by something that the employer did or said, that merely indicates that the employee would only be discharged under certain conditions. This type of agreement can override any contract if proven in court, and therefore would be a breach of contract in the eyes of the court. This usually will take the power away from the employer to terminate "at will" even if it were not done in writing.

By protecting those who are discharged for reasons of financial gain by the company or for reasons of not wanting to pay the employee his due commissions, the Good Faith and Fair Gains Law will see that these employees will get their monies.

The laws dealing with employers' freedom to terminate an employee are changing and growing at a rapid pace. For this reason, it would be wise for management to consider the assistance and advice from a competent lawyer before proceeding with the termination procedure for any employee. In this way, an employer can make sure that the reason(s) for terminating the employee are legal.

AVOIDING LITIGATION

Terminating the employment relationship is as likely as not to result in lawsuits, and a wise employer would take action to prevent them. Sandra Rappaport suggests six steps of preparation for an employer to take in order to reduce the chances of a lawsuit.

1. Employee Handbook. Management should carefully review each handbook provision, preferably with an experienced professional or lawyer.
2. Employee Evaluation. Managers should review their evaluations of each employee. Too many times businesses have been sued on the grounds of good cause and were hung by their own evaluations of the employee.
3. Document the Misconduct. Laying the groundwork for a "good cause" defense must begin before an employee is actually let go.
4. Voluntary Resignation. This angle can insulate the employer from future lawsuits by having the employee resign instead of being terminated. This usually works by sweetening severance pay in return for a signed release of all claims from the employee.

5. One-Year Contracts of Employment. This is an alternative that an employer may want to try. Termination can be in effect at the end of the one-year contract simply by not renewing the agreement.
6. Incorporate an Employee Grievance Policy. Some experts feel that this is a critical step for employers to take in order to preserve good labor-management relations.

Along with these steps, the process starts the minute a potential employee comes in for an interview. An employer should not imply limitations on her own power to discharge, or imply that the applicant is being hired for a certain period of time, and remove all references in company documents of "probationary" or "trial" periods of employment. Some courts have found that using such terms suggests that an employee becomes a "regular" after the probationary period is over and cannot be discharged except for provable good cause.

If a contract exists, examine its terms. Usually the terms of employment are not in writing; however, on occasion there is a letter of agreement in effect or even a detailed contract. If so, examine it carefully for potential problems. If the person being fired has signatory power, revoke it immediately. It is important to know your state employment laws. Pay the terminated employee all money due. Check the employee benefit plan. You can encounter a great deal of difficulty if you fail to observe the requirements of the Employee Retirement Income Security Act laws adopted a few years ago.

When an employee is fired, he has the right to know what the reasoning is behind the termination. Once the decision has been carefully weighed to get rid of an employee, the terminating manager needs to have a well-thought-out plan to deal with the feelings of a person being fired. In today's society, employment contributes a great deal to the person's sense of self-worth. A manager must take these feelings into consideration and do all that is possible to avoid destroying the employee's self-esteem.

Typically managers make reasons for termination vague and interject false ones into the process. Whatever reason or reasons a manager may have for firing an employee, they must be documented. The documentation of infractions against any policy should be ongoing in an organization. Results of evaluations and any communication that has been done regarding the performance of the employee should be updated and secured in the personnel file. An updated and thorough personnel file may be utilized to build a case against the employee.

Of the utmost importance in a business environment is the utilization of honest and frequent performance appraisals. It is important to fight the urge to make evaluations a "whitewash" and the necessity of saying what should be said in the evaluation. By having relevant and truthful information available in a personnel file, an employer can have these "truths" to fall back on, and strengthen his position.

Of course, as a part of the evaluative process, an employee should constantly know where she stands so she will know where to improve. Formal, written guidelines for discipline should be rigorously followed. Reasoning for termination should be more of a process than a rash decision. By working in advance to establish an honest relationship with the employee and by consistently stating expectations and deficiencies, a manager will be dealing with problems as they arise, rather than trying to come up with reasons for the termination. When expectations are made clear from the beginning and performance is frequently evaluated, the termination may not be completely unexpected.

One preventative measure against termination is a progressive discipline system. This system incorporates an oral warning for a first offense, a written warning for the second, suspension for the third, and termination for the fourth. One problem with this implementation is that by giving warnings, the management may be implying that it will not fire until the fourth infraction. Therefore, the system must also provide for immediate dismissals under certain conditions, such as embezzlement, unprofessional conduct, or some other gross infraction.

Management has a responsibility to maintain credibility. An employee who is not satisfactory must be disciplined and held accountable for his actions. Most experts recognize five criteria that are almost always grounds for dismissal. The first is poor job performance. If an organization is getting no return on their investment (the employee), and job performance criteria have been explained repeatedly, the employee is a prime candidate for dismissal. Incompetence is the most common reason for dismissal of employees in lower-level positions. Failure to let an inept employee go could lead to decreased trust and respect for management by other employees in the organization.

A second reason for elimination is the failure to comply with company rules and regulations. If policies are established, they need to be followed by everyone.

Thirdly, another type of employee who should be let go is the ambitious worker who is consistently passed up for promotions. If the person is not quite good enough to be moved up, she may eventually experience feelings of discontent that may affect her performance. Letting an ambitious person go is often a benefit to both the company and the employee.

The fourth reason listed is the case of a personality conflict between the employee and his boss. If the boss cannot deal with the employee, and the employee does not live up to the expectations of the boss, dismissal is a relevant option.

The fifth is that organizations go through cycles, and sometimes the strength of a particular employee is no longer needed. When an employee reaches the career stage of stagnation, and possibly even maintenance, dismissal may be the best option.

Once all the reasons for termination are organized logically, the time comes to meet with the employee in a termination interview. Many issues need to be taken into consideration: first and foremost, the timing of the dismissal should be analyzed. An employee should not be fired on a holiday, birthday, or anniversary. An employee should be spared from being terminated late on a Friday afternoon and having the entire weekend to become upset. Rather, a termination should take place early in the week, so the person can at least think about applying for another job. Termination interviews should also be conducted in a private setting such as an office.

A manager should expect an array of reactions to the termination and be prepared for outbursts. Recently in Houston, Texas, a produce company worker who was fired shot his boss to death with a semi-automatic pistol and wounded a coworker before fatally shooting himself. In Santa Fe Spring, California, a social worker fired from her job walked into the Los Angeles County office and shot her former boss in the head. Regardless of the reaction of the employee, she should still be treated with respect and dignity. The termination interview should be respectable and professional, and the personal feelings of the manager must not be discussed. A manager who tells an employee how terrible he feels about the dismissal will probably be viewed as a hypocrite. A manager should also take great care not to say anything that might result in a lawsuit on racial, sexual, age, or some other grounds.

The bad news should be broken in the least traumatic way. One theorist suggests trying to appreciate the employee's strengths and understand why you, as a manager, could not harness them. In admitting his lack of ability to ensure success for the employee despite many *documented* attempts of assistance, the manager tends to alleviate some of the blame the employee may feel. By shifting part of the blame from the employee on to himself, the manager can preserve a sense of worth in the person. Of course, the person being fired should know what she has done to lose the job. One way to anticipate this query is writing out comments beforehand as part of the preparation to fire an employee. Listing problems that have been discussed previously with the employee and the lack of resolution will also help the employee see where the problem lies.

To reduce the shock of termination, a continuation of some benefits that the employee had on the job is often appropriate. These should be defined in the termination interview. When a separation occurs, many employers offer severance pay and insurance for a specified time as a measure of goodwill. The type of severance agreement should be tailored to the individual, and it should be put down in writing.

To help a person adjust to termination, assistance techniques are often utilized when the case permits. These could include counseling, or typically an exit interview. A counseling interview to facilitate emotional adjustment can serve as a buffer between the termination interview and re-entry to the outside world. It is suggested the 72 hours following dismissal are crucial for the departing employee. The attitude carried through the three days following termination can be seen as an indicator of the psychological health and adjustment potential of the terminee.

In some situations, employers may assist the employee in a job search and allow the terminated employee the use of the office facilities to look for a new job. Sometimes an arrangement is made for messages to be taken for the employee in his job hunt. An organization may also assist in outplacement that provides further counseling, career development, résumé assistance, interview training, and other applicable services that would help the terminated employee develop skills needed to be successful in the workplace.

A very important and often neglected phase of termination is breaking the news to other workers. While the conditions of the termination should be kept confidential, the manager has the responsibility to address the issue honestly with remaining employees. The goal of this step is to stabilize the workforce and try and eliminate the rumors that may start after a firing. If the employee was terminated because of the inability to perform or for breaking rules, most remaining employees will have already realized that there was a problem and will understand the dismissal. If the action was justified, there will be little sympathy. But if the remaining employees are bitter toward the dismissal, the efficient function of the organization could be in jeopardy. Whatever statements are made regarding the situation, they must be genuine and should alleviate fears and concerns of the employees who remain.

As unpopular as firing employees is, it would seem logical to take steps to avoid the confrontation of a dismissal. But regardless of how successful an organization is, it is almost certain that at some point one of its managers will face the unpopular task of dismissing an employee. An employer needs to look carefully at the situation that forced the dismissal and learn from it. A company may need to analyze how well it utilizes efficient hiring techniques, develops up-to-date job specifications, utilizes performance appraisals, and clearly explains and communicates expectations to the employee. Although termination and the consequences of it are never enjoyable, they should and can be managed. Organizations should establish policies that allow termination to be conducted in a fair, honest, and prompt manner, which will cause only a slight amount of upheaval in the organization.

SUGGESTIONS FOR TERMINATING EFFECTIVELY

- Before you inform the employee about his termination, write down in advance what you wish to say. Also prepare a written explanation of the severance benefits. Have notes that deter you from saying something you could regret later.
- Be aware of all the organization's policies regarding terminations before confronting the employee.
- Do not ask the terminated employee to clean out his desk immediately and leave the office. After hours or on weekends is more appropriate moving time, unless the person has been fired for cause.
- Do not allow the termination notice to be delivered by one individual. You should deliver the message with another person present, preferably from the human resources department, to serve as a witness.
- Employees should not be notified of termination on a Thursday or Friday, nor the day before a holiday, when they would have time to feel depressed about the occasion. Avoid discharging an employee on his birthday, wedding anniversary, or service anniversary. Also avoid firing when an individual is pregnant, going through a personal crisis, facing a serious medical problem, or when there is a death in the family or of a close friend.
- Be prepared for the possibility that terminated employees may behave emotionally and sometimes violently. This may take place immediately or within a short time after the termination.
- Be sure to seek legal advice on potential problems. Court decisions are changing constantly and it is very difficult for a manager or a supervisor to stay abreast of these changes.
- If an individual has been terminated because of a workforce reduction, organization downsizing, or department consolidation, he should be told that is not as a result of poor performance.
- When firing an employee for poor performance, try not to sound too righteous. Avoid saying things like "How could you...." Start with phrases like "I don't think it right that...."
- Keep the discussion to about 15 minutes or less. A short session will be less emotionally exhausting for both you and the employee. You will be more likely to say what you planned.

SUGGESTED READINGS

Coulson, R. (1981). *The termination handbook.* New York: The Free Press.

Gilberg, K. R., & Voluck, P. R. (1987, May). Employee termination without litigation. *Personnel,* 31-37.

Grandholm, A. R. (1991). *Handbook of employee termination.* New York: John Wiley & Sons.

Morin, W. J., & Yooks, L. (1982). *Outplacement techniques: A positive approach to terminating employees.* New York: AMACOM.

Shepard, I., Heylman, P., & Duston, P. (1989). *Without just cause: An employer's practical and legal guide.* Washington, DC: The Bureau of National Affairs.

Chapter 39

VOLUNTEERISM

Recruiting volunteers may be easier than you envision. More than half of the nation's adults and teenagers donate time to some volunteer activity.

The first step in establishing a volunteer program is to determine the purpose for introducing new personnel into the organization. At a time that demands more and better services with fewer financial resources, the answer to the above statement seems obvious. If the primary reason for establishing a volunteer program is to save money, those in charge will be greatly disappointed. From a perspective of effectiveness and efficiency, volunteers offer an organization the ability to utilize existing funds better and more efficient use of existing personnel.

Over the last 20 years, volunteering has become big business in the United States. However, in the early 1970s, volunteering expanded dramatically and became much more organized. Volunteer work has included services to health and medical organizations, welfare agencies, recreational groups such as garden clubs, ski clubs, youth-serving organizations and issue-related groups such as Sierra Club, National Organization for Women, professional organizations, business and commercial associations, labor unions, artistic and cultural organizations, fraternal groups, legal and criminal justice system, and many others.

Prior to 1970, not much was known about the extent and size of volunteer programs. Since that time there have been a number of studies that give some perspective of the growth of these programs. One of the most recent surveys revealed:

1. Nearly half of all Americans over 14 years old (approximately 89 million) volunteer to some organization.
2. Volunteers contribute an average of 3.5 hours per week.
3. The major areas of volunteering were religion (23%), informal volunteering (19%), education (13%), general fund-raising (11%), and recreation (10%).
4. Most volunteers (80%) contribute time to charitable organizations, 18% contribute time to governmental organizations, and 3% reported contributing time to for-profit organizations.
5. Volunteers do a variety of jobs, including assisting the elderly, performing janitorial work, and being an officer of an organization. The most popular form of volunteer work was assisting the elderly, the handicapped, or a social welfare recipient.
6. The primary reasons for volunteering were wanting to do something useful to help others (52%), having an interest in the work or activity (36%), or enjoying the work (32%).
7. People who volunteer their time are much more likely than nonvolunteers to donate money to charitable organizations.

CHARACTERISTICS OF VOLUNTEERS

There has been a great deal of research on the characteristics of volunteers. Research has been done on a national basis and provides information in the following areas:

Age: The prime years for volunteering are from about age 27-29 through retirement with the peak being 35-49 years of age. It is commonly thought that retirees are a prime target group for volunteer recruiting because they have the most free time. However, studies have revealed that retirees who were not volunteers when they were younger are not likely to volunteer when they retire.

Gender: More women volunteer than men. A survey by the American Volunteer reported women made up 51% and men 45% of volunteers. However, studies show men participating more, particularly in groups relating to their occupations. It should be pointed out that as women enter the workplace their own voluntary associations activities change, becoming more like their male coworkers and less like female full-time "homemaker" patterns.

Education, Income, and Occupation: These three characteristics of all demographic indicators of volunteering are the most predictive of volunteer activity. Studies indicate that education, income, and occupation are highly positively correlated with voluntary organization membership, with education having the most predictive power.

Marital Status: Married individuals participate more extensively in volunteering than any other marital status group. They are followed by widows and widowers, single people, and divorced and separated people. Higher rates of volunteering by married people is a function of their greater integration into community institutions and their sharing of home responsibilities, which give them more free time.

Children in the household: Having children is associated with higher rates of volunteering, and having children all of school age produces even greater involvement in volunteering. Parenthood probably more than any other role ties adults into the community by creating an interest in the availability of youth-oriented services and overall quality of life in the community.

Race: The majority of volunteers are white, but they are also the majority population in the United States. Very little research on volunteer participation has been conducted along ethnic lines. One study reported that among blacks and whites of lower socioeconomic status, blacks participated more.

Myths about Volunteers

Although many agencies have benefited from using volunteers, there are still many myths about volunteers. These myths are counterproductive to the effective and full utilization of volunteers. Some of these myths include:

- Volunteers are just not dependable.
- Recruiting, training, and supervising volunteers takes too much time.
- Volunteers lack commitment to the program.
- There are few jobs that can be turned over to volunteers.
- Volunteers may be more accepted by participants than staff members.
- A volunteer program takes valuable resources away from other programs that are more beneficial to the agency.

These views are out of date and if subscribed to by full-time, paid personnel, will act as a detriment to a successful volunteer program.

Recruitment of volunteers

"Where do you find capable volunteers?" This question is often repeated as a major concern by agency officials. There is no one simple solution to the problem of finding any recruiting volunteers. Recruitment is a continuous task since new volunteers must be constantly recruited in order to bring in the fresh energy needed to sustain a good program. In urban areas there is intense competition among many diverse groups for potential volunteers. Although the pool of possible volunteers is much larger in urban areas than in rural areas, the difficulty in attracting them is also greater. There are many more volunteer opportunities and situations from which the interested volunteer may choose.

It is important to make the volunteer program and its recruitment campaign as appealing as possible. Volunteers do not usually "walk in off the street" to say they want to give their time and energy to a volunteer organization. When asked how did you get started in volunteering, the most common answer is "Because someone I knew asked me." Volunteer administrations indicates there is no substitute for being asked. One of the most important keys to success in volunteer recruiting is being asked by someone the prospective volunteer already knows and respects. When discussing volunteer involvement with prospective candidates, the following areas should be mentioned:

Background of the agency. Be prepared to tell what your agency does and whom it serves. Some facts of interest may be how many people were served last year, what were the agency's accomplishments, and what are the long-range goals.

Why you are being asked. Be able to tell exactly why the individual is being asked to serve in a particular capacity. The general appeal, "We need your help," is becoming less effective because there are so many causes that need help. A specific statement of why the prospect was selected accomplishes several things at once: It tells the prospect that he was selected based on certain important qualifications, it conveys the impression that recruiting is a dignified and purposeful activity in the organization, and it sends the message that the job is meaningful and important.

Have a job description available. Use a job description to guide the discussion explaining the actual content of the volunteer role. Do not read the job description to the prospective volunteer. Have a copy to give the person. Let her mull it over on her own time. Do not ever minimize the task or its importance. On the other hand, do not overemphasize the task difficulty. Just discuss the actual job responsibility.

Benefits of volunteering. Discuss with the prospect what he will get out of volunteering. Meeting new people, gaining experience, learning a new skill, advancing one's career, helping others, gaining recognition, and having fun are a few of the more often-mentioned rewards for volunteering. When potential volunteers are convinced that volunteering can be personally beneficial to them, they are more likely to take the job and work hard at it.

Answer questions. Recruiters should be prepared to anticipate questions. If questions arise that the recruiter cannot answer, she should find the answer and recontact the prospect.

Volunteer Recognition

Obviously volunteers are not in it for the money. They bring to your agency a desire to be useful. They also have certain expectations and needs that must be met. These needs and expectations are best met by using their time effectively and showing appreciation for their efforts. Even though volunteers work for no pay, remember that "nobody works for nothing." Supervise volunteers as you would paid staff, but compensate them through even more recognition and attention for work well done. In dealing with volunteers, recognition is often an underused administrative tool. A well-placed commendation acts as an incentive to the recipient to continue to perform well or to strive to perform even better. It is a morale booster for all involved, an intensely humanizing gesture that costs little in monetary terms. Specific forms of recognition may vary according to your imagination. Certificates, mentions in professional publications, bulletin boards, special events such as luncheons, banquets, picnics, and breakfasts, commemorative gifts, T-shirts, pins, public praise at organizational ceremonies, yearly volunteer day ceremonies, and hiring the volunteer when a staff vacancy occurs are all examples. Consider the following points when developing a volunteer recognition program:

- **Recognize and express appreciation for volunteer contributions on a day-to-day basis.** Simple gestures such as a smile, handshake, or a sincere thank-you are still the most appreciated ways of showing appreciation.
- **Recognize volunteers by asking their advice and following it.** Volunteers may have excellent suggestions on ways to im-

prove your program. If your operation is small and you only have a few volunteers, simply encourage them to express their feelings and concerns about the program on an informal, regular, ongoing basis. For larger organizations, consider using written evaluation forms to get more formal input from your volunteers.

- **Recognize volunteers on a more formal basis.** Many agencies have found it useful to reinforce the daily forms of recognition with formal recognition events. Formal events can be a meeting, a luncheon, a tea, a picnic, a barbecue, or dinner.
- **Present special awards.** Some agencies have found that plaques, books, insignia clothing, and speciality items such as belt buckles, watches, rings, etc., are often prized by volunteers. The Roosevelt-Vanderbilt National Historic Park rewards its volunteers with an annual picnic on the grounds of the Eleanor Roosevelt home. All volunteers receive certificates; special awards go to the most improved young volunteer and outstanding achievement. The teenage volunteers receive a belt buckle, while the adult volunteers receive a print of one of the National Park sites.
- **Provide recognition through public praise.** Publish articles in professional magazines and newsletters as well as write letters to the editor of the local newspaper. Take a group picture of the volunteer staff that has worked on special projects. Constantly praise your volunteers and, by all means, involve your paid staff at ceremonies and other events.
- **Do not forget paid staff.** A job well done by the volunteer staff means that a good job has been done by the paid staff as well. Paid staff can be recognized in commendation letters to their personnel files and to their supervisors by special mention at staff meetings and through performance appraisals.

Orientation and Training of Volunteers

The purpose of orienting and training volunteers is to ensure the highest possible degree of satisfaction with and the contribution to the program that they are to implement. Orientation helps volunteers become acquainted with one another, and the staff learn the organization culture and learn about their own volunteer role in relation to the entire organization. Training introduces new skills, knowledge, and abilities, or reinforces existing ones; can be used to plan and manage program changes; and provides opportunities for self-renewal and growth. Orientation differs from in-service training in that orientation usually occurs at the beginning of the volunteer's commitment and in-service training at various times during a volunteer's involvement with the organization.

Orientation can be conducted by paid staff or by current volunteers. Experienced volunteers usually enjoy the opportunity to share their accumulated wisdom with newcomers. Often they will have excellent ideas about what to include in an orientation manual or on the orientation agenda. There are many ways to orient volunteers based on the type of organizations, skill level, and personal style. Orientation often occurs in a group meeting and provides opportunities for volunteers to ask questions.

Another method is a personal, one-on-one orientation conference between agency staff and each new volunteer. Unfortunately, time and financial constraints usually work against this approach. For some types of volunteer roles, especially when the volunteer is heading up a major project and a successful outcome is essential, the personal conference is especially effective.

Orientation information can be presented in a variety of ways. Many organizations find a volunteer's handbook a useful way to present information. The following information should be included in such a manual: the philosophy and conceptual framework of the volunteer's role, the program of the agency, the organizational structure of the organization, history of the agency, policy and procedures, by-laws and constitution of the agency, volunteer benefits, identification of key personnel, telephone numbers and key offices, emergency procedures, and job descriptions.

Once the volunteer has the benefit of orientation and has made the initial adaptation to the new role, his or her needs for more information can be satisfied by in-service training programs. Training programs may be used to reinforce or introduce skills, knowledge, or attitude and are ideal vehicles for enhancing volunteers' growth and development. For example, through a leadership-training program, an employee who has never held a leadership position may learn skills that help him to be a successful leader. The skills may also have relevance in the volunteer's paid employment and lead to promotion or recognition in the workplace.

Another purpose for training is the creation of quality-control systems. Training helps ensure that the program standards are communicated. Informing volunteers of procedures and policies is part of the agency's risk-management plan. Through training, the agency can ensure the smooth implementation of programs and activities, set up reliable communication patterns, and outline procedures. Training is also a valuable motivational tool. When an organization spends time and money to train personnel, it sends the message, "You are important to us—you count." Training helps volunteers reach their maximum potential; as a motivational tool, training may help to reduce volunteer turnover.

SOME SUGGESTIONS TO IMPROVE YOUR VOLUNTEER PROGRAM

- Be systematic and comprehensive when you implement a volunteer program. Do not use the "shotgun approach." Do not just rush out and recruit a few volunteers and put them to work. Look at the big picture first and be sure you have the commitment necessary to run an effective volunteer program.
- Deal with paid staff attitudes and reservations about volunteers. Build a consensus on appropriate roles for volunteers in your organization.
- Develop and use effective volunteer application and evaluation procedures that will allow you to screen and accept only qualified volunteer candidates.
- Evaluate the volunteer's work performance. Resolve problems as quickly as possible.
- Review your program at least annually. Make sure that what you are doing is necessary and that the volunteer program fits the future needs and opportunity before you continue the program unchanged.
- Educate paid staff to the advantages of working with volunteers and reward them for effective collaboration.
- When possible, appeal personally to prospective volunteers by asking them face-to-face. Surveys indicate that the ideal way to approach volunteers is by personal contact.
- Establish high standards of conduct and expectations for your volunteer staff. Volunteers should feel as important as paid staff.
- Identify at least one prospect for each volunteer position. This matching process should involve two considerations: first, which job can best meet the needs and interests of the

> prospective volunteer, and second, which prospective volunteer can best fill the job.
> - Do not always be in a rush. Let volunteers see you have time for them—they have time for you.

SUGGESTED READINGS

Adams, D. (1980). Elite and lower volunteers in a voluntary association: A study of an American Red Cross chapter. *Journal of Voluntary Action Research 9*, 95-108.

Brundney, J. L. (1990). *Volunteer programs in the public sector.* San Francisco: Jossey Bass.

Phillips, M. H. (1982). Motivation and expectation in successful volunteerism. *Journal of Voluntary Action Research, 11*, 118-125.

Vizza, C., Allen, K., & Keller, S. (1986). *New competitive edge: Volunteers from the workplace.* Arlington, VA: The National Volunteer Center.

Volunteers and nonvolunteers: A profile. (1988, August). *Employee Services Management,* 20-21.

Chapter 40

WORK-TIME OPTIONS

Even though many of today's work-time options are primarily designed to make workers happy, countless employers have discovered that the benefits go both ways.

Our changing workforce—and the special needs of the individuals in it—have created a society where work and leisure often clash. Ever-increasing numbers of women, single parents, and older employees are working harder and longer to support themselves and their families, thus sacrificing their personal time. As a result, today's employees are changing the way they look at job schedules, and they are looking to management to help them balance their work and family responsibilities.

Managers have been forced to change their way of thinking, too. Concerns about productivity and quality of work life have prompted managers to explore various work-time options, including creative alternatives to the traditional five-day, 40-hour work week. In the process, employers have found that they also benefit from these scheduling options through improvements in cost effectiveness, quality of work, morale, and customer service.

FLEXTIME

One of the most popular and widely accepted schedule alternatives is the flexible full-time schedule, often called flextime. Begun in the early 1970s, flextime allows employees to vary their starting and finishing times to suit their own needs, provided that they are all present during a mandatory core period. The core hours—typically the middle hours of the day, with a flexible lunch break—are maintained to provide continuity within the organization.

A similar system is the staggered fixed schedule. Workers choose a starting time and adhere to it for a set period, usually a month or a quarter. The schedule can then be changed when a new month or quarter begins.

Employers and workers have found that flextime schedules can deliver great benefits. Even the most limited forms of flextime—those that provide only a one-hour latitude in starting and finishing times—can reduce stress and help employees fit their non-work activities into the day. It frees workers to use their full potential on the job instead of worrying about taking a child to the doctor or getting to the bank before it closes. Many companies have found that flextime reduces sick leave because workers do not have to make up imaginary illnesses to handle personal matters during working hours.

Flextime can also contribute to improved job performance and quality of work. Employees can adjust schedules to their own "big clocks" and work the hours they feel most productive. Job skills improve because of the need for employees to cover for other workers on different schedules. Tardiness is almost eliminated, in part because people can set their schedules to reduce commuting time and avoid rush-hour traffic.

There are, however, disadvantages to flextime. There are significant limitations on the number of workers who are eligible to participate because of the types of jobs they perform. Also, people who are dependent on public transportation may find it difficult to fit work into bus or train schedules.

One of the biggest challenges in designing a flextime program is to provide adequate coverage during non-core hours. Occasionally, customer service problems can occur because of unusual scheduling and lack of supervision during certain periods. Communication among employees can be interrupted when scheduling meetings, and it can be difficult to coordinate work among employees on different schedules. If supervisors have different schedules than workers, it may be difficult to keep track of the actual hours worked by each employee.

COMPRESSED WORK WEEK

Another option is the compressed work week, which allows employees to work fewer days each week, without reducing their total hours. One of the most common schedules is four 10-hour days each week, rather than five eight-hour days. Others vary the schedule to provide four 12-hour days each week, which allows alternating three- and four-day weekends. A similar approach is the "5-4-9" schedule, where employees work five nine-hour days one week, and four nine-hour days the next.

The compressed work week was developed primarily to achieve maximum use of expensive facilities and to provide greater leisure-time opportunities for employees. It is not as widely used as flextime, but is most common in entertainment and recreation services, the health-care industry, and government.

The obvious advantages to the employee are extended weekends and reduced commuting time and costs. The employer benefits from improved employee recruitment, lower turnover rates, reductions in absenteeism, more accurate time keeping, and higher employee morale. In addition to decreasing production costs, the shorter work week also increases employee output and job performance. This has prompted some employers to reduce the number of total hours in the compressed work week—they have found that employees accomplish more in a 36-hour, four-day week than they do in the traditional 40-hour, five-day week.

The compressed work week can create some problems. There have been mixed reviews from employees, especially from women or older workers who generally dislike the scheduling because they want to feel a sense of stability in life. Although most employees may be more productive in a longer work day, a long schedule will cause fatigue in some, which can result in lower productivity and more accidents. Supervision can be more difficult, and some employers have problems with overtime costs.

JOB SHARING

Job sharing is the practice of having two or more employees share the responsibility for one full-time job, while maintaining the status, promotion prospects, security, and pro-rata pay levels that would be attached to the full-time job. It is an attempt to develop a flexible work schedule that best suits the needs of the individuals and the organization.

Job sharing takes many forms, depending on the type of job and the preferences of the workers involved. The most popular pattern is the split week, where one job sharer works full days the first half of the week and the other works the second half. Alternate weeks are also used, with each employee working a full week at a time. The least-used schedule is the split-day, where one person works mornings and the other works afternoons, with some overlap during a hand-over period. Job sharing can also be set up with no fixed hours, but this requires great cooperation from supervisors and job-sharers.

Job sharing arose in response to increases in unemployment and the belief that employees would prefer to have half a job rather than no job at all. It was also explored as a way to accommodate women who wanted to reduce their work schedules to handle the responsibilities of their growing families.

Job sharing is often confused with part-time work. The main difference is that job sharing is based on the premise that there is a full-time job with the attendant salary and benefits. Part-time workers have skills that usually do not match up with what management expects a job sharer to be able to do. Also, the pay and benefits are usually much greater for those who job share.

There are many benefits to the practice of job sharing. A shared job offers higher status, more fulfilling work, and better pay than the jobs that are typically available to people who wish to work part-time. Because job sharing emphasizes cooperation rather than competition, employees are more comfortable and happy with their jobs. Also, it offers employees a way to accommodate work and family responsibilities by opening alternative times to work.

These advantages to the employees translate into a number of benefits for management. One of the big pluses is the gain in potential manpower flexibility, which helps maintain continuity when one job sharer leaves. It helps retain experienced staff and generally allows expanded ranges of skills and experience to be incorporated into a particular job title. Employers see reduced turnover and absenteeism, much greater continuity and effort, the availability of wider ranges in skills, the tapping of a wider employment pool, better training for younger people, and easier transition into retirement for older workers.

The success of job sharing depends largely on the people involved and the type of work performed. Supervisory and management jobs are difficult to divide, so companies may choose to limit job sharing to routine tasks. Management and the job sharers must cooperate to determine salary levels, resolve seniority issues and divide vacations, holiday pay, and other benefits. Management may incur higher Social Security tax burdens. The job sharers may be affected by the reduced chance of advancement within the company.

Communication and continuity problems are some of the most difficult problems found in job sharing. Management may have trouble communicating with employees because of staff changes and lack of consistent supervision. Delays can occur, or the job sharers may duplicate their counterparts' work because they do not sense a clear division of responsibility.

WORK SHARING

This relatively new work-time option was designed to counter job shortages while maintaining rather than increasing employment. It encourages employees to reduce the length of their work weeks to compensate for temporary business slowdowns, such as a recession. It is based on the premise that most employees would rather sacrifice hours instead of risking the possibility of being laid off or terminated.

As an example, work sharing could be used when a company is facing a 20% reduction in business and may need to lay off 200 employees. Under work sharing, the company could avoid any layoffs by reducing every employee's work week by 20% (one day). In addition to their regular paychecks being 20% less, employees would be eligible to receive 20% of their normal unemployment compensation payment, thereby minimizing income losses.

The attraction of work sharing is that it helps employees keep their jobs and most of their usual income. Employers enjoy the benefit of reduced employee turnover. When layoffs occur, many people begin to look for new jobs. Work sharing alleviates staffing concerns because employees are not forced to leave. The most interesting benefit is that it preserves Affirmative Action profiles, which assume "last hired is first fired." This is important in eliminating perceived discrimination if the most recent hirees are female or minorities.

Critics of work sharing claim that blocking unworked hours into new jobs is impractical. Businesses will have difficulties finding additional skilled employees who are willing to make up for the time relinquished by the preexisting work force.

When work sharing is advocated, the employer loses the flexibility of its workforce, especially if overtime is restricted. This practice is not legal in all states, probably because of the cost of making up the difference of wages lost through unemployment compensation, which in turn costs the state and the employers insurance. This explains why only a small number of U.S. employers implement work sharing.

V-TIME

V-Time is the combination of the scheduling options of full-time flextime, part-time work, and work sharing. Like work sharing, it operates on the idea that part of a job is better than no job at all. It was created to keep employees from being laid off or losing their jobs to part-time employees during a business downturn.

V-Time allows full-time workers to volunteer to reduce both their hours and their pay, but still maintain some benefits. Employees can choose among several alternatives, ranging from reductions of a few hours per week to as many as 15-20 hours. Then, when the business climate improves, employers can return workers to full-time status.

The benefits to management are reduced costs and the ability to maintain the current staff. Employees like V-Time because it allows them to keep their jobs and some of their benefits. They also enjoy the opportunity to relax and spend more time with their families.

PEAK TIME

Peak time—also called prime time, premium time, preferred time, or right time—is the practice of hiring individuals to work only during high-volume periods. The periods could range from a few hours to 25 hours a week. These schedules are widely used by banks, where workloads are greatest during lunch hours and Friday afternoons when people have just been paid. Peak-time hiring is also being introduced into the retailing, computer programming, accounting, and airline industries.

Employees are fond of peak time because it offers higher wages per hour than the average full-time wage for the same position. This helps entice people who would regularly be unwilling to work part-time. Peak time has been successful with housewives, young singles, retirees, and those looking to add to their income.

Employers benefit because these jobs attract quality people who are eager and willing to work. Other advantages include drastic reductions in turnover rates and training costs. And because peak-time workers have part-time status, many employers realize great savings because they do not pay benefits for these employees.

JOB BIDDING

Although these options are most widely used for full-time staff, similar programs can be offered to part-time workers, too. Some companies have also had success with job-bidding programs, which allow part-timers to bid for open positions, extra work assignments, and special projects.

Opportunities are posted for permanent and temporary work assignments. Interested employees submit proposals for the number of hours they can commit to the job or project and the schedules they want to work. Priority goes to qualified bidders with the greatest seniority.

Management benefits because open jobs are filled by already-trained employees who really want to do the work and can fit it into their existing schedules. As part-timers increase their hours, they develop a stronger commitment to the organization. The system also improves recruiting because part-time employees know they will have the opportunity to increase their hours.

As our society undergoes constant change, so do our attitudes about work schedules, job-performance issues, and conflicts between our personal and professional lives. The options discussed here are just some of the ways managers are attempting to meet the changing demands of today's workforce. Our whole approach to work life and productivity is being transformed in other ways, too, as normal, full-time work weeks are getting shorter, vacations are getting longer, and people are retiring earlier than ever before.

These trends present special challenges to today's employers, and meeting them requires management to be more creative, flexible, and generous than in the past. Even though many of these work-time options are designed primarily to make workers happy, countless employers have discovered that the benefits go both ways. By giving people more choices about how they schedule their daily lives, management creates a more positive work environment and better morale. The final payoff is satisfied, committed employees who perform more efficiently and deliver higher-quality service and products.

SUGGESTIONS FOR DESIGNING WORK-TIME OPTIONS THAT WILL BE SUCCESSFUL FOR YOUR EMPLOYEES

- Solicit workers' ideas about all the possible options. Use a simple survey and/or interviews with selected managers and employees to determine need, interest, and potential participants' willingness to adapt to others' schedules.
- Avoid thinking of time spent at work as the main measure of performance. Flexible schedules are most successful when managers can evaluate employees on their ability to get tasks done, not on the number of hours at the office.
- Consider adopting more than one scheduling option. The greater the flexibility within the system, the greater the impact on employee attitudes.
- Variability of scheduling is one of the most critical design features. Most satisfying to employees is the ability to allow daily variations at the individual's discretion. If this option is not feasible, consider a staggered fixed schedule, where the employee can make changes in her schedule at the end of every month or quarter.
- Extend options to as many employee groups as possible. Offering flexible schedules only to certain groups may create the appearance of a class system, which can foster resentment.
- If some groups cannot be included in a new program, meet with them to explain why. Consider offering an alternative benefit (an extra half-day of personal leave or a longer break).
- One of the ways to reduce stress (and tardiness) is by minimizing travel time and costs. When setting up schedules, consider local traffic, the availability of public transportation, and the commuting time of individual employees.

- When setting up a job-sharing arrangement, provide at least one to two hours per week when both participants are on the job. This overlap will improve communication, efficiency, and teamwork. It also provides a convenient time for both job sharers to meet with supervisors.

Chapter 41

WRITING A RÉSUMÉ

The variety of résumés crossing a busy personnel desk is infinite. There are short ones and long ones, cute ones and straightforward ones, clear ones and illegible ones. There are even résumés without names and addresses.

Employers are faced with a flood of job applicants annually, but they have been forced to become highly selective when hiring employees. Since many businesses are downsizing and closely monitoring their budgets, managers have placed an emphasis on extensive staff recruitment. Smart managers have realized that proper staff selection can eventually lower turnover rates and decrease the expensive costs of training.

Changes in careers or jobs can be traumatic and unsettling. With each such change, an individual is introducing himself to new people, selling his skills to prospective employers, and creating an impression of himself and what he can bring to the organization. Often the entire effort is contained in one or two pages that must tell it all for you—your résumé. If your résumé does not get you in the door for that all-important interview, you may never get a chance to introduce or sell yourself in person.

Managers in parks and recreation administration are continually faced with the problem of shrinking budgets. But, due to the popularity of these career fields, managers often have the advantage of choosing from numerous qualified candidates to fill position openings. So, like employers in other competitive areas, they must build an effective staff recruitment program to ensure the hiring of the best candidates.

WHAT IS A RÉSUMÉ?

A résumé is a written communication that clearly demonstrates an individual's ability to produce results in an area of concern to potential employers in a way that motivates them to want to meet the person. The résumé is not designed to get the individual the position. The best that a résumé will do is to get one an interview. Employers do not hire on the basis of a résumé alone. But a good résumé will get interviews with employers who count by demonstrating that you have a valuable, potential contribution to make you move the job search process to the next level.

WHO NEEDS A RÉSUMÉ?

Everyone. Years ago résumés were assembled primarily by people of distinctly professional class—educators, lawyers, and professional managers. Today with increased specialization and mobility, the use of résumés extends through all white-collar levels from executives to hourly workers in lower-paying blue-collar and unskilled occupations. The test for using or not using a résumé is whether a person would typically show up for an interview, get hired, and start work the same day, or become involved in a longer hiring procedure in which the résumé would precede or follow the interview. Regardless of the level or scope of the work targets, a résumé can be a valuable tool. By putting yourself through the discipline of preparing a résumé, you will have greater clarity about your work-life purpose and will increase your ability to present yourself in a way that motivates employers.

WHY WRITE A RÉSUMÉ?

A résumé serves to introduce you to the organization you are seeking employment with and help you gain a personal interview. You need a résumé because, for most advertised jobs, a résumé is required before an interview will be granted. Employers require résumés because they are executive time savers. A personal interview takes hours; a résumé can be read in a few minutes. There are other reasons for writing a résumé:

1. A résumé forms the basis for a mail campaign about yourself using either the actual résumé or a summary of it in the form of a general letter.
2. A résumé can be used to test your marketability while you remain safely employed.
3. A résumé helps you to organize the facts of your accomplishments, clarifying what you can or wish to do in the future. Many advisors recommend that a résumé be updated every six months or once a year. Thus, you remain continuously aware of your progress (or the lack of it) and of your up-to-the-minute responsibilities and achievements.
4. A résumé may be used when buying a business to impart information about yourself to the seller.
5. A résumé can serve to solicit business if you are a consultant, freelance artist, writer, or a part-time worker.
6. A résumé prepares you for your job interview by forcing you to think about and express yourself in an organized way.

TYPES OF RÉSUMÉS

There are several types of résumés that can be used in your job search, and the right one depends on the situation. The first type is the *chronological* résumé. This form lists your experience in the order it was received and generally is divided into sections such as experience, education, and possibly achievements. This style is easy to read since descriptions of work experience are done in a brief manner.

The following rules should be adhered to when writing a chronological résumé:

1. Start with the present or most recent position and work backwards with most space allotted to recent employment.
2. Detail the last four or five positions or employment, covering the last 10 years. Summarize early positions.
3. Use year designations, not month and day. Greater detail can be given in the interview or application.
4. Do not list every major position change with a given employer. List the most recent or present and two or three others.
5. Do not repeat details that are common to several positions.

6. Stress the major accomplishments and responsibilities within each position listed that demonstrate your competency to do the job. Once the most significant aspects of your work are clear, it is generally not necessary to include lesser achievements.
7. Keep your next job target in mind. As you describe prior positions and accomplishments, emphasize those that are most related to your next move up.

The second type of résumé is the *functional* résumé. This type is used when the job seeker has limited experience in the area and wishes to use her experience to present a background of skills learned rather than from the actual job experience in the area. An example would be an individual who has a background in management, sales, communications, or possibly all three and is interested in a position requiring one or all of these skills. The individual may have no knowledge or experience in a particular industry but is trying to draw attention to her abilities and not the position she presently holds.

By selecting a functional résumé format, you have chosen to highlight your basic area of ability and potential rather than your work history. In doing this you will be able to organize and highlight information in a particular career target direction and play down possible gaps or inconsistencies in past work. If you are changing careers, entering, or reentering the job market, you have chosen an approach that will also allow you to talk easily about nonpaying work experience and community activities. Some rules to follow are:

1. Use four or five separate paragraphs, each one handling a particular area of expertise or involvement.
2. List the functional paragraphs in order of importance, with the area most related to your present job target at the top and containing slightly more information.
3. Stress the most directly related accomplishment or result you have produced within each functional area.
4. Know that you can include any relevant accomplishment without necessarily identifying with which employer or nonemployment situation it was connected.
5. Include education toward the bottom, unless it is within the past three years. If it is an unrelated field, include it at the end, regardless of how recent.

6. List a brief synopsis of your actual work experience at the bottom, giving dates, employers, and titles. If you have had no work experience or a very spotty record, leave out the employment synopsis but be prepared to talk about the subject at the interview.

The third type of résumé is the one considered *industry-related*. This is the style that would be helpful to the individual who has an identical or closely related experience to the position he is seeking. An example of its use would be the individual who has worked in a few positions with different agencies and is applying for a similar position. Here, the applicant would want to highlight these experiences. These experiences may be listed under a specific section that would describe them to the reader. This form becomes a more customized résumé and should not be used in a general search.

Unlike the chronological and functional résumés that are geared toward an affirmative picture of past history, the industry-related features a series of statements concerning what you can do whether or not you have actually had directly relevant experience. You may want to use the industry-related résumé when you are clear about a particular job and want to focus on it alone. Some rules that need to be followed when developing an industry-related résumé are:

1. You must be clear about a specific job target or targets if you plan several versions. A job target is a clear description of a particular title or occupational field that you want to pursue.
2. The statements of capability and accomplishments must all be directly related to the job target. This may require some reading or research in the field.
3. Both capabilities and accomplishments will be short statements of one or two lines, generally written in an active style.
4. Listed capabilities will answer the underlying question, "What can you do?" Listed accomplishments will answer the underlying question, "What have you done?"
5. Experience and education are listed but not openly stressed.

NO-NO'S IN RÉSUMÉS

- Religion or church affiliations
- Race
- Color
- National Origin
- Political Preferences
- Previous Salaries
- Anticipated Salaries
- Reason for Leaving Position
- Opinions of Previous Employers

THE COVER LETTER

The individual cover letter that accompanies the résumé is important in the job search and is a personal introduction to a potential employer. The purpose of the cover letter to is communicate to a prospective employer a personalized message about the applicant's value to the organization. It should be written with great care, for it adds a powerful element to your résumé. Cover letters should do precisely what they are intended to do—provide cover for an enclosure. If you want your reader to examine your résumé, your cover letter must have impact. This letter advertises your résumé. It should *not* regurgitate or substitute for it. The cover letter captures the reader's attention, stresses the employer's needs, and your value to the company. It invites her to read the résumé in depth. The résumé, in turn, repeats the process of getting attention and stressing value. Moreover, the résumé sustains and heightens the reader's interest. It provides additional credibility by detailing your value in relationship to your goals and the employer's needs. Your cover letter should follow certain general rules:

1. Type on good-quality bond paper.
2. Address to a specific person and title. If you are uncertain whom to address, look in the library reference material or call the organization and ask the receptionist for an appropriate name and title.
3. Writing style should be direct, powerful, and error free. Edit the letter to eliminate extraneous words and to check gram-

mar, spelling, and punctuation. In addition to stating your purpose, the letter tells the reader how well you communicate.
4. No more than *one* page. Do not overwhelm the résumé with a lengthy cover letter or excessive repetition of the résumé content.
5. Keep the letter short and to the point. Three paragraphs are sufficient.
Paragraph 1: State your interest and purpose. Try to link your purpose to your interests and the organization's needs.
Paragraph 2: Highlight your enclosed résumé by stressing what you will do for the organization.
Paragraph 3: Request an interview and indicate you will call for an appointment.
6. Use appropriate language. Use the organization's jargon. Use active verbs. Do not try to be cute or too aggressive.
7. Always be positive by stressing your past accomplishments and skills, as well as your future value.

The cover letter should communicate to the prospective employer a specific personalized message about your potential value to the organization. It generates interest in you from the person who counts. It is not difficult to add a powerful element to your résumé.

USEFUL WORDS TO BE USED IN RÉSUMÉ WRITING

accomplished	extensive	qualified
analyst	experienced	reliable
certified	innovative	responsible
communicator	inventive	skilled
complete	leadership	solid
conceptual	motivated	successful
contribute	negotiator	systematic
creative	organizer	talented
delegates	planner	traveled
economizer	problem solver	writer
effective	progressive	

SUGGESTIONS FOR IMPROVING YOUR RÉSUMÉ

- It is not necessary to list references in your résumé, although if your references contain a person who is very well known in the field, listing their name can be a plus. "References available upon request" is sufficient.
- The best way to emphasize key words is by using capital letters, underlining, or if you are sharing your résumé, typeset or printed out by a computer with the use of boldfaced or italic lettering. Headings should be easy to locate, which can be done by leaving white space surrounding them.
- As you put together your résumé, keep in mind that it is an advertisement for you. It is designed to sell your potential, not just your past experience and education. How it looks and what it says should be evidence of your ability to express yourself and your concern with appearance.
- A shorter résumé that features an abundance of white space offers several advantages. It looks cleaner and more appealing, it is easier to read, and it allows the personnel manager to use the white space for notes during the course of the interview.
- An effective way to make a concise résumé is by using bullet copy rather than full sentences. You can keep it simple by staying with bullets. Owners and managers are adamant about their preference for short résumés.
- Résumés should have as few as six and as many as nine sections, depending on the format used and how one plans to highlight his background. The most commonly used include: identification (name, address, phone), job, career objective, education, leisure activities, work experience, skills and capabilities, achievements and accomplishments, personal interests, and references.
- Your name, address, and phone where you can be reached during business hours should always be placed at the top of your résumé. Avoid putting it in the upper left-hand corner, where a staple or paper clip may be used.
- Always indicate or imply your job objective early in the résumé. If pertinent, indicate or imply also your career objective and specialization.

- Remember, flashy approaches and gimmicks in résumés get noticed, get talked about, and passed around the office for a few laughs, but they seldom get the applicant called in for an interview.
- Employers have identified several factors that can lead to instant résumé rejection. These include:
  (1) admitting being fired,
  (2) showing gaps in job history,
  (3) having less education than the job requires,
  (4) showing no experience in the field for which you are applying and,
  (5) trivia in the personal section.

SUGGESTED READINGS

Angel, J. L. (1980). The *complete résumé book and job-getters guide.* New York: Pocket Books.

Kelley, J. (1985). *Résumé writing: A comprehensive how-to-do-it guide.* New York: John Wiley & Sons.

Rulek, R. E., & Suchan, J. A. (1988, November/December). "Application letters: A neglected area in the job search." *Business Horizons*, pp. 70-75.

Weinsteir, B. (1993). *Résumés don't get jobs.* New York: McGraw-Hill.

Wendover, R. W. (1989). *Smart hiring: The complete guide for recruiting employees.* Englewood, CO: Management Staff Press.

Part III

Executive Development

Chapter 42

BUSINESS ETIQUETTE

Manners govern how people treat each other, whether in the coal mines or in a boardroom. When people who work together in either place adhere to the rules of social behavior, their workplace becomes efficient.

Business etiquette is becoming a very important element to success in the business world, whether it be for an individual or an organization. Without proper etiquette, clients, coworkers, and employees may be offended by certain behaviors or may receive an undesirable message about the organization. Content of the message portrayed is important, of course, but other variables, including etiquette, influence reactions.

Proper etiquette is important because it portrays a positive image of the company and the individual employee. It also improves employee morale, increases the quality of life in the workplace, and embellishes the image of the organization. Many believe it helps generate profit. Inappropriate etiquette, on the other hand, puts the employee's job and possibly an organization's success on the line.

Problems with employee etiquette can be very difficult for management to deal with. Managers often find it difficult to confront an individual about issues that are so personal. The issues are often avoided and either lead to problems for the employee or for the organization.

Employee etiquette encompasses many different areas of behavior, from mode of dress to dining habits. For a new employee, it is important to grasp these unspoken rules of behavior quickly. This will ensure membership into the employee work group and will help with an easier transition to the work habits of the organization. When an employee first enters an organization or a new department, it may be helpful to keep a low profile. After an individual feels more comfortable, it is important to ask a number of questions and pick up on the unspoken rules of the organization.

The most important thing to remember about business etiquette is that everyone is important. Everyone, from the mail clerk all the way up to the president, deserves equal respect and fair treatment. As the old cliché says, "Whatever goes around, comes around," and this is especially true in a professional situation. Positive, helpful behavior may not be immediately rewarded, but the payoff is often found later on in the employee's career.

INTRODUCTION

One of the first issues that an employee is faced with when meeting a prospective client, a fellow employee, or colleague is the issue of introductions. These brief phrases that orient two people who were otherwise unacquainted beforehand should always be succinct and relevant. Extraneous information that may be interesting but not pertinent should be avoided. Interesting facts or tidbits about the individual that may help initiate a conversation are acceptable. When making an introduction where a male and a female are present, the male is always introduced first. For example, "Miss Green, may I introduce Dr. White?" When two people of the same sex are present, the younger of the two is introduced first. For example, "Professor White, have you met my brother, Joe Brown?" When introducing someone to a group of people in a meeting, it is often helpful to hand out background information beforehand. If an individual has been neglected in the introductions, the individual may introduce himself to the individuals in the group by adding that they have not met yet.

Often when making introductions, it is difficult to know how to address someone being introduced or refer to someone being spoken about. The best advice, when in doubt, is to ask the person what she feels most comfortable with. Someone named Charles may prefer to be called Charlie. Showing sensitivity by asking beforehand is far better than making an introduction and being incorrect. Following other employees' leads when it comes to proper titles is also acceptable. When referring to executives, it is almost always safe to assume the use of last names, unless told otherwise. When dealing with subordinates or equals, it is often common to use first names. In a small company, it is common for everyone to be addressed by his or her first name. If still confused after using all of the above methods, noting the way an individual signs his correspondence can help to discover his proper introduction name. Using nicknames for introductions is usually not preferred, unless they are short versions of a name (e.g., Bobby from Robert).

MEETINGS

Meetings also have their own set of etiquette rules to follow. Timeliness is of the utmost importance. By being tardy, the individual is implying that the other person's time is not as important as her own. This message can be detrimental when trying to forge employee-client relations and can cause an employee to make a terrible impression on upper management. If being on time is almost impossible, a call to the party or parties involved to explain the situation and to suggest a possible meeting time is recommended. If the meeting will be delayed too long, the other party or parties are expected to suggest another meeting time at their earliest convenience.

Adequate preparation is another important factor to remember when meeting with an individual. If no preparation has taken place, it is expected that the meeting will be postponed and rescheduled for the next most convenient time. This eliminates the problem of having unproductive and unsuccessful meetings. During a meeting, if issues have been discussed that need further attention, a course of follow-up action should be agreed upon. This may be a continuation of research done prior to the meeting, a new course of research, or a future meeting. Be sure to emphasize these follow-up points and ensure that proper action is taken to complete them.

TELEPHONE ETIQUETTE

Telephone etiquette is becoming more and more important in the age of highly technological business deals. It is not just important for secretaries to have good telephone manners, but also for the entire organization to know how to handle the telephone properly. The first task is to answer the phone professionally. When on the phone, it is important to focus all of your attention on the person on the other end. If it is inconvenient to speak with him or he is undesirable to speak with, it is acceptable to take a message and call him back at a later time.

The importance of taking complete phone messages should be easily learned by all employees. It is important to return all messages, whether personally or by a fellow colleague, in a timely manner. Adequate listening techniques are important to understand the message and the content of the conversation. It is often easy for people to tell when the person on the other end of the telephone is not focusing completely. Because body language cannot be a factor in phone conversations, positive tone of voice and listening quality are essential.

DRESS FOR SUCCESS

Modes of dress emphasize commitment and level of seriousness in relation to a job. Each company has its own set of rules for dress. The important thing to learn is what is required by the organization. This does not mean that every employee must dress exactly as the executives. Anything that may draw unneeded attention to the employee may be detrimental to her future with the company. Observing other individuals in the company may be helpful in gauging the normal mode of dress. In conservative industries, such as banking and finance, employees should always dress conservatively. Men in all industries should have at least four suits. Men should keep their jackets on in most situations, unless alone. Shirts wrinkle much easier than jackets and may look unprofessional. Women should have at least fourteen business outfits and should never wear pants to the office. Accessories should complement rather than destroy the effect of the outfit. Along with appropriate clothing, employees should be presentable: clean-shaven, showered, clean-looking, and well-groomed. It may also be necessary to have emergency supplies handy, such as a sewing kit, a small assortment of toiletries, and a few extra items of clean clothes.

BUSINESS MEALS

Business meals of any type are some of the most important and most feared types of business etiquette. Proper manners in these situations help to create a more comfortable, social relationship between the two individuals. Whether breakfast, lunch, or dinner, each has its own set of social rules. In any situation, the person who has extended the invitation is responsible for paying the bill at the end of the meal. Rank is the deciding factor in payment of a meal, while gender is not so much a factor anymore. Inviting the boss out to lunch or dinner may be awkward when the issue of payment arises. For this reason, it is wise for the employee to wait to be asked out to a meal by the boss.

When inviting someone out for a meal, it is considerate to inquire about his preference in restaurants. If he has no suggestions, it is wise to choose a place that has been tried before. Inadequate service or terrible food is not an issue that needs to be dealt with during a meeting of this sort.

Again, prompt arrival is imperative. At this time, it may also be helpful to take charge of the seating arrangement so as to ensure the most political arrangement. When a meal is meant to be a business date, general conversation should take place during the appetizer and business should be discussed during the main course. If dessert is being eaten, it is acceptable to wait to discuss business just before dessert has been served. When in doubt as to when to commence discussion of business, it is helpful to take cues from the individual who initiated the lunch or dinner meeting.

When ordering, the less worry, the better. Always offer suggestions to the invitee about good dishes and specials. Ordering something because of what other employees may think of eating habits or tastes is foolish. Price should be considered and dishes should always be from the middle price range of the menu. One area that needs attention during a meal is the consumption of alcoholic beverages. As a general rule, alcohol should not be consumed at luncheon meetings. If alcohol is ordered, it should never be consumed in excess and the person invited should never consume more than one drink more than the person who has invited her. The average duration of breakfast and lunch meetings should be about one to one and a half hours; for dinner, two to three hours is acceptable. Careful attention to details may signal to the employee or potential client that business will be handled in the same manner.

FOREIGN TRADE

The world is becoming smaller and smaller as technology increases. Business deals between cultures are becoming a more usual occurrence. To ensure success in the international business world, executives must be aware of the proper rules of etiquette in the various cultures that they will be dealing with. In order to do this, executives must research and prepare before they interact with these different cultures. It is often very apparent when an individual has attempted to prepare for an international meeting but has only done rudimentary studies of the different culture. This lack of effort can detract from the presentation and may show that the individual has a less than adequate level of commitment to the individual and to the detail of the presentation.

Many American business people have the attitude that the American way of doing business is the proper way. To achieve positive results, it is imperative for American businesspeople to dispose of this attitude and adopt a more encompassing, accepting attitude. Americans should learn to be more flexible and understanding in their responses to other cultures but should be careful not to forfeit their individuality as Americans.

It is not enough for Americans dealing in foreign countries to adopt a "foreign" etiquette. Every country has its own set of etiquette rules and these must be learned prior to interaction with the individuals. Some of the major differences when conducting business in foreign countries are off-hours business, holidays, verbal messages, and nonverbal cues.

The following are general guidelines to follow when dealing in certain countries. Before entering these countries, more thorough research *must* be conducted. The area of the world that is most compatible with American business practices is Western Europe. Each country does have variations, though. In Germany and other German-speaking countries, business is practiced with seriousness and rigidity. Decisions are reached systematically and often lack flexibility. Meetings are often efficient and well planned. Language problems will most likely not exist, as many Germans know fairly good English.

In France and Belgium, the subject of the meeting may be approached in a more general manner and will eventually focus in on the details. Rhetoric and the art of speaking is very important to the French-speaking population, and this may help slow down negotiations. Remember to be patient and take every point in consideration. It is important to decide beforehand if the meeting will be held in English or French; reliable translators must be present if needed.

In the United Kingdom, businesspeople are more often underprepared than overprepared. Their negotiation style is friendly, respectful, and flexible. It is important to take as much time as necessary, as negotiations that do not take a lot of time are not always the most productive.

In contrast to Western Europe, business deals in the Mediterranean and Central and South America rely heavily on the quality of personal relationships. Often more time is spent on establishing a strong relationship than on the actual deal. Businesspeople from these areas do not see this effort as a waste of time but see it as vital to the success of their ventures.

In the Arab world, customs surrounding culture and religion are extremely important. Religious holidays may interfere with negotiations, which must be taken into consideration when establishing meeting times. Bargaining is also important, and many Arabs feel negotiations are open-ended and never truly finalized.

Negotiations in the Far East are all very dependent on the various countries involved. They cannot be dealt with like a solid block of nations with similar rules of conduct. In China, personal relationships are extremely important. Small courtesies and extra efforts make an incredible impact. Because of language differences, communication is often very difficult. Preparation for these meetings should be detail oriented. Chinese businesspeople will notice the small details of the presentation techniques and any positive interaction that took place. Informal communication is vital, as the Chinese will often open up more during these informal meetings.

In Japan, the long-term relationships of the individuals and of the projects are often more important than the actual negotiations. Lots of planning and time should go into the preparations for these meetings. The language barrier will again pose a problem and must be dealt with effectively. Agreements in this country are often seen as flexible and may be continually negotiated.

Eastern European countries, including the Soviet Union, have very similar problems in relation to their business dealings. The infrastructure is often poor, the products are poor quality, the productivity is low, and the manufacturing facilities are run down. At first glance, Westerners would seem to have nothing to gain from these relationships but that is not the truth. Western businesses can hope to gain low labor costs, knowledge of the local history, and the current political system and possible expansion into new marketing and distribution opportunities. Negotiations in these countries are often extremely bureaucratic. Decisions must be ironed out completely to ensure the desired results have truly been obtained. Ensuring success in international negotiations is often very difficult, but adequately preparing for such meetings can help to push the negotiations in a positive direction.

Business etiquette is extremely important at all levels of an organization. Without proper attention to such details, crucial business deals and relationships can be lost. These rules often seem tedious to follow but are important in ensuring success in the business world.

SUGGESTIONS FOR FOLLOWING PROPER ETIQUETTE

- Knowledge that social skills are important in a professional environment. Realize the importance of emphasizing this to employees.
- All levels of the organization must practice proper etiquette, as they all reflect the company's image.
- Remember that everyone who comes in contact with the organization must be treated with equal respect and must feel equally important to the organization.
- Even the briefest business encounter can leave a potential client with an impression about the organization.
- Make sure that the first impression an individual has of an organization is a positive one.
- Make an effort to learn office norms and etiquette of a new department or organization.
- Superficial impressions made by fancy clothes or expensive dinners will not be effective in the long run.
- Be aware of additional rules of etiquette for business meals above and beyond the regular table manners that were taught by Mom and Dad.
- Respect cultural differences and make an effort to learn and assimilate the habits of different cultures.

- Employees who have problems with etiquette must be confronted by management to assure that a positive corporate image is maintained.

SUGGESTED READINGS

Blanchard, K. (1992, December). Business etiquette. *Executive Excellence.*

Caudron, S. (1993, February 1). Doing business with Miss Manners. *Industry Week.*

Frank, S. (1992, May). Global negotiating: Vive les differences! *Sales & Marketing Management.*

Thomsett, M. C. (1991). *The little black book of business etiquette.* New York: American Management Association.

Voss, B. (1991, January). Eat, drink, and be wary. *Sales & Marketing Management.*

Chapter 43

EFFECTIVE PRESENTATIONS

Words work wonders. Words mean power to inform, influence, impress, entertain, and move. The person who can command them has an "unfair" advantage. People notice and listen to a good speaker, while many individuals may be overlooked because they can't express themselves.

I'm sure you are familiar with the situation—the speaker for the occasion has received a resounding introduction. It probably was said the speaker needed no introduction because of her successes and her well-known reputation. As the speaker stood behind the podium and thanked the master of ceremonies for the kind remarks, a couple of humorous jokes were told to "loosen" the audience or perhaps the speaker. Then it happened. The speaker put on her glasses, opened her folder, and began reading about the topic she was requested to talk about. After a few short minutes the audience looked bored. Before the speech was over some of the audience had left and many who stayed did so out of politeness.

Situations such as this are not unusual. Why is this so? In most cases, successful executives are skillful in the techniques of marketing, sales, administration, production, and distribution. Yet the skills needed for the above duties are the very ones needed to give an effective presentation. Many presentations fail long before the speaker appears before the audiences.

In today's business world, it is essential to be articulate. Today more than ever, executives must be able to put over their points of view, whether in the boardroom, at social gatherings, or at public meetings. Many find this skill difficult. Few individuals are born public speakers; there is a world of difference between casual conservation and addressing an audience. As one expert points out,

"A person's brain is a wonderful thing. It starts out working at birth and only stops when you stand up to make a speech."

Are you nervous when you have to speak in public? You are? Good. Like actors, speakers should feel nervous. Those who do not lack sensitivity and are likely to bore an audience. Some of history's greatest orators have been nervous before, during, and after their speeches. Remember, the audience wants you to do well. They are not waiting for you to flounder and make a mistake. They like to be led, roused, and dominated. Those who study the subject of public speaking will tell you the chief desire of an audience is to be part of a successful occasion.

Let us say you have been asked to give a speech. Before you do anything else, find out who the audience is, what it is interested in, what the setting is, and what precisely you are supposed to talk about. Too many speeches fail because the speaker does not bother to check these points. Remember, the key to an effective speech is selecting a topic that will enable you to offer material that is different from what anyone else can offer.

The first words you speak should "turn on" your audience. If you appeal to people's self-interest, they will become interested. Do not apologize for yourself or say how hard it is to speak on a subject. The audience wants to be led by someone authoritative, not by an individual who displays a lack of self-confidence.

No good speech was ever made without sound planning, so prepare thoroughly. Start by writing down in one sentence the exact message you want to convey. Without a clear purpose, a speech is likely to ramble over a variety of points. The audience can find this unsettling. But when the purpose is clear, the speaker can think more clearly about the subject. A speech that moves in an orderly way makes the audience more comfortable; the audience likes to feel that the talk has a definite destination. Jot down facts, ideas, and examples concerning your presentation. Firsthand thoughts are often the most powerful. Every point you are trying to convey should be carefully planned. The best advice I ever received came from a preacher I once heard: "I tell 'em what I'm going to tell 'em, I tell 'em what I'm telling 'em, and then I tell 'em what I told 'em."

The foundation for an effective presentation begins with one important rule: learn exactly who the audience will be and what its interests and objectives are. This means if you are making a presentation to a board of directors about a new voicemail system, do not tell them the details of the technology that makes it happen. Tell them why the new system will save personnel and financial resources. Explain the benefits of the system, not how it operates. The proceeding are key steps to follow when developing a presentation: setting objectives, determining audience needs, explaining the benefits of your proposal, knowing the audience attitudes, protecting your presentation, and final delivery of your presentation.

Before the presentation, decide what your objectives are. Ask yourself a series of basic questions: "What is the purpose of the presentation?" "What message am I trying to communicate?" "What results am I trying to achieve?" "What point of view do I want to convey?" Prepare notes concerning your delivery. The function of notes should not be so much to remember information as to remember to include activities and ideas stored in the memory or to direct yourself to do something that you might forget to do. The ideas will be most effective if they come out of your own experiences. Seasoned speakers are compulsive newspaper clippers and note jotters. They make their own collections of anecdotes that will help put the shine on some future speech. Such personally discovered illustrations also help make the speech unique. Books are full of humorous anecdotes, but experienced speakers avoid "canned" stories. Many audiences have been turned off by hearing the stale anecdotes over and over again from different speakers. Preparing podium notes is a highly personal matter. Good technique is whatever works for you in delivering your message. Delivery notes should contain the entire sequence of what you do and say during your presentation. This includes:

- The subject matter of the presentation in an abbreviated form.
- Directions for movements, gestures, readings, passing out your material, and use of blackboard.
- Cues for the use of visual aids, charts, slides, videos, etc.
- Any information to be written on a blackboard, easel, or chart.
- Podium samples of each individual aid to be used in the presentation.

At some point you will have to decide whether your speech will be more effective with visual materials. Be careful not to use these materials as crutches. If they do not add to your presentation, do not use them. If you decide to use this material, it should be remembered that visuals must be first class. Charts, graphs, slides, and videos should be sharp and clearly visible to everyone in the audience.

Once your objectives are set, select three or four major points you want to make. Every presentation has three points: the introduction, the body, and the conclusion. The introduction should be general and establish the presentation's direction and objectives. The closing statement or conclusion should summarize what was covered and ensure that your objectives have been met. Next learn who will be in the audience and, if possible, their current attitudes about the subject you will be discussing. Do they have concerns or skepticisms that might be addressed in your presentation? Issues that are troublesome should be addressed head-on, not avoided. If you wait for serious concerns to be addressed during the questions and answers, you may be put on the defensive.

SUGGESTIONS FOR SUCCESSFUL PRESENTATIONS

- Know the outline of your presentation thoroughly. Make it a mental picture or pathway in your mind.
- Do not argue. Never engage in a personnel argument or a personality conflict with an audience member, regardless of who is right.
- Be committed to the content of your presentation. It will mean you will be interesting, energized, and real.
- Try not to read your presentation. Reading from a prepared text usually results in being boring as well as lacking eye contact with the audience.
- If you use audiovisual material, do not rush it. Give everyone time to see and understand the point you are highlighting.
- If you respond to questions from your audience, be sure to tie your answers to your presentation objectives and the point your are trying to make.
- Be sure your audience understands the benefits of the ideas you present. The main purpose of expressing benefits is to gain attention and acceptance for the ideas and information contained in your major points.
- Concentrate on developing visual aids that are clear, concise, and, most important, relevant to the audience.

- Learn who will be in the audience and, if possible, their current attitudes about the subject you will be presenting.
- Avoid any reference that suggests to your audience you are insecure, naive, or unfamiliar with your subject matter.
- Use gestures to underscore your important points. Avoid gestures that imply nervous mannerisms and have nothing to do with the message you are trying to convey.

SUGGESTED READINGS

Cavanagh, M. E. (1988, March). Communications: Making effective speeches. *Personnel Journal*, pp. 51-55.

Franco, J. (1985, March). Making effective presentations. *In Flight Magazine*. Piedmont Airlines.

Gorkscheit, G. M., Cash, H. C., & Cressy, J. E. (1981). *Handbook of selling* (Part 2). New York: John Wiley & Sons.

Saunders, D. S. (1985, November). Selling systems ideas. *Journal of System Management, 36*, pp. 29-33.

White, J. R. (1991, January). Suggestions on public speaking. *The Appraisal Journal, 50*, pp. 71-75.

Chapter 44

THE IMPORTANCE OF WRITING CAPABILITY

Gobbledygook
Part of an Office of Management and Budget memorandum entitled "Impact of Conversion of Interyear Comparability":

The period July 1, 1975, to June 30, 1976, will be comparable to the preceding years. The period from July 1, 1975 to September 30, 1976, can be compared with the period from July 1, 1976, to September 30, 1977, by addition of separately published transition period data to fiscal year 1976 and the 1977 date.

—Washington Star, Friday, March 26, 1976

The ability to write effectively should be a concern of every manager. If you are unable to express yourself on paper, your employees as well as your colleagues may misunderstand your instructions. You should try to develop guidelines that give your messages the advantage of being easy to handle and pleasant to read. Successful executives try to put themselves in the position of the persons who will read their communiqué. An attempt should be made to develop communications that appeal to the minds and emotions of the receivers. They should also anticipate and try to answer the reader's "what's in it for me" questions.

The distinction between literary and business style is created by the time available to the writer and the reader. The busy manager rarely looks for an entertaining literary tone in his communications. He is more concerned with accuracy, brevity, clarity, digestibility, and empathy—and he should be.

Studies in the dynamics of reading indicate that maximum efficiency in reading speed and comprehension is achieved when the matter is expressed in clear, simple language. Simple language can be rich and communicative. Lincoln's Gettysburg Address contained only 265 words. Almost three-quarters of them are one-syllable words, yet there is variety, vitality, and drama in this classic message.

Professional writers search for key words and phrases to convey their story with vigor. Insecure people who are fearful of committing themselves in writing are frequently guilty of using abstract and technical jargon, clichés, and platitudes as a cover-up. They may find it necessary to communicate to "impress" rather than "express" when they have little to say.

Business writing need not be a complicated assignment. One of the reasons why it is considered difficult by some managers is that they do not understand the character and function of their memo or report. They are not familiar with its purpose in relation to the conditions from which it arises. Understanding these fundamentals is an absolute requirement for anyone who wants to write an effective business report that will get results.

A business communication is nothing more than a written presentation of useful data or information. But if this information is to be used, it must be more than vague ideas, opinions, prejudices, or feelings. Since the report is directed to a specific reader or audience for business purposes, its information must be verifiable facts or at least the conclusion of acceptable authorities. In essence, a business report is a highly specialized form of communication. It is flexible in subject content as well in organization, form, and use.

Regardless of how capable and knowledgeable a manager you may be, the inability to communicate by means of the written word is a great handicap. The ability to write adds an important skill to the manager's repertoire. One of the major reasons for writing failure at the executive level is a lack of preparation. By keeping a few basic rules in mind, you can improve your writing.

1. **Write an outline.** Before you put anything on paper, think through what you want to say. Review the purpose of your communication, do the required research, and make an outline to help structure the flow of your thoughts.

2. **Revise your writing.** The first draft of your writing should not be locked in stone. After you have finished your first draft, go back and make corrections. If you have time, let it sit overnight and then review it. Revision is an essential element of all good writing. No matter how competent you may be, you will find passages that will need to be changed. It is a good idea to show what you have written to a colleague for her reaction.
3. **Get to the point.** In the first sentence, tell you readers your reason for writing. Do not make them search for the meaning of your message.
4. **Be clear.** Everything you write must be explicit and not misleading. Use simple, short words (Remember Lincoln's address): your goal is to write only what you need to say, using the fewest words possible.
5. **Stay focused on your subject.** Do not wander to irrelevant issues that weaken your point and confuse your readers. Your message should flow logically and smoothly from one point to the next.
6. **Each sentence should discuss only one main idea.** Each paragraph should deal with a single, clearly stated theme. The rest of the paragraph should strengthen the idea with supporting facts.

A WORD ABOUT LETTER WRITING

Business letters can take different forms and deliver different messages. According to the magazine *Today's Office*, there are four basic kinds of letters: those that convey good news, those that impart bad news, those that are neutral or informative, and those that try to persuade the reader to take action. Before deciding which approach to use, ask yourself how the recipient is likely to react to your letter. The answer will guide you in how to present your material.

The order in which the information is analyzed can dictate the tone of your message. A letter that contains good news is usually written in direct order. The first paragraph (the most emphatic) imparts the good news. The middle section explains the situation, and the final paragraph usually has an action closing.

However, the bad-news letter is presented in indirect order. It starts with a buffer statement to soften the blow. The second section provides an explanation or analysis of the situation. This leads to the third section, which contains the decision. The final paragraph is usually a friendly, routine closing.

When writing neutral or informative letters, you should use direct order. This format is preferable because it is easier to read.

The fourth type of letter attempts to persuade the reader to take some action. It uses a modification of the indirect arrangement. You start with a paragraph that acts as an attention-getter. The second tries to create interest in whatever ideas, products, or services you are promoting. The third section attempts to generate desire or conviction in the reader, and the final paragraph has an action closing.

As the letter writer, put yourself in the other party's shoes. How would you feel after receiving the letter? If this is a letter that you would want to receive under the existing circumstances, then your hard-to-write letter will have been worth the effort you took to compose it. The following is a checklist you may want to review to evaluate your letter-writing ability.

THE LETTER WRITER'S CHECKLIST

The questions are so worded that checkmarks in the "No" column may indicate your correspondence trouble spots.

		Yes	No
1.	Are most of your letters less than a page long?	☐	☐
2.	Is your average sentence fewer than 22 words?	☐	☐
3.	Do you try to keep paragraphs short—fewer than 10 lines?	☐	☐
4.	Do you avoid beginning a letter with *Reference is made* or *This office is in receipt of your letter?*	☐	☐
5.	Do you know some good techniques for beginning letters naturally and conversationally?	☐	☐
6.	Can you think of four words that will take the place of *however?*	☐	☐
7.	As a rule, do you paraphrase laws and regulations instead of playing safe and quoting them?	☐	☐
8.	Do you know what's wrong with phrases like these: *makes provision for, held a meeting, gave consideration to, meets with the Bureau's approval?*	☐	☐

9.	Are your letters free of pat phrases like *the records of this bureau indicate* and *this office has no jurisdiction over. . .?*	☐ ☐
10.	Do you use personal pronouns freely, particularly the personal pronoun *you*?	☐ ☐
11.	Are your letters written in the first person *(we /I) shall appreciate)* rather than the third person *(they will appreciate)*?	☐ ☐
12.	Do you prefer active verbs *(the manager read the letter)* to passive ones *(the letter was read by the manager)*?	☐ ☐
13.	When you have a choice, do you choose little words *(pay, help, mistake)* rather than big ones *(remuneration, assistance inadvertency)*?	☐ ☐
14.	Whenever possible, do you refer to people by name *(Mr. Jones, Miss Smith)* rather than categorically *(the claimant, the veteran, the applicant)*?	☐ ☐
15.	Compare your letters with your talk. Do they sound as you do when you talk in a careful manner?	☐ ☐
16.	Do you answer a question before explaining the answer?	☐ ☐
17.	Do you encourage your stenographer to correct obvious errors in your letters?	☐ ☐
18.	Have you an urge to use a red pencil on phrases like *attention is called to the fact, it is to be noted,* and *it will be apparent?*	☐ ☐

SUGGESTIONS FOR IMPROVING YOUR WRITING SKILLS

- Keep your sentences to 20 words or fewer. More than 20 words usually are difficult to understand. Sentences must vary in length and in structure if the reader is to be saved from boredom. Effective writers maintain a balance between long and short sentences.
- Avoid unnecessary words. A great deal of business writing is diluted with words that do not count. Surveys have shown that most letters can be cut almost by half and still say the same thing.
- Write to express, not to impress. Inexperienced writers often try to impress rather than to express. The results are usually confused, fail to communicate, and probably will irritate the reader. Put your message in clear, concise, and simple language.

- Put your purpose in writing. Before trying to write, do some visualizing of your result. Developing a concrete, visual picture of your target forces you to create a positive expectation that your efforts will lead to a good performance.
- Find a comfortable place to write. To many, that is not the same office desk where you conduct regular business. Eliminate the distracting influences of other work waiting for you, phone calls, and other interruptions.
- Do not feel that every business letter needs a "finishing touch" such as "Please let me know if I can be of further assistance." That is probably what you are there for.
- Do not overwhelm your reader with intensives and emphatics. Intensives include such adjectives and adverbs as *highest, deepest, very much, extremely,* and *undoubtedly*.
- When you receive a business letter that has good form, is clear and concise, and meets all the requirements of good letter writing, share it with your staff so they can learn from the correspondence.
- Before sending a business letter, ask yourself three questions: (1) Does the letter respond to the other party's feelings, perceptions, and needs? (2) Does the letter indicate that you looked carefully and thoroughly into the situation? and (3) Does the letter report on the action you have taken or plan to take?
- In writing business reports, a solid page of print may be made more digestible to your reader if it is broken up with headlines and graphics. Other suggestions include underlinings, spacing, italicizing, capitalizing, enumerating, excerpting, indenting, boxing, summarizing, and illustrating.

SUGGESTED READINGS

La Fler, S. A. (1989, February). Guidelines for writing effective management letters. *The practical accountant,* 43-54.

The right way to write a business letter. (1991, March). *Today's Office.*

Seekings, D. (1987). Effective writing. *Management Decisions, 25*(3), 11-15.

Smeltzer, L. R., & Glesdorf, J. W. (1990, Nov/Dec.). How to use your time efficiently when writing. *Business Horizons.*

When writing words fails you. (1984, November). *Nation's Business.*

Chapter 45

LEADERSHIP

Leadership is always directed toward achieving goals desired by both the leader and the group being led, and control is exercised by all.

There is no such thing as a stereotypical leader. Leaders come from all backgrounds: some are male, some are female; some are black, some are white; some are tall, some are short. George Washington was a wealthy planter of English ancestry. Abraham Lincoln was a poor rail-splitter from Kentucky. Martin Luther King, Jr., an African-American minister, was the greatest civil rights leader of the 20th century, and Susan B. Anthony was a pioneer and leader in the women's rights movement.

Education is important, but many of our past leaders had little. Lincoln, Edison, Carnegie, and Ford had little grade school education and no high school education. Einstein was called a "dumm kopf," Eisenhower was 66th in his class at West Point, and Custer was last in his.

Leadership is a difficult term to define. Scholars who study it also find it difficult to agree on a definition. When I am asked to explain it, I indicate three answers. The first is one word—*influence*. By leading from behind, suggesting, persuading, and guiding others, you exert influence on them. Subordinates are more likely to listen and follow.

The second definition is *leadership*—getting people to do what you want them to do. It is your ability to motivate others by their consent without the use of authority on your part.

The third definition is more detailed. A leader is a person who has influence with people, which causes them to listen and agree on common goals, to follow that person's advice, and to go into action towards these goals. John Gardner stated in *Excellence: Revised edition* (1984), "Leadership is the process of persuasion and example by which an individual or leadership team induces a group to take action that is in accord with the leader's purposes or most likely, the shared purposes of all."

In order to define leadership, we must identify those characteristics that distinguish leaders from followers. In a review of more than 100 studies of personal traits of leaders of various groups, researchers found the leader was above the average of his group in intelligence, scholarship, dependability, and social participation. There was less agreement among researchers that leaders possess greater initiative, persistence, self-confidence, cooperativeness, adaptability, and verbal facility. Although leaders are reported to possess these traits in different measures from followers, researchers found that the differences cannot be too great in either direction. Many of these traits are undoubtedly desirable in any leader, but it is almost impossible to define them to everyone's satisfaction. The definition of a given trait varies from situation to situation. There is little agreement as to the meaning of the trait and how it can be measured. Personality tests can assist in this assessment but, unfortunately, such tests often have low reliability and validity. Furthermore, the trait approach does not include examination of technical skills and specific abilities that may be necessary for leadership in specific work organization, nor does it recognize that different situations may require different personal characteristics. This approach does not deny the value of seeking a cluster of personal characteristics that would form the nucleus of leadership. However, we must recognize that this search is only a part of this problem.

Since a study of the personality traits of leaders seems inadequate, many researchers have turned to analysis of the situational factors in leadership. Situational analysis assumes that leadership emerges in a group when one member's characteristics are perceived by others enabling the individual, more than others, to contribute effectively to the attainment of group goals. The characteristics necessary for leadership, then, would vary from situation to situation, depending upon the nature and function of the group.

Another characteristic often considered in the discussion of leadership is charisma. Charisma has many labels: personality, star quality, magnetism, and charm. Charisma, coming from the Greek language meaning "a gift from the gods," is that special something that attracts an individual to certain people even if we cannot understand why we are attracted. We all recognize extraordinary power in world figures of the past such as Gandhi, Joan of Arc, Napoleon, Roosevelt, and Kennedy. Many of us admire contemporary personalities such as Martin Luther King, Jr., former President Reagan, Jacqueline Onassis, Phil Donahue, Oprah Winfrey, and Robert Redford.

These "stars" are widely different, but they all have the same unique ability to capture and hold the spotlight. Charisma in public figures manifests itself in many different forms: former President Reagan exudes political charm; Pope John Paul II and the late Mother Teresa epitomize spiritual force; Henry Kissinger, at the height of his influence, held people spellbound with his aura of power.

A word of caution about charisma: we should be careful not to "hitch our wagons" to an individual who possesses only charisma. History recalls that Hitler had charisma, yet he was responsible for the annihilation of 6 million Jews. Reverend Jimmie Jones was able to influence over 800 individuals to end their lives in the South American jungles, and Fidel Castro, through his tirades on television, has been able to suppress democracy and freedom of the Cuban people for over 30 years. Charisma is certainly a desirable trait in a leader, but honesty, integrity, and knowledge must also be a part of the leader's makeup.

Recently, Warren Schmidt of the University of Southern California conducted a study of 1,500 managers relative to their views concerning the personal traits one looks for in leadership positions. The most frequent responses were integrity, competence, and decisiveness. A follow-up survey sponsored by the Federal Executive Institute Alumni Association, involving over 800 senior public sector administrators, replicated these findings. In a four-year series of executive seminars conducted by Santa Clara University, the Tom Peters Group/Learning Systems asked 5,200 top-level managers what they thought were important characteristics of a leader. In virtually every organizational setting, the four leadership characteristics regularly selected as most important were honesty, competence, foresight, and inspiration.

Leadership is an exceptional quality of personality characterized by specific attitudes, talents, and abilities that combine to produce outstanding performance. The performance is measured by results. The results are interpreted in terms of goals. It is this performance, not status or position, that will identify leaders of the future. Future leaders cannot be made—they must be discovered. Once recognized, potential leaders must be developed by challenging them with a work environment that will allow them to use their talents optimally. The future of American enterprise depends on finding and developing a new breed of management leadership. These must be individuals who are willing to "fight against the tide." Leaders of the future will be philosophers as well as entrepreneurs. They need to be oriented toward discovering and obeying reality rather than manipulating and controlling people. The formal education they have acquired will be only a beginning to an ongoing, continuous program of self-study in their chosen areas of expertise.

Leaders of tomorrow will need to be tough physically, intellectually, and spiritually, and be able to withstand conflict. They must be self-starters motivated by the opportunity for achievement. Individuals who assume positions in management must be independent thinkers, relying on their own assessments of the issues rather than accepting mandates from others. Being able to examine specialized fields and correlating and generalizing the results will be another requirement for future leaders. Most importantly, they will be required to be decisive, objective, knowledgeable, and prepared to speak out on major issues of our time. Tomorrow's leaders will need to stand alone, whether representing government, labor, or the public.

SUGGESTIONS FOR SUCCESSFUL LEADERSHIP

- Be sure you know as much as possible about the person or persons you are trying to influence.
- Set goals *with* your followers and not *for* your followers. Individuals often accept new ideas quicker if they feel they, rather than the leader, generate the ideas.
- People like to be praised, especially in front of others. Praise is one of the most important techniques a leader can use. Everyone needs to experience the thrill of success.

- Challenge your followers to something new and different. People get bored and lose interest if they are not introduced to new challenges.
- Time must be made available for you to listen to your followers, to consult, think, organize, and plan with them and receive their ideas.
- Act as a facilitator rather than a teacher to your followers. Encourage them to plan, organize, and carry out their own ideas.
- Give your followers as much responsibility as they can handle. Watch for leaders within the group. Put them in charge when appropriate. Provide them with opportunities to acquire skills to solve their own problems.
- Lead with your head, not with your emotions. Decisions made while you are angry are usually not rational and are ineffective.
- Leaders must be both thinkers and doers. They must be as competent with philosophical concepts as they are with administrative and operational skills.
- Create work rules and policies that will not stifle initiative and creativity. Delegate authority and responsibility when it is appropriate and deserving.

SUGGESTED READINGS

Bass, B. M. (1985). *Leadership and performance beyond expectations.* New York: Exec Press.

Batter, J. D. (1989). *Tough-minded leadership.* Chicago: AMACOM.

Bennis, W., & Nanus, B. (1985). Leadership. New York: Harper & Row.

Byrd, R. E. (1987, Summer). Corporate leadership skills: A new synthesis. *Organizational Dynamics, 16,* 34-43.

Vroom, V. H., & Jago, A. G. (1988). *The new leadership.* Englewood Cliffs, NJ: Prentice Hall.

Chapter 46

LISTENING

As the sign in Lyndon Johnson's office proclaimed: "You ain't learning nothing while you're talkin'."

Today's executives spend nearly 45% of their workday just listening. Research indicates that the average person spends 9% of her time reading, 16% writing, 30% speaking, and 45% listening. The higher she rises in the management hierarchy, the greater that percentage is apt to be. This is as a result of more meetings, as well as interviewing, counseling, exchanging information, and decision making. Most executives simply talk about listening, but do little to actually enhance listening skills. Many top executives say they are aware of the high cost of poor listening but do not seem to be committed to providing the training necessary to improve listening skills. The executive must not only listen, but also must try to *hear* what is being said and what is not said.

Most of the time the manager is not talking should be spent in creative listening. Talking is one phase of communication. Listening is the other. The purpose of this listening should be to get information, to develop a positive relationship with the speaker, and to make the speaker feel comfortable in his position.

What do managers who are good listeners do? They maintain eye contact, which shows attention and interest in the speaker. Good listeners concentrate attention on what the speaker says and does, never faking attention. This is quite an achievement because speed of thinking is much faster than talking speed. Good listeners note changes in the speaker's tone and inflections and try to interpret what the speaker does not say. Individuals who listen well do not hesitate to ask questions for clarification and confirmation, but do not interrupt, do not run out of patience, and do not reveal any lack of interest in what the speaker is saying. The goal of creative listening is to assure effective two-way communication.

One basic barrier to effective listening is the inability to concentrate, which causes facts and ideas to be lost. Lack of concentration may have several roots. Most of us speak about 140 words per minute, but we can comprehend at a much faster rate. This speed permits us to wander into other areas as we listen.

Another obstacle to the art of listening is that most of what is heard is soon lost. Retention of what is heard is approximately 50% immediately after a 10-minute oral message and about 25% 40 hours later. Given these facts, it is not surprising that remembering oral information correctly is a problem.

Opinions and prejudices can also cause poor concentration. When a statement the listener does not like is made, he turns off the communicator. Or he may concentrate on a statement with which he disagrees, allowing other statements to go unheard.

The way a communicator is dressed, the look on her face, her posture, accent, skin color, mannerisms, and past experiences may also cause a listener to react emotionally and tune out the communicator. The listener should try to put aside his preconceived ideas and prejudices. The person speaking may have a new idea that is worth putting into practice. If you want to concentrate, do not try to do something unrelated to the discussion while the speaker you wish to listen to is talking. The good listener makes every effort to listen to everything said and tries to determine facts from perceptions. If you listen carefully, you can distinguish important ideas and separate them from minor ones. It should be emphasized that good listening is hard work.

Effective listening produces many positive results. There will be more effective listening on the other person's part. When the individual notes that you are sincerely and carefully listening and not just waiting for her to pause so you can interrupt, she will not feel threatened. Also, after the individual has had her say, she is more ready to listen to you.

Effective listening encourages the speaker to present more information that may benefit you. Your careful listening will motivate the speaker to reveal more facts about the subject. This will help you make more intelligent discussions.

When you demonstrate good listening techniques you can improve your relationship with the speaker. This results in a better understanding in what the speaker is saying. You may also recognize that one individual requires frequent praise, while another does not; that one responds favorably to counseling while another resents it.

Another result of careful listening can be easy solutions to problems. When the speaker is permitted to express himself in an unthreatening environment and he feels that he has the listener's complete attention and respect, he may hear himself more clearly; as a result, solutions may come through to him or you more clearly.

Take time to listen to the speaker's entire message. As the receiver of the information, it is your responsibility to understand thoroughly what the speaker says. This means that you must listen to the entire message without interrupting or allowing yourself to jump to conclusions about what the speaker means. Some experts call this "focus listening." You stay with the speaker rather than cut the conversation short or turn it around so it centers on your concerns.

The listener gains a great deal when the speaker becomes involved in a two-way communication process. Some of the benefits include additional information that can be used to your advantage because the speaker will likely reveal thoughts and feelings of the moment. The listener can recognize the motives and objectives, if there are any, of the speaker and can proceed accordingly. Speakers who are actually involved in the conversation are more apt to lead themselves through the stages of interest, desire, and conviction. Speakers who are involved in a two-way communication will feel more comfortable in disclosing what they want, how they actually feel, and how the listener can help them achieve it.

It should be emphasized that the speaking and listening process is not over when the conversation ends. An important part of the process is feedback to those involved. Do not let listening be something that takes place infrequently in your organization. Give your employees an opportunity to learn the listening-speaking process. If they do, your messages will be more clearly and quickly understood. Active listening won't eliminate all your problems, but it is a beginning.

SUGGESTIONS TO IMPROVE YOUR LISTENING SKILLS

- Keep your mind open until you have heard all sides of the issue. By being open and attentive as a receiver, you will be able to pick up and receive many more *transmissions* much more clearly. When people learn you are honest with them, they will come to trust you and in time will reciprocate with honesty.
- Take notes during the speaker's presentation. This will help you retain ideas and facts. Never become so absorbed in the task of taking notes that you lose the ideas being transmitted.
- Look directly into the eyes of the speaker. Do not slouch in your chair. You may listen better when you are relaxed, hands clasped behind your head, and feet propped on the desk; however, the speaker may see this posture as a lack of interest and perhaps an uninterested listener.
- Consider the speaker's background and experience. This will help you analyze the speaker's point of view and more accurately interpret the meaning of his words. Most effective communication occurs when you completely understand the other person's position and needs.
- Give the speaker enough time to complete her entire message. Do not interpret or allow yourself to jump to conclusions too quickly.
- Remember, hearing is not listening. Pay attention to what is being said, take mental and written notes, think along with what is being said and look for meanings and messages in everything that is being said.
- Prepare to listen. Clear your mind, sharpen your senses, and tear down the barriers to the introduction to new ideas. Do your homework if you are not familiar with the subject to be discussed.
- Always summarize what is being said. This will increase retention and reduce the likelihood of misinterpretation.
- Be sure to distinguish and clarify the main points the speaker is trying to make. To be an effective listener, you should ask yourself, "What is the purpose of the communication?" "What is the speaker really trying to say?"

- Be aware of nonverbal as well as verbal messages. Attention to nonverbal cues enables you to "hear" feelings as well as words. Nonverbal indicators often carry 60% of the meaning of a message.
- As a listener, analyze the logic of what is being said. Test the assumptions that are being communicated. Be sure the words being stated are representative, current, complete, and relevant to the issue at hand.
- If you are not hearing the messages, you may have a hearing problem and should seek medical advice. Tremendous advances have been made in the last decade and these can be of great help in improving your hearing capability.

SUGGESTED READINGS

Glatthorn, A. A., & Adams, H. R. (1983). *Listening your way to management success.* Glenview, IL: Scott, Foresman.

Maidment, R. (1985). Listening—the overlooked and underdeveloped half of talking. *Supervisory Management, 30*(8), 397 10-12.

Margerision, C. J. (1987). *Conversational control skills for managers.* London: W. H. Allen.

Montgomery, R. L. (1981). *Listening made easy: How to improve listening on the job, at home and in the community.* New York: AMACOM.

Steel, L. K. (1980). *Your listening profile.* Minneapolis, MN: Sperry Corporation.

Chapter 47

MANAGING STRESS

Remember, we all have to face and live with occasional states of stress. We can either ignore them or we can turn the situations of stress and pressures into an opportunity for further emotional growth.

It has been called the disease of the '90s. It has been blamed for everything from heart attacks to rising medical costs and the increase in homicides in the workplace. It is stress. While much is understood about stress, its causes and effects, much is still a mystery.

The medical experts are aware of many things that can cause stress, but it is different for each individual. The same situation may create great stress for one person, while the other person is motivated and not affected at all by stress. Some stress can even be good. It can help people perform their best or motivate them. It is when that level of stress is sustained over a period of time that it can cause problems.

Stress is a condition with which every human being is familiar, yet the term is so widely misused that it is often subject to confusion and ambiguity. Most people automatically assume that stress is bad; as a matter of fact, it may or may not be harmful, depending on the circumstances.

It is often useful to distinguish between what causes stress and what it is. The various pressures or demands from the external environment that could stem from your family, job, friends, or the government are called *external stressors*. The various pressures or demands from your internal environment are called *internal stressors*. They include the pressures you put on yourself by being ambitious, materialistic, competitive, and aggressive. In most of us, these internal stressors have far more intense an effect than do the external stressors or demands.

The U.S. Department of Health and Human Services has ranked the inability to deal with work-related stress as one of the five most critical threats to American health. In 1990, Northwestern National Life Insurance released survey results of a random sampling of 600 workers in the United States. Forty-six percent of those surveyed responded that their jobs were "highly stressful," while 34% felt so much stress that they were actually thinking of quitting their jobs. And stress appears to be increasing each year.

Also in 1990, Sirota & Alper Associates, a New York firm that researches employee attitudes, released findings from 18 years of research for over 170 corporations where they monitored more than one million employees. The number of managers who say they have too much work to do had increased from 34% to 46% over the previous five years. And the number of non-managers who complained of having too much work increased from 30% to 39% in those same five years.

Stress has been blamed for high turnover rates, absenteeism, job dissatisfaction, down time, grievances, high medical costs, and a plethora of other problems. It is estimated that the cost of stress, including factors like absenteeism and rising stress-related compensation claims, is $150 billion annually. A Gallup survey showed that on average, 25% of an organization's employees suffered from stress-related disorders, and typically such an employee misses about 16 days of work a year. Many employees are actually suing their organizations, and winning, for harm that they have suffered due to stress while working. States like California have been awarding damages to employees who have filed worker's compensation claims for these types of "mental" injuries.

Like other management professionals, those in recreation and parks experience a great deal of negative stress. Although the profession is geared to help others enjoy leisure as a healthy balance to work or study, many of our own people do not practice what they preach. If they engage in recreational or leisure pursuits, too often they play as hard as they work, replicating the same competitive challenges and stresses found in most work. Signs of stress include insomnia, restlessness, irrational irritation and anger, difficulty concentrating, loss of interest in usual recreational activities, prolonged or excessive fatigue, needless worry, poor or failing health, difficulty meeting deadlines, excessive working, confusion about what is important in job and family life, and habitual reliance on tobacco, alcohol, or sedatives.

Although physical exertion causes a certain degree of stress, mental effort usually produces more. The mind and emotions are far less resilient than the body and thus more likely to show unquestionable signs of strain. Thus, it is important to understand and resolve mental and emotional stress to the extent these can be divorced from the body.

We live in stressful times, in a highly tuned society and culture. In fact, Americans have been called "the most electric of people." To get at the sources of personal stress, researchers of the human psyche have concentrated on the individual, especially in the context of organizations and societies. Change is with us daily; routines, traditions, expectations, and plans are often modified or assailed, causing managers to feel beleaguered and burdened, rather than challenged and fulfilled by their assignments. Added to the uncertainty of change are the constraints of organizational life: competition for high-paying executive jobs, erosion of decision-making power, subordinates' claims for power or autonomy, lack of career opportunities, inadequate rewards or feedback on job performance, in-house politics, budget and job cuts, heavy workloads, and the ongoing conflict between a manager's personal values and those of the organization. And, to make matters more difficult, few people today can claim their homes as respites from the stresses of work. It is not that a manager takes his work problems home, or the concerns of home to work, but that these are often inseparable—both worlds manifesting stressful interrelations.

Stress Quiz

		Often	Sometimes	Seldom	Never
1.	During the past three months, how often were you under considerable strain, stress, or pressure?	☐	☐	☐	☐
2.	How often do you experience any of the following symptoms: palpitations or racing heart, dizziness, painfully cold hands or feet, shallow or fast breathing, restless body or legs, insomnia, chronic fatigue?	☐	☐	☐	☐
3.	Do you have headaches or digestive upsets?	☐	☐	☐	☐
4.	How often do you experience pain in your neck, back, arms, or shoulders?	☐	☐	☐	☐
5.	How often do you feel depressed?	☐	☐	☐	☐
6.	Do you tend to worry excessively?	☐	☐	☐	☐
7.	Do you ever feel anxious or apprehensive, even though you don't know what has caused it?	☐	☐	☐	☐
8.	Do you tend to be edgy or impatient with your peers or subordinates?	☐	☐	☐	☐
9.	Do you ever feel overwhelmed with feelings of hopelessness?	☐	☐	☐	☐
10.	Do you dwell on things you did, but shouldn't have?	☐	☐	☐	☐
11.	Do you dwell on things you should have done, but didn't?	☐	☐	☐	☐
12.	Do you have any problems concentrating on your work?	☐	☐	☐	☐
13.	When you're criticized, do you tend to brood about it?	☐	☐	☐	☐
14.	Do you tend to worry about what your colleagues think of you?	☐	☐	☐	☐
15.	How often do you feel bored?	☐	☐	☐	☐
16.	Do you find that you're unable to keep your objectivity under stress?	☐	☐	☐	☐

		Yes	No
17.	Of late, do you find yourself more irritable and argumentative than usual?	☐	☐
18.	Are you as respected by your peers as you want to be?	☐	☐
19.	Are you doing as well in your career as you'd like to?	☐	☐
20.	Do you feel you can live up to what top management expects from you?	☐	☐
21.	Do you feel your spouse understands your problems and is supportive of you?	☐	☐
22.	Do you have trouble with any of your associates?	☐	☐
23.	Do you sometimes worry that your associates might be turning against you?	☐	☐

24. Is your salary sufficient to cover your needs? ☐ ☐
25. Have you noticed lately that you tend to either eat, drink, or smoke more than you really should? ☐ ☐
26. Do you tend to make strong demands on yourself? ☐ ☐
27. Do you feel that the boundaries or limits placed on you by top management regarding what you may or may not do are fair? ☐ ☐
28. Are you able to take problems in stride, knowing that you can deal with most situations? ☐ ☐
29. Do you seldom "lose your cool" and stay productive under stress? ☐ ☐
30. Do you feel neglected or left out in meetings? ☐ ☐
31. Do you habitually tend to fall behind with your work? ☐ ☐
32. During the last year, have you or anyone in your family suffered a severe illness or injury? ☐ ☐
33. Have you recently moved to a new home or community? ☐ ☐
34. During the last three months, have any of your pet ideas been rejected? ☐ ☐
35. Is it difficult for you to say no to requests? ☐ ☐
36. Do you generally work better when under pressure? ☐ ☐
37. Are you able to focus your concentration when under pressure? ☐ ☐
38. Are you able to return to your normal state of mind reasonably soon after a stressful situation? ☐ ☐

Scoring: Add up your points based on the following answer key.

	Often	Some-times	Seldom	Never		Yes	No		Yes	No
1.	7	4	1	0	17.	4	0	33.	3	0
2.	7	4	1	0	18.	0	3	34.	4	0
3.	6	3	1	0	19.	0	4	35.	3	0
4.	4	2	0	0	20.	0	5	36.	0	3
5.	7	3	1	0	21.	0	5	37.	0	3
6.	6	3	1	0	22.	3	0	38.	0	4
7.	6	3	1	0	23.	4	0			
8.	5	2	0	0	24.	0	3			
9.	7	3	1	0	25.	5	0			
10.	4	2	0	0	26.	4	0			
11.	4	2	0	0	27.	0	3			
12.	4	2	0	0	28.	0	3			
13.	4	2	0	0	29.	0	3			
14.	4	2	0	0	30.	4	0			
15.	4	2	0	0	31.	3	0			
16.	6	4	1	0	32.	6	0			

What your score means

90-167—A score in this range not only indicates that your troubles seem to outnumber your satisfactions, but that you presently are subjected to a high level of stress. You are, no doubt, already aware of your pressures, and you are rightfully concerned about your own psychological and physical well-being.

By all means, you should do everything possible to avoid as many stressful situations as you can until you feel more in control of your life. It might be a good idea for you to go over the quiz to pinpoint the major sources of your present stress.

You also might need to develop more effective ways to manage your response to stressful situations. Your vulnerability to stressful events shows that you are, perhaps, overreacting to problems or it may be that you are not as willing to cope with adversities as you could be.

You might want to consider seeking professional help. Sometimes even a few hours of counseling can be of great help. You also might want to pay heed to the wise words of a cardiologist, who offers the following three rules for combating stress: "Rule No. 1: Don't sweat the small stuff. Rule No. 2: Everything is small stuff. Rule No. 3: If you can't fight it, or flee from it, flow with it."

45-89—A score within this range either indicates that your stress seems to be moderate, or that you probably are handling your frustrations quite well. You should, however, review various aspects of your daily life and try to relieve stress before it starts building up. Because you may have occasional difficulties coping with the effects of stress, you might want to consider adding some new methods of dealing with disappointments.

Remember, we all have to face occasional states of unwellness. We can ignore them, or we can turn situations of stress and pressure into opportunities for further emotional growth. Life either can grind us down or it can polish us up—and the choice is largely yours.

0-44—A score in this range indicates that your stress is relatively low and you probably are in great shape. In spite of minor worries and concerns, stress does not seem to be causing you any serious problems.

You have, no doubt, good adaptive powers and are able to deal quite well with situations that make you temporarily uptight. You seem to have been able to strike a good balance in your ability to cope with and control stress.

Recognizing Stress

Parks and recreation managers—either full-scale recreational agencies or educational institutions with a variety of physical and cultural outlets—are in an ideal setting to reduce stress. However, few have the time or hardiness to go to the root of a complex problem, addressing such issues as "Why am I working this hard?" or "What pressures led me to participate unquestioningly in this race for wealth and success?" Instead, most persons seek practical outlets for these pressures, trying to dispel stress through variety and distraction, leaving the contemplation of life to intellectuals and philosophers.

SIX SOURCES OF JOB STRESS

Most executives suffer from job stress for one of six reasons.

1. **A Sense of Failure.** Like a student who is getting bad grades in school, an individual will not feel good if she believes she is not doing a good job. Often such a sense of failure is a misconception on the individual's part. It can occur whenever (1) you feel you do not measure up to the standards you have set for yourself, (2) you are doing a good job but have the misfortune to be working among others who are unusually superior performers, (3) you have a critical supervisor or board member who takes excellent performance for granted and only comments when something does not go perfectly.
2. **Too Much to Do.** You may be doing a good job but still feel the stress of not having enough hours in a day. This sense of always being behind schedule, never caught up, and rushing from task to task takes a psychological toll. Everyone can survive this kind of pressure for a short period of time, but if such a pattern continues unabated, the pressure can turn into emotionally draining long-term stress.
3. **Personality Conflicts.** Disagreements between family members are frequently dealt with immediately and openly, but people are often advised by peers to keep office conflicts under the surface, where they remain unresolved.

    We spend eight to 10 hours a day with members of our staff—probably more time than with our families. This amount of time creates a closeness that affects your emotional well-

being no matter how we try to ignore conflicts. We need to tend to our relationships with work associates with the same level of concern as with our family.
4. **Lack of Money.** If we are experiencing financial pressures in our personal life, it is natural to blame our job. Financial frustration is not technically a job stress. If you do not have enough money to keep the lifestyle you want, you may feel you are suffering job stress. Other personal conflicts can also spill over to affect our attitude toward our jobs. Health problems, marital concerns, parent/child conflicts, or any other personal dissatisfactions can be carried into the workplace, and unless you are aware of what is happening, you may tend to blame your work for the negative emotions.
5. **A Career Mismatch.** We have a unique blend of interests, abilities, temperaments, aptitudes, and personality qualities. If these patterns come close to matching the demands of your career, the odds that you will find fulfillment are greatly increased. Likewise, if our personalities vary markedly from the personalities of people who are successful in our careers, we may be mismatched. The odds of making a positive adjustment to the demands of your job are thus diminished. If it is too much of a mismatch, we may always experience some level of stress and may want to consider another position.
6. **Job Security.** The fear of losing one's job can create a gnawing, persistent stress. The reasons behind such a fear can range from a slowdown in the economy that raises the possibility of budget and staff cutbacks to a proposed shift in organization focus that includes major personnel to an unexplainable sense of anxiety with no logical basis. No matter what the origin, a lack of security can play a significant role in job stress.

WHO COPES WITH STRESS BEST?

Individuals who are comfortable with change, involvement, and control tend to cope effectively with the strain of their careers. Those who are not struggle harder.

Change
What is your attitude toward change? Do you enjoy change in your life? Do you encourage it? If change is not taking place, do you

make it happen or do you like things to remain constant and stable? According to research, the more positive your attitude toward change, the more stress-resistant you are. Economic conditions will fluctuate, new board members will arrive and others will move on, staff will resign, and even an organization's purpose can be altered by outside events. People who have a positive attitude toward change will respond more comfortably to the inevitable changes that occur in life.

Involvement

Involvement means how close you get to others. Some people like emotional closeness. They enjoy talking, listening, and sharing experiences. They strive for intimacy in their professional as well as their personal lives.

Other people are more removed. They believe that getting too close to fellow workers complicates the work environment. How close do you get to people? Do you reveal your true emotions; or do you use your public persona? Consider this question whenever you have a social exchange with your staff, board of directors, colleagues, or fellow members. Which executive is more stress resistant? It is the one who constantly strives to drop the walls and move closer to other people. Involvement meets your basic need for emotional closeness. Granted, there is some additional stress in being close to other people, but for the most part, closeness is energizing.

Involvement keeps dialogue alive. Many managers use the anchor of personal interest and involvement to stay with a conversation or confrontation until a misunderstanding is corrected, a misconception clarified, and any ambiguity reduced.

Control

To what extent do you feel you are in control of your job and of your life? If you have a job skill in high demand or a job where *you* call the shots, you tend to feel in control. If, on the other hand, you are living from paycheck to paycheck, have a job that is valuable only to your present organization, or work for a very autocratic supervisor, you may feel less in control. Control is often mentally achievable no matter what the circumstances.

Organization Support

For many stress management experts, a stronger approach to stress management is advocated. Many organizations are using Stress Management Interventions (SMI) to help address stress in the workplace. *Stress Management Interventions* are any activities, programs, or opportunities initiated by an organization that focus on reducing work-related stress by reducing the presence of work-related stressors or assisting individuals to minimize the negative outcomes of exposure to these stressors.

Stress Management Interventions can target three different points in the stress cycle. They can attempt to (1) change the situation by determining and reducing the number of stressors in a situation, (2) help employees modify their appraisal of the situation, or (3) help employees cope more effectively with the consequences of stress.

Many stress experts agree that any organization serious about stress management should start with a stress audit. Questionnaires can be used to determine where conditions at work promote stress, either by overwork or by boredom. The answers may illuminate areas or needs that are causing stress to employees that can be altered. Employees may also have information on how tasks can be done better or more efficiently that may also help to reduce stress. Not only is all this information helpful for organizations that are seriously trying to reduce their employees' stress, but by asking employees for feedback and information and listening and responding accordingly, employees will feel more in control of their situations and that they have the ability to change them when necessary. This feeling of control can also help to reduce stress.

Another way of addressing situations in the workplace that can cause stress is to respond to the situation before it causes stress. Management personnel should try to prepare employees for potentially traumatic organizational changes. This measure includes giving them any information available. Employees cannot feel in control of a situation if they lack information. The current economy is rife with these kinds of situations, layoffs, downsizing, reorganizations; rumors of these without solid information can create a great deal of stress in an organization. By giving employees information about what is going to happen, how long it will last, and what their parts in it will be, organizations can attack the situation and alleviate much of the stress that employees might normally feel.

SUGGESTIONS FOR IMPROVING YOUR STRESS MANAGEMENT

- Strengthen your personal and professional qualifications. If you are underused, undercompensated, or underappreciated, you have basically three options: (1) You can try to get your superiors to see you differently, (2) you can work for a transfer within the organization, or (3) you can leave the organization and find a less stressful position.
- Recognize the pressures associated with the leadership role. These include decision making, overwork, motivating others, and job security.
- Talk with a friend or confidant as a way of relieving stress. Verbalizing problems helps clarify your thoughts and gives you an objective perspective of the situation.
- Keep a balance between work and recreation. The first step is an assessment of work habits. Every individual requires a different balance of fun and work, and overdoing either one will impair efficiency.
- Get physical exercise. A regular program of physical activity or other leisure pursuits is imperative.
- Begin your day by ordering your priorities. Make a plan of what you want to accomplish and the priority in which these goals need to be reached.
- Do not allow other people to waste your time. This means cutting off nonproductive conversations quickly or closing the door on unnecessary interruptions.
- Accept change. Realize that while some change is good and some is bad, the world is changing. Develop the attitude that says, "The world is not moving too fast; simply people are simply moving too slowly."
- Find yourself a mentor. Seek out someone who can advise you at critical times in your career. A mentor usually is one with experience and good judgment who can act as a sounding board for your ideas.
- Avoid being a perfectionist. Put your best effort into whatever you are doing, then relax. By continually striving for perfection, you create tension for yourself. Perfection implies unrealistic expectations for yourself.

SUGGESTED READINGS

Everly, G. S., & Smith, K. J. (1992, April). Stress management. *CPA's Journal of Accounting, 80*, pp. 82-85.

Ivancevick, J. M., Matteson, M. T., Freadman, S.M., & Phillips, J. S. (1990, February). Worksite Stress Management Interviews. *American Psychologist*, pp. 252-260.

Jones, J. W., & DuBois, D. A. (1989). A review of organization stress instruments. In T. Murphy & Schoenborn (Eds.), *Stress Management in the Workplace Settings*. New York: Praeger.

Matteson, M. T., & Ivancevick, J. M. (1987). *Controlling work stress*. San Francisco: Jossey-Bass.

Sutherland, V. J., & Cooper, C. L. (1990). Understanding stress: A psychological perspective for health professionals. New York: Chapman and Hall.

Chapter 48

PERSUASION

When we persuade people, we do not merely change their thinking on a subject, we cause them to do something. Their actions are modified, as well as their thoughts.

We constantly negotiate, persuade, and sell. We sell our ideas, products, services, and often ourselves. To persuade effectively we must use all the resources at our command. When persuading individuals to our point of view, we communicate at two levels: the content level and the relationship level.

The content level of communication deals with the subject, product, topics, or problem at hand. The relationship level of communications involves the way two individuals define their relationship. Either when we accept ideas, products, or services from another person or when we resist his attempts to persuade, we are usually responding to relationship messages. In any sort of persuasion transaction, the relationship between the persuader and the person being persuaded is determined in advance. Most experts on persuasion feel that real communications difficulty develops not in the content part, but rather in the relationship part.

One rarely has a second chance to sell a product, service, or idea; persuade a board to enter a new venture; convince colleagues that one's idea is best; or even ask the boss for a raise. Persuasion requires clear goals and a sound strategy. Every audience consists of individuals, each whose perceptions and motives are unique. The same information will strike different people in different ways. Some will be motivated to act; others will be impelled to resist.

The fundamental objective of persuasion is to modify an attitude, to change a person's predisposition to behave in a certain manner. But attitudes are difficult to change for three reasons: (1) attitudes accentuate certain aspects of an individual's environment. (2) Attitudes simplify life. (3) Once a person has an attitude, he will do everything possible to prove himself right, finding reasons that never before existed. It is vital, of course, to know the kind of person you are dealing with. Does the individual prefer a summary first and then the details? Or the details first? Does the individual respond best to a "folksy" approach or to a "sophisticated" explanation?

Never attempt to set yourself up as an authority in too many areas or you may find yourself losing credibility in areas where you may rightfully be an expert. If you must make decisions and persuade people in areas in which you cannot operate with authority and assurance, admit it. Preface your attempt at persuasion by saying, "I'm not an expert in this area, but I have talked with experts. Here's what they say and I'm inclined to agree," then continue. You are very likely to be considered a credible and highly regarded source of information, which is almost as good as being an expert.

We must be prepared to "sell" our ideas. Becoming familiar and comfortable with persuasion tactics is a must. This means presenting your ideas clearly, listening to responses attentively, and persuading others to accept your ideas. You must learn to be assertive without being obnoxious, and always to remain patient with those who object to your ideas. These skills do not come automatically or even naturally to most of us. However, we can acquire them with time, knowledge, and a solid commitment.

Before you can put your idea to work, you have to plan how to achieve successful implementation. You can combine one or several approaches. These approaches usually involve several aspects for selling the idea: inform others, persuade them of the idea's value and effectiveness, convince them it is the best idea under the circumstances, let them know how much consideration went into selecting this particular idea over others, overcome their resistance to change, and be prepared for emotional and cultural blocks.

TECHNIQUES OF A SUCCESSFUL PERSUADER

Certain techniques can help you become a successful persuader. All of them serve but one purpose, which is to get your point of view across effectively. Listed below is a brief description of some of these techniques:

- **Be careful in challenging the power structure in the organization.** In meetings especially, avoid direct or public confrontations with the "power people." You usually come out on the short end of such confrontations even if you have the right answers. In those situations where you do win in a direct confrontation with a power person, you usually create a dangerous situation for yourself in the future.
- **Stay on the content level.** While avoiding public confrontation with the power people is wise, the same does not hold true for others involved in the situation. If you have powerful persons supporting your point of view, go ahead and confront the opposition head on. It is important, however, to use tact and stay on the subject at issue. Always stay on the content level rather than attacking or addressing your opponent personally.
- **Stay positive and work your strong points.** When persuading, do not be excessively negative or start from positions of weakness. In sales, this is often interpreted as "Don't knock the competition," since this form of persuasion is usually interpreted by the buyer as a reflection of the seller's lack of confidence in her own product or service. Presenting your point of view in a negative light gives the buyer a choice between two "worsts." Instead, present ideas in a positive way and respond positively to comparisons.
- **Use your eyes.** Use constant and direct eye contact without staring. Keep your eyes focused and direct but not cold and expressionless. Watch for subtle changes in your opponent's eyes, such as pupil dilation or increased blinking. These changes might signal an emotional state is changing. Also watch for and evaluate facial expressions for similar signs of emotional state.
- **Keep your voice controlled, low, but well modulated.** Most importantly, keep it relaxed. When lowering your voice, you often see listeners moving in towards you to hear better what you are saying. Keep them from becoming bored with a monotone voice by raising and lowering it. At the other extreme, avoid becoming soft and wavering in your tone because these characteristics are considered too passive and nonauthoritative.

- **Stand still.** When seated, keep your hands in sight but not folded across your chest or behind your head. Do not rock or swivel in your chair. Keep both feet firmly and flatly on the floor. Crossed legs tend to create more barriers and sometimes look sloppy. When standing, keep your body as still as possible, slightly forward. Maintain a balanced stance, erect but not rigid. Avoid slouching, leaning, and stooping. Try to avoid excessive head nods. All of these movements and postures might be interpreted as signs of weakness. Others might signal aggressiveness. Hands on hips, arms folded across the chest, or feet wide apart are postures that can be taken as domineering signals.
- **Do not let your fingers do the talking.** The amount of hand movement each of us uses is controlled somewhat by culture and learning factors over which we have little or no control. Although some hand movement is natural and very effective as a means of emphasizing your points, in general avoid excessive hand fluttering. When persuading, avoid playing with beards, mustaches, and long hair. Avoid clenched fists, finger pointing and table pounding. All of these gestures can be, and frequently are, interpreted as dominant gestures and are seen as threats to a person's position.
- **Personalize your appeal.** According to most persuasion studies, the most successful persuaders pitch their presentations to the specific needs of the person being persuaded. You are better off not using the standard line or appeal when trying to persuade. Some people are susceptible to direct appeals and others must be persuaded using specially designed presentations.
- **Listen for needs, then respond.** Persuasion is not simply a matter of talking to someone about the many benefits, advantages, and features of your idea, product, or service. Too often talking becomes a high-pressure situation in which the persuader takes charge in attempts to overcome any and all objections. The major problem with this approach is that it can easily turn off persons being persuaded. Instead of high pressure, try to listen for their needs and hidden objectives. Use questions to probe them, to learn what may be on their minds. Once you have a clear idea of the feelings and attitudes of your "buyers," you are in a better position to make a successful persuasive appeal.

- **Do not be "flip," be genuine.** Make individuals feel you sincerely care about them and their problems and that your suggestions are designed to satisfy their major needs. Unless you are with close colleagues or friends, avoid sarcasm or flippancy. These messages often imply an equality position that usually is not present in a persuasion transaction, even after the point of view has gotten across. Do not overdo "superiority messages" and exaggerated shows of strength or knowledge.
- **Make your conclusions clear.** A common misconception is that people are much more likely to accept conclusions if they develop them themselves. Yet according to the best research on persuasion, people are more likely to change their opinions in the directions you desire when you give them a clear indication of that direction. Unless you give such signals, you cannot be really sure that your persuasive efforts will yield the results you desire. You should never force others to accept your ideas. Any acceptance you get from this strategy is likely to be in response to pressure, not to the value of your idea or to the persuasiveness of your logic. Gentle guidance, along with a clear but subtle indication of your position, is best.
- **Do not be too dramatic.** Some people feel highly sensational or dramatic appeals are the best ways to produce a long-term attitude or opinion change or to persuade someone not presently in favor of accepting your viewpoint. The most sensational or dramatic forms of persuasion can often be the least effective of all types of appeals in producing significant attitude changes, especially over a long period.
- **Do not oversell your idea.** The most difficult job in any presentation is knowing when to stop pushing your point of view. This may mean asking an individual for an opinion, or commitment towards the idea or proposal you are advocating. Knowing when to quit is a matter of listening for subtle signals. A long period of silence, a lowering of the voice, fewer questions, or leaning back can be signals that it is time to end the persuasive encounter. Knowing when to "close" takes time and experience, and even experts agree that such a skill develops only as you do more and more persuading.

PREPARING TO PERSUADE OTHERS

Much of the work in getting an idea accepted is done before it is submitted to others. In fact, it is at this stage that you will spend the most energy on behind-the-scenes preparation. You will have to be as alert as possible to the many contingencies that might develop.

Being prepared is absolutely essential. Your preparedness will show others that you have given much consideration to your idea. You must thoroughly understand your idea, be able to defend it, explain how objections to it can be met, and discuss how you intend to implement it. The following tips may help you sell your ideas.

- Make the advantages of the idea obvious to others.
- Satisfy the needs of those who will benefit from the idea.
- Indicate the economic value of your idea.
- Build on the support and enthusiasm you may already have from those in favor of your idea.
- Prepare audiovisual materials.
- Be convinced of your own idea before trying to sell it to others. If you do not buy it, no one else will.
- Back up your idea with good facts, research, and know-how. There is no substitute for information. Others will want to know just what you based your idea on, and why they should commit their energies to it.
- Do not criticize those who object to or resist your idea.
- Do not get excited or emotional if anyone resists your suggestion. Calm, reasoned responses will convince others of your confidence more quickly than angry rebuttals.
- Do not debate your idea, sell it. Do not become so involved in refuting objections that you forget to emphasize your solution's positives.
- Do not distort the worth or impact of your idea. Be realistic about what it can achieve, and others will be more likely to trust your assessment of the idea's real worth.
- Be willing to compromise on how your idea is put into action.
- Do not be in a hurry for agreement or approval. Give others time to mull your idea over and get used to it.
- Be ready to change your mind about your idea. Be willing to discard it, if necessary.

Timing

In preparing to persuade someone to accept your ideas, consider the best time for presenting it. Unless you are facing a critical situation, avoid times when others are busy or tired. Timing is even more important in urgent situations. Remember that just because you are ready for change does not mean everyone else is.

Understanding Resistance

Change is likely to be resisted, even by those who may benefit from your idea. Individual needs and desires for recognition can become barriers to implementing your idea. These needs can be physical, emotional, intellectual, creative, psychological, economic, social, and cultural.

Individual needs should be anticipated and met if possible. Ignoring them will only hinder your efforts. You can offset resistance and respond to misconceptions by discussing both the advantages and disadvantages of your idea. Of course, you will emphasize the advantages to encourage acceptance of your idea.

Objections will depend on the status, experience, perceptions, and self-concepts of the people raising them. Sometimes those raising objections have been influenced by what they have already heard about the idea. Respond to these prejudgments by answering them, evading them, replacing them with substitute objections, or agreeing with them. Respond in some way.

Most experts agree that objections must be dealt with if an idea is to be successfully implemented. Mastering this skill will enable you to handle difficult situations. In handling objections, two principles apply. First, every objection should be viewed as an opportunity and an advantage. The reason is that you will usually have an opportunity to analyze the objection, counter it, and reply in such a manner as to turn the objection into an advantage.

The second principle is that every objection should be seen as an opportunity to analyze the doubter's viewpoint. Recognizing others' viewpoints is essential for success. For example, when you are attempting to get a smoker to become a nonsmoker, you must try to understand why smoking is important to him. Some smokers fear they will gain weight if they quit their habit. Some rely on smoking to cope with stress. A teenager may be responding to peer pressure.

By understanding others' points of view, you can develop better strategies for overcoming their objections. Figure 48.1 assists in clarifying this concept.

Figure 48.1
The Perspectives of Others

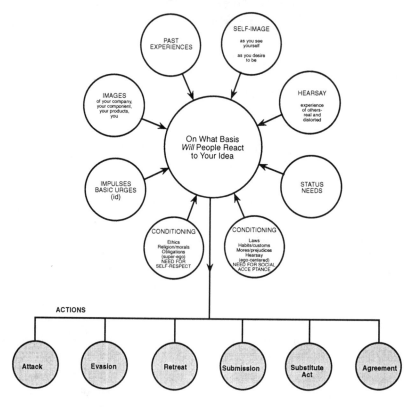

Discovering Smoke Screens

Objections to your idea may be raised for valid reasons, or they may be part of a hidden agenda. An objection may be a smoke screen that covers a real objection. Some may want to introduce ambiguity, uncertainty, and confusion. Because the objector or doubter is not stating her real objection, you will have difficulty discerning what that objection is and how to handle it.

Reasons for establishing smoke screens vary. Unknown motives can present roadblocks to implementation that may be impossible to overcome. If the smoke screen cannot be removed, selling your idea may be impossible.

Responding to Objections

Several tactics can be used to respond to objections. The most effective is to anticipate them and raise them yourself! When you do this, you are fully prepared to respond. If you have answers in advance, you will gain acceptance more easily than if you seem unprepared and surprised by doubters' questions. Use your imagination to anticipate all possible objections. You will probably discover that you anticipated many of the objections.

There can be at least 11 kinds of objections to any new idea. These include:

1. The idea actually costs too much. You need to reevaluate your estimates of required human and capital investment.
2. The idea is believed to cost too much, based on misconceptions or misinformation. You need to share your estimates of cost and how you arrived at them.
3. Others do not see the urgency of agreeing to a solution and putting it into effect. You need to convey to them a sense of urgency.
4. Others have fears about making the wrong decision. You need to convince others that your idea is the best, based on a good deal of analysis and investigation, and that you are prepared to take full responsibility for it.
5. Others seem dissatisfied with the idea. If you have argued your case poorly, your argument needs to be strengthened.
6. The idea is not practical. This occurs when others are not convinced by your evidence, regardless of its worth.
7. Others believe the idea is not practical. This may be due to poor presentation on your part or not handling objections properly.
8. Others may be strongly committed to their own ideas, regardless of their merits. This may simply be a lack of information on their part. If so, you can provide more detail.
9. The idea is full of weaknesses you did not anticipate. This means that it was not handled properly during the analysis process.
10. Few want to commit themselves to future action. Again, you need to create a sense of urgency and its need for a solution, as well as to demonstrate your willingness to put it into action.
11. Not everyone is convinced of the value of your idea. You need to prepare for ways to stimulate others' excitement and to enhance their interest in your idea.

SUGGESTIONS FOR IMPROVING YOUR PERSUASION SKILLS

- Adopt a positive attitude. Look at each situation as a challenge rather than a chore. Enthusiasm and confidence are contagious. So are uncertainty and pessimism.
- Be professional. Be knowledgeable, poised, businesslike, and open. Professionalism comes from hard work, study, asking for critiques of your performance, and learning from past mistakes.
- See the viewpoint of the person you are trying to persuade. Remember, you are selling benefits, not a product or a service.
- Do not stress negatives. Do not argue nor knock the competition.
- Do not get discouraged. You will not persuade all the people all the time.
- Confidence is critical and essential. Many convincing arguments are made solely because the person being convinced has such confidence in the presenter that little actual convincing needs to be done.
- In preparing for objections from individuals you are trying to persuade, consider the best time to present your ideas. Unless you are facing a critical situation, avoid times when others are busy, tired, or short of funds.
- Build on the support and enthusiasm you may already have from those in favor of your position. You should not become so concerned about objections from those opposed to you that you forget to bolster existing support.
- Do not get excited or emotional if anyone resists your ideas or suggestions. Calm, reasonable responses will convince others of your confidence more quickly than angry rebuttals.
- Do not be in a hurry for agreement or approval of your idea or position. Give others time to mull your position over and get used to it.

SUGGESTED READINGS

Cohen, W.A. (1990). *The art of the leader*. Englewood Cliffs, NJ: Prentice Hall.

Freedman, J., & Sears, D. (1985). Effects of expected familiarity with arguments upon opinion change and selection. *Journal of Personality and Social Psychology*.

Kuykendall, D., & Keating, J. B. (1990, Spring). Mood and persuasion: Evidence for the differential influence of positive and negative states. *Psychology and Marketing, 7*(1).

Mancuso, J. (1993). *Winning with the power of persuasion*. Dearborn, MI: Enterprise Publishers.

Silver, D. A. (1992). *How to close a deal*. Englewood Cliffs, NJ: Prentice Hall.

Chapter 49

PROCRASTINATION

Procrastination does more than any other habit to deprive us of satisfaction, success, and happiness. It does not solve any problem when we toss it into the tray marked "pending."

Everyone living in the fast pace of today's society suffers under the feeling of being pressured for time. Many of us feel we are behind in those tasks that we place importance upon. We often feel guilty when we just sit with our thoughts because those thoughts inevitably turn to something we should be doing. People who put things off do not gain peace of mind or leisure as a result. If they gain free time, it is invariably haunted by all that needs to be done—a load of work or decisions that become monsters in their own right. People who put things off can be quite busy with other matters. In fact, one common delay tactic to justify procrastination is to claim that you have other, more pressing responsibilities. Being busy does not mean you are being effective, especially if you have avoided doing something that is essential to the performance of other tasks.

Why is it we procrastinate? Some claim it is because of fear of failure. Others say we may not have confidence in our skills to do the task. Many develop lack of interest in what they are doing. In some cases, we procrastinate because we become angry with our supervisor or our business colleagues and social contacts. But whatever the reason, the fact is it creates excessive stress on the individual. One thing is for sure—if you put off a task that is important to you or your organization, you will do an inferior job.

Are you a procrastinator? J. D. Ferner developed an exercise that will assist you in determining if you are guilty of this devastating weakness. Rate yourself by completing the questions below.

		Strongly agree	Mildly agree	Mildly disagree	Strongly disagree
1.	I invent reasons and look for excuses for not acting on a tough problem.	☐	☐	☐	☐
2.	It takes pressure to get on with a difficult assignment.	☐	☐	☐	☐
3.	I take half measures that will avoid or delay unpleasant or difficult action.	☐	☐	☐	☐
4.	There are too many interruptions and crises that interfere with my accomplishing the big jobs.	☐	☐	☐	☐
5.	I avoid forthright answers when pressed for an unpleasant decision.	☐	☐	☐	☐
6.	I have been guilty of neglecting follow-up aspects of important action plans.	☐	☐	☐	☐
7.	I try to get other people to do unpleasant assignments for me.	☐	☐	☐	☐
8.	I schedule big jobs late in the day, or take them home to do in the evening or weekends.	☐	☐	☐	☐
9.	I've been too tired (nervous, upset, preoccupied) to do the difficult tasks that face me.	☐	☐	☐	☐
10.	I like to get everything cleared off my desk before commencing a tough job.	☐	☐	☐	☐

Total responses X Weight (Strongly Agree=4, Strongly-Disagree=1) = Score

Total score   Procrastination quotient

If your procrastination quotient is below 20: You are not a procrastinator; you probably only have an occasional problem.

If your quotient is 21-30: You have a procrastination problem—but not too severe.

If your quotient is above 30: You probably have frequent and severe problems of procrastinating.

From J. D. Ferner. (1980), *Successful Time Management*. New York: John Wiley & Son.

The tendency to put off action needed to accomplish a task is common. In many instances, the task delayed is considered distasteful or will require much effort, part of which may be unknown or at least not completely identified. Procrastination, of course, is closely related to the reluctance to make decisions and to move ahead on some course of action. Without implementation, the best plans and hoped-for accomplishments are worthless.

Procrastinators do not like to look back, because that is where the crumbling foundations of their present lives are revealed. Occupying ourselves with small matters does not get at the poison of procrastination. We are simply creating a smoke screen when we engage in mini tasks to the point of exhaustion with "no time" for pressing concerns. Because most of our daily actions and inactions are interrelated, procrastination can impede the smooth functioning of the rest of our lives. You cannot get the laundry done if you have not fixed the washing machine. You cannot lead an orderly life if your clothes are dirty.

Procrastination serves many purposes. We do not procrastinate against our wills. Its cancerous development is encouraged because it allows us to avoid risk, uncertainty, and change in our lives. No matter how restless we are, too often we remain with the familiar and put off what should or could be done. Our jobs, family lives, and friendships may have gone stale long ago. Yet we put off having to deal with these issues directly to avoid risk of failure, not realizing that the biggest failure is procrastination. When others do for us what we have neglected, we are reinforced in our procrastination.

Even our leisure is eaten into by procrastination. So many people complain that they have no time for leisure. They are constantly driven. Life for them is either a steady grind or a dream. These are people who do not organize their time and energy. They are the sort who find themselves nervously unfit to deal with immediate things, to stand the pressure of an urgent job.

What causes procrastination and why does it persist? No behavior continues unless it serves us in some way. Procrastination is another word for protectionism. In most cases, the procrastination is protecting themselves from something else that feels far more difficult to face. Procrastination may also be an indirect way for people to rebel. They may think that if they delay beginning a task they dislike, they are saved from saying that they did not want to complete the task in the first place.

Procrastinators pretend that things will get better through no effort of their own. Because of this fallacy, there is no reason to alter behavior, and action is postponed. Fantasies are easier to cope with than taking the risk to change. Procrastination can be a serious block to an individual's problem-solving and decision-making ability. Its symptoms may include boredom and dependency on drugs, alcohol, tobacco, or other distractions. Other symptoms of procrastination include:

1. Continuing in an unchallenging job.
2. Adopting bad habits: smoking, drinking, and overeating.
3. Avoiding a necessary confrontation.
4. Always being late for work, meetings, and social events.
5. Beginning a project so late in the day, you have no chance of doing a good job.
6. Putting off menial or unpleasant tasks.
7. Becoming ill on the day you were going to accomplish something.

Young people need to be aware of "putting off." When things are deferred until the last minute and nothing is prepared beforehand, every step finds an impediment. It becomes harder to do things. We are pushed into blundering through on hasty judgments. By trying to take things easy, we do not make things easy. It is possible to spend more energy in figuring out ways to escape a task than is necessary to accomplish it. Our available energy is lowered by inward conflict between "do it now" and "put it off." We lose our poise because we are always catching up, always in a hurry to do today what we should have done yesterday. Not only is procrastination a blight on a person's life, but an irritant to individuals we must come in contact with. Everyone with whom the procrastinator must deal is thrown from time to time into a state of disarray. Everyone else has to work harder to take up the slack of the procrastinator.

The practice of "putting off" has a way of inching upon us without realizing it. So what if we do not make that important phone call today, or pay our insurance before it is due or talk to one of our employees about a serious mistake he has made or postpone that report that is due. We say to ourselves we can always do it tomorrow. The consequences of procrastination can be disastrous and overwhelming. Many individuals have found that by putting things off, they have created a crisis for themselves as well as those around them.

What can we do to begin to eliminate our procrastination habits? To start, analyze why you procrastinate and write down several of the last experiences in which you procrastinated. What were the reoccurring reasons? Write what led to the incidences, then list the kinds of consequences that occurred. You may find that you experienced guilt, depression, anxiety, or embarrassment. Try to look at activities you tend to substitute for what you should be doing. Becoming aware of your style of procrastination may help you better recognize when you are doing it. At the moment you decide to begin the unpleasant task, do not ask yourself the questions, "Am I really too busy or too tired? Do I really not have the time right now?"

An excuse sometimes made by individuals when they begin to procrastinate is that they are waiting for the special motivation or inspiration. However, motivation and inspiration do not simply happen for the procrastinator as much as they do for busy and goal-oriented individuals. The best way to avoid procrastination in your life is to never let it get started. If it creeps into your life, recognize it and admit it. As long as you allow it to exist, you will never be able to overcome it. Procrastination, like any problem, can be eliminated, as long as you are willing to take definite, decisive action.

SUGGESTIONS FOR ELIMINATION OF YOUR PROCRASTINATION HABIT

- Be orderly and systematic in handling your mail. Lay aside only such letters as really need further thought and then take them up immediately after routine mail has been handled.
- Make out a complete and honest statement of what you wish to accomplish each day, each week, each year, and determine what obstacles may be in the way that prevent you from accomplishing them.
- Schedule your time. Write down the various jobs you must do or would like to do. Estimate the time needed for each. Number each in order of importance and then do it.
- Break your tasks down into smaller, more manageable units. The whole project may seem overwhelming, but completing a smaller task is workable. To do this, lists can be helpful in

- breaking down a job and making sure that each step is completed before going on to the next phase. Completing the highest priorities first will make the whole project seem less overwhelming.
- Do not look too long at a job before starting it. Even if progress seems slight and futile, the act of starting is a strong motivator to continue toward successful completion.
- If you are a procrastinator, admit it. Once you acknowledge it, you can examine your situation to determine why and then do something about it.
- Schedule your most unpleasant tasks first. Do the distasteful first and get it behind you rather than dreading it and continually putting it off.
- Consider the cost in delaying a task. When you are tempted to procrastinate, think of the problems you will create for yourself. If you do not want to deal with those problems, do not procrastinate.
- Reward yourself when you complete a task. By rewarding yourself, you feel motivated to get on with the job on schedule and help conquer the problem of procrastination.
- Personal disorganization is a trap and increases your potential of becoming a procrastinator. Examine your daily routine for habits that may delay the starting of a task. Ask yourself which habits would help you become a more organized person.

SUGGESTED READINGS

Douglass, N. E., & Douglass, D. N. (1980). *Manage your time, manage your work, manage yourself.* New York: AMACOM.

Engle, H. (1983). *How to delegate: A guide to getting things done.* Houston, TX: Gulf Publishing Co.

Ferner, J. D. (1980). *Successful time management: A self-teaching guide.* New York: John Wiley & Sons.

Mays, C. A. (1991). *Strategy for winning.* New York: The Lincoln-Bradley Publishing Group.

Quality Digest. (1989, January). Outlook: Up to 94 days a year spent on paperwork, p. 3.

Chapter 50

SELF-ESTEEM

Many employers have found that improving employees' self-esteem can be a more effective motivator than raises. Sorting out the practicalities, though, can be a difficult challenge.

Self-esteem—or more specifically, the lack of it—seems to have become a national preoccupation during the last decade. The "feel-good" movement, which got its name from its emphasis on thinking positive thoughts about ourselves, established the current popular view of self-esteem. It also spawned an entire commercial industry, with hundreds of books, self-help programs, and consultants now specializing in building self-esteem. Meanwhile, a lack of self-esteem has been blamed on everything from alcoholism to teenage pregnancies to serial murders.

Businesses have also become interested in self-esteem and the role it plays in the workplace. Many have found that self-confidence does have a significant effect on morale, job satisfaction, and performance, and that improving employees' self-esteem can be a more effective motivator than raises. Sorting out the practicalities, though, can be a difficult challenge.

WHAT IS SELF-ESTEEM?

The concept of self-esteem and how it affects our lives goes back to Freud and his "ego ideal." Since then, more than 10,000 scientific studies have focused on its effect and more than 200 tests have been used to measure it. Although there is still no universally accepted specific definition, research has identified meanings that can be useful in a work setting.

In its broadest sense, self-esteem can be thought of as an innermost sense of confidence, competence, self-worth, and value. One simple explanation is that it is the self-reflective perceptions people have of themselves. These perceptions affect every aspect of a person's life, from getting up in the morning to making decisions about a project on the job.

There are several theories about the different factors that contribute to self-esteem. One is that people's feelings of their worth are determined by how well they function within their own perceived domains of importance. For example, if John feels that athletic ability is important and he is quite a skilled athlete, then his self-esteem is high. If John is a bad athlete but athletic ability is not important to him, then his self-esteem will not be greatly affected.

Another theory is that self-esteem is influenced by societal constructs, and that people who do not have "socially acceptable" traits are at a higher risk for low self-esteem. If athletic ability is highly stressed by society, then good athletes will be more accepted. And if society emphasizes appearance or achievement, people who are beautiful or smart will find it easier to feel good about themselves.

Gender also accounts for some differences in an individual's self-esteem. Girls particularly seem to be on the low end of the scale. In a 1990 survey conducted by the American Association of University Women, boys of all ages showed higher levels of self-acceptance than girls. When asked to respond to the statement "I am happy with the way I am," 42% of high-school boys answered *yes,* as opposed to only 23% of the high-school girls.

Studies have also found that girls sometimes gain the impression that if they are intelligent or excel in school, they are not "cool" or popular. In order to fit in, many of these girls become withdrawn in class and try to mask their intelligence. Girls are also less likely to excel in science and math because society perceives these as "male" areas, and their teachers and peers offer them little encouragement. Oftentimes, girls settle for careers they are not happy with because the career of their choice was not viewed as appropriate.

Boys are not exempt from this self-esteem slide. Boys grow up with the social message that it is not appropriate to display sensitivity and nurturing. As a result, many boys try to repress these traits, with negative results. Studies have shown that boys are more likely to be grade repeaters, dropouts, hyperactive, and/or drug and alcohol dependent.

"Working" Self-Esteem

People who possess a high self-esteem tend to be less affected by threatening situations because they have the coping mechanisms to build themselves back up. Markus and Worf (1987) call this ability to self-actualize and reaffirm oneself a "working self-esteem." When faced with a stressful or anxiety-producing situation, a person can use working self-esteem to go through a series of self-affirmations that helps him perceive negative feedback as negative opinion.

On the other hand, when people with low self-esteem are faced with a stressful situation, they resort to rationalizing and making social comparisons (comparing themselves to persons with less ability to make themselves look good). They do this because they do not possess the working self-esteem that would allow them to disregard the situation.

Working self-esteem also influences how a person responds to praise or ridicule. People without a working self-esteem are more sensitive to threats and praise because they have no means of self-affirmation to balance the two. These differences in how people react to praise and criticism affect consistency of moods and job performance.

Stability and Careers

The question of stability and self-esteem has often been controversial. Some researchers believe that self-perceptions are stable and that adults and children with low self-esteem are slow to change their opinions of themselves. Others argue that although early childhood years are critical in the formation of self-esteem, children can quickly change their perceptions of themselves if their competence and/or support changes.

These two conflicting views are prevalent in theories of how people view career choices, attitudes, and behaviors about work. When a person selects a vocation, she must consider her personal interests, competencies, and overall attitudes about work. Self-esteem dovetails these decisions because it influences how an individual perceives her interests and how she searches for jobs.

There is debate, however, about whether a low self-esteem can directly influence a career choice. One theory is that low self-esteem causes people to perceive themselves as less suited for a particular career, and therefore they do not pursue it. Other studies have found that people with high self-esteem are more likely to choose a career to satisfy their desires than people with low self-esteem, and that the former were more likely to believe they possess the abilities needed to succeed.

Self-esteem can also affect how people search for jobs. People with high self-esteem are more likely to use their personal contacts and send applications and résumés to employers. People with low self-esteem have less faith in their own efforts and are more likely to depend on employment agencies to find jobs for them.

Once a person selects a career and a job, his personal self-esteem may also be responsible in part for his overall attitude about work and his level of job satisfaction. Some studies have shown that people with high self-esteem are more satisfied with their jobs than people with low self-esteem. Others argue that there is no way to determine whether or not job satisfaction leads to higher self-esteem or if high self-esteem leads to higher job satisfaction.

SELF-ESTEEM AND PRODUCTIVITY

Managing people with various levels of self-esteem can be a very delicate and complicated task. To work effectively with people, a manager must understand the concept of self-esteem, how it influences career choice, and how it affects work performance.

The work environment itself is one of the most important factors in productivity. In a positive work environment—one in which supervisors support workers' decisions and are generous with praise—employees feel freer to work to their potential. The positive reinforcement of praise encourages them to increase their productivity to please their supervisor.

An individual employee's inherent self-esteem can also affect productivity, although there is an ongoing debate about the exact relationship. One theory is that a person with low self-esteem constantly struggles with whether or not she is contributing anything to the work environment. If such a person experiences a low productivity time, she attributes it to her own inadequacies, considers herself a failure, and becomes more unproductive. When productivity is high, she questions whether or not it were really her contributions that led to the success or if it were luck. She will not be temporarily motivated again until another low productivity time, and the cycle begins again.

Conversely, when someone with high self-esteem experiences a low productivity time, he may look for ways to improve his performance. He chalks up this drop to experience, learns from it, and goes on. Regardless of the level of productivity, he performs his best, and his work is usually more consistent than the work of a person with low self-esteem.

These cycles can also be affected by the number and amount of extrinsic rewards from a manager. A supervisor's praise can increase a person's positive self-perceptions, just as criticism can decrease them. If an employee with low self-esteem is doing inconsistent work and a manager talks to her about it, this person may begin to question her usefulness to the organization, feel unimportant, and work less because she feels that her contributions do not matter.

In the same situation, however, a person with high self-esteem is more likely to realize that his manager's criticisms or suggestions are not related to his worth as a person. He will work harder to please his boss and begin a new cycle of high productivity.

The two cycles described are, of course, extreme cases. Factors outside of work, such as family stress, economic factors, and coworker dynamics, play into this picture. The manager needs to remember that she has the ability to maintain a positive work cycle and enhance her employees' positive perceptions of themselves.

Stress in the Workplace

Some job-related stress and pressure to get a project finished can be healthy in the proper amounts. There is a fragile balance, however, between too little and too much pressure. If standards are set too high, workers can become frustrated and production may suffer.

Performance goals, production incentives, deadlines, and other requirements of a job can be very effective motivators. They can build rewards into the work and provide for competition among employees. It is important to remember, though, that individuals of varying levels of self-esteem may react very differently to identical work settings. If a worker perceives himself as being unimportant, he will often feel helpless when he is challenged by a deadline. He will not have the working self-esteem to convince him that he can finish the task, and he may give up. This failure may prompt him to compare himself to people with lower abilities, and his self-esteem falls even lower.

In a similar environment, an individual with high self-esteem will face the challenges and find the motivation to complete the task. Her working self-esteem reassures her that although the task at hand may be challenging, it is not impossible, and it can be done.

Management Skills

Employers of all sizes have recognized the benefits of building their workers' self-esteem. In their efforts to do so, however, they often face a dilemma—how to balance the personal needs of employees with the needs of the group and the employer. Managers must find a way to fulfill the organization's objectives and provide people with a positive, nurturing work environment.

One way to accomplish this is for the manager to set specific, realistic goals for workers. Workers should know what they are expected to do and what types of incentives or penalties they can expect if they properly or improperly complete certain tasks. Once these expectations are established, the manager must communicate with his employees to get a clear picture of how they perceive his approach and expectations. This will help the manager select strategies that will make workers feel they are making important contributions—and thus increase their satisfaction and productivity levels.

An effective manager will also know how to balance recognition and motivation—and that different types must be used for different types of people. She must set challenging (but not daunting) goals, and provide both formal and informal recognition for meeting them. An employee with low self-esteem will often need to experience some success before he can develop confidence in his abilities, and the manager may have to bring him along slowly by giving him easier-to-attain goals at the beginning, providing recognition for meeting them, and then offering more challenging tasks.

Another consideration is the individual employee's gender and age. The relationship between gender and self-esteem is often more evident in older workers, and managers should be aware of the problems this can create. Compared to younger generations, today's older men and women grew up with a much stronger message about their "proper" roles in the social structure and the importance of excelling in work-related activities. Older women are especially subject to self-esteem problems because of the difficulty in overcoming a lifetime of conditioning that they are supposed to be less able and less intelligent than men.

THE DELICATE BALANCE

Even with all the unanswered questions about the relationships between self-esteem and job performance, it is clear that management plays a critical role in determining how employees feel about their jobs and themselves. It is important to remember, though, that managers cannot "fix" a person's self-esteem. They can only foster it. The programs and policies that will be successful depend on the unique personality and abilities of both the employee and the manager.

The effective manager will learn to assess each worker's level of self-esteem, anticipate problems, and develop techniques to build the individual's self-confidence. In making her decisions, though, she must also consider the overall goals of the organization.

Finding the proper balance is the key to developing a motivated workforce that will move the organization in a positive direction.

SUGGESTIONS FOR ENHANCING EMPLOYEES' SELF-ESTEEM

- Build some form of regular, structured recognition into each employee's job.
- Provide clearly defined objectives for each task an employee performs. Create a challenge and a way for workers to measure their success.
- When setting objectives, focus on the task, not the person performing it. Structure the wording of a goal so it emphasizes a positive, work-related result instead of personal abilities.
- If possible, match production quotas and incentives to the abilities of the individual worker. If goals are set for a diverse group of employees, consider the interactions of the personalities and maintain regular, frequent communication with the group members.
- When evaluating an individual employee's work performance, use the PIN technique. Structure your thinking (and your discussion with the worker) so you begin with what is *positive*, then what is *interesting* or *innovative*, and last, what is *negative*.

- Remember that praise is much more effective than criticism, especially with workers who have low levels of self-esteem. Look for opportunities to give positive feedback—watch problem workers closely to try to catch them doing something right.
- Efforts to build employees' self-esteem should come from multiple sources, not just the employee's direct supervisor. Involve high-level management in the formal and informal processes of recognizing and rewarding employees.
- Appreciate the value of recognition from coworkers. Take advantage of every opportunity to publicize an individual's achievements and awards in company memos and newsletters and in local newspapers.

SUGGESTED READINGS

Adler, J. (1992, February 17). Hey, I'm terrific. *Newsweek*, pp. 46-51.

Bass, B., & Barrett, G. (1981). *People, work, and organizations.* Boston: Allyn and Bacon, Inc.

Baumeister, R. (1993). *Self-esteem: The puzzle of low self-regard.* New York: Plenum Press.

Brockner, J. (1988). *Self-esteem and work: Research, theory and practice.* Lexington, MA: Lexington Books.

Dobson, J. (1988). *Hide or Seek.* Pomona, CA: Focus on the Family Publishers.

Hale, R., & Maehling, R. (1993). *Recognition redefined: Building self-esteem at work.* Exeter, NH: Gray Media, Inc.

Lewis, M. (1969). *Self-concepts and the level of occupational aspiration.* Unpublished doctoral dissertation, University of Illinois, Urbana-Champaign.

Rinke, W. J. (1988, March). Maximizing management potential by building self-esteem. *Management Solutions*, pp. 11-16.

Sadker, M., & Sadker, D. (1994). *Failing at fairness: How America's schools cheat girls.* New York: Charles Scribner's Books.

Walsh, E. (1974). *Job stigma and self-esteem.* Unpublished doctoral dissertation, University of Michigan, Ann Arbor.

APPENDICES

APPENDIX A

GLOSSARY

A

Acceptance theory of authority. The theory that subordinates will accept orders only if they understand them and are willing and able to comply with them.

Accountability. The practice of holding subordinates accountable for exercising delegated *authority* in such a way as to fulfill assigned responsibilities and of rewarding or punishing them on the basis of their performance.

Accounts payable. A liability that includes obligations owed to creditors, usually arising from purchases of goods and services on credit.

Accounts receivable. An asset that includes obligations owed, usually arising from its sales.

Activity. One of the elements in a PERT network. It represents the work done to complete a particular *event* and is usually represented by an arrow.

Ad hoc committees. Committees that function similarly to a task force in seeking to accomplish a specific purpose and are then disbanded; they do not have permanency.

Administration. Those activities that have as their purpose the general direction, execution, and control of the affairs of the organization.

Administrative activities. One of four groups of activities engaged in by managers. They include processing paperwork, preparing and administering budgets, monitoring policies and procedures, and maintaining the stability of operation.

Administrative model. A model that is descriptive and provides a framework for comprehending the nature of the process that decision makers actually use when selecting among various alternatives.

Advertising appeal. A theme intended to trigger buying decisions or to project a better organization image in the target market.

Affirmative action. Performance required to ensure that applicants are employed, and that employees are treated appropriately during employment, without regard to race, creed, color, or national origin.

Affirmative-action program (AAP). An employer's program to actively seek out women and minorities and promote them into better positions as required by the *Equal Employment Opportunity Act of 1972*.

Analysis. The breaking up of study subjects into manageable elements for individual evaluation.

Analyst. A person skilled in the definition of problems and the development of logical procedures and data structures for their solution.

Arbitration. An attempt to settle labor-management conflicts through the intervention of a third party neutral to the dispute whose decision is binding. *Voluntary arbitration* takes place when the parties in a dispute agree among themselves to submit the dispute to arbitration. *Compulsory arbitration* takes place when the parties are forced into arbitration by an outside organization, usually the government.

Area cluster sampling. A type of probability sample that involves the random selection of a number of geographic sections from a mapped area on which a grid has been superimposed. After row and column coordinates are selected from the grid at random, interviews or samples are done at those locations.

Assessment center. A systems approach to diagnosing individual potential or assessing individual performance based on group techniques. When used for *diagnosis*, a battery of exercises, games, or tests is administered to judge current abilities and future potential. Some portion of individual performance is judged by observation and evaluation by a panel of experts numbering perhaps four to six people. The results of individual performance and judgments of potential are then fed back to the person for further discussion.

Attitude surveys. Eliciting responses from employees through questionnaires about how they feel about their jobs, work groups, supervisors, and/or the organization.

Audit trail. A system of providing a means for tracing items of data from one processing step to the next, particularly from a machine-produced report or other machine output back to the original source data.

Autocratic leader. One who dictates decisions down to subordinates.

Autonomous work teams. Groups that are free to determine how the goals assigned to them are to be accomplished.

B

Balance sheet. A statement of the overall financial condition of a business at a given date. A formal statement of assets, liabilities, and fund balance as of a specific date.

Balance theory. A theory used to explain how people react to change. In essence, the theory places primary attention on the consideration of three relationships: (1) the attitude of the worker toward the change, (2) the attitude of the manager toward the change, and (3) the attitude of the worker toward the manager.

Behavioral approach to leadership. A theory based on the assumption that leaders are not born but developed. It focuses on what leaders do rather than what they are.

Body language. A nonverbal method of communication in which physical actions such as motion, gestures, and facial expressions convey thoughts and emotions.

Brainstorming. An idea-generation process that specifically encourages any and all alternatives, while withholding any criticism of those alternatives.

Break-even point (BEP). That point on a chart at which total revenue exactly covers total cost, including both fixed and variable costs.

Budget. A statement of planned allocation of resources expressed in financial or numerical terms.

Bureaucracy. A prototype form of organization that emphasizes order, system, rationality, uniformity, and consistency.

Business goals. In addition to making a profit and providing employment, business goals are the special contributions a business makes to its environment.

C

Career management. A process that integrates the individual's career planning and development into the organization's personnel plans.

Career planning. The process by which an organization helps employees to choose their career goals and to identify the means of attaining them.

Cash flow. The total funds available to a business during a given period. Cash flow is approximately equal to the firm's earnings *plus* the depreciation charges included in its accounting statements. Depreciation is included because it is not a "real" expense of a business in the sense that there is an outflow of funds to another party.

Centralization. The degree to which authority is retained by higher-level managers within an organization rather than being delegated.

Centralized data processing. Data processing performed at a single, central location on data obtained from several geographical locations or managerial levels. Distributed data processing involves processing at various managerial levels or geographical points throughout the organization.

Chain of command. The line along which authority flows from the top of the organization to any individual.

Civil service. A permanent government staff whose positions are based on merit as measured by entrance examinations or recommendations and by periodic examinations for promotion.

Coaching. An on-the-job method of *employee development* whereby superiors provide guidance and counsel to subordinates in the course of their regular job performance.

Coercive power. Power that is based on fear.

Collective bargaining. The process of settling disputes between unions and management.

Committees. Groups of persons from more or less the same level whose purpose is to exchange information, advise top management, or even make decisions themselves.

Communication network. A complex of data communication equipment, data links, and channels that connects one or more data processing systems.

Conflict management. The methods and program that managers use to deal with excessive conflict between either individuals or departments.

Consumer behavior. The processes individuals use to arrive at purchase decisions and the factors that influence those purchase decisions and product usage.

Contingency theory. A management theory that refers to a manager's ability to adapt to meet particular circumstances and restraints a firm may encounter.

Controlling. One of the basic management functions; devising ways and means of assuring that planned performance is actually achieved.

Coordination. The process of ensuring that persons who perform interdependent activities work together in a way that contributes to overall goal attainment.

Corporate culture. The system of shared values, beliefs, and habits within an organization that interacts with the formal structure to produce behavioral norms.

Cost accounting. A method of accounting which provides for the assembling and recording of all the elements of cost incurred to accomplish a purpose, to carry on an activity or operation, or to complete a unit of work or a specific job.

Critical path method (CPM). A planning and control technique that involves the display of a complex project as a network, with one time estimate used for each step in the project.

D

Data. A general term used to denote any or all facts, numbers, letters, and symbols that refer to or describe an object, idea, condition, situation, or other factors. It connotes basic elements of information which can be processed or produced by a computer.

Data dictionary. A cataloge of an organization's data bases, including names and structures.

Debenture. A long- or intermediate-term fixed obligation of a business.

Debt. Any legally binding obligation of an organization to pay a fixed amount of principal or interest for a specified period.

Decentralization. A principle of organization that states that decision making should be moved to lower levels of an organization that are independent enough to have their performance measured objectively.

Decision making. The process of generating and evaluating alternatives and making choices among them.

Delegation of authority. The process by which managers allocate *authority* downward to the people who report to them.

Delphi process. An intuitive methodology for eliciting, refining, and gaining consensus from individuals within an organization regarding a given issue.

Direct costs. Those elements of cost which can be easily, obviously, and conveniently identified with specific activities or programs, as distinguished from those costs incurred for several different activities or programs and whose elements are not readily identifiable with specific activities.

E

Efficiency. The ratio of *outputs* to *inputs*; an efficient manager is one who achieves higher outputs (results, productivity, performance) relative to the inputs (labor, materials, money, machines, and time) needed to achieve them.

80-20 rule. States that 20 percent of any firm's customers account for 80 percent of the firm's sales volume, gross margin, and profits.

Eminent domain. The right of government to confiscate private property, usually real estate, for just compensation as determined by private negotiations or by the courts. Some privately owned utilities have been granted the right of eminent domain.

Employee development. The improvement and growth of abilities, attitudes, and personality traits that make a person a more productive worker.

Entrepreneur. A person who conceives of, gathers resources for, organizes, and runs a business; entrepreneurs tend to be risk takers who are motivated by the profit motive.

Environment. Those factors outside an organization which influence the organization. Included in the environment are items such as competitors, geographical considerations, market status, and customer attitudes.

External audit. Verification of financial records conducted by outside agencies such as bank examiners or certified public accountant (CPA) firms.

Extrinsic rewards. Rewards received from the environment surrounding the context of the work.

F

Feedback. The degree to which carrying out the work activities required by a job results in the individual obtaining direct and clear information about the effectiveness of his or her performance.

Fixed costs. Costs that would be the same even if the production level were changed. These include some utilities, rent for the building, and salaries for supervisors.

Flextime. A technique that supposedly increases employee motivation; it permits the employee to choose when to start work between designated hours (e.g., between 7:30 and 9:30 a.m.) and when to leave work between designated hours (e.g., 4:30 and 6:30 p.m.), as long as the normal work day is completed (e.g., as long as a total of eight hours is worked).

Flowchart. A series of symbols connected by lines to demonstrate a sequence of events and decisions.

Fringe benefits. Employer contributions to workers in addition to basic wages or salaries.

G

Gantt chart. A method of showing scheduling requirements in chart form. The method was developed by Henry L. Gantt prior to World War I. In a typical Gantt chart, the time required for completion of each step of a process is expressed as a bar across a sheet, with the length of the bar denoting the total time required for the step. This method quickly shows overlapping steps.

Geographic segmentation. Creating marketing segments according to the region of the country, city size, market density, or climate.

Gobbledygook. Written or verbal statements containing so much technical jargon that they fail to communicate the message effectively.

Grapevine. Transmission of information by word of mouth without regard for organizational levels, *departmentation,* or *chain of command;* the best-known type of *informal communication* within an organization.

H

Heuristic method. Any exploratory method of solving problems in which an evaluation is made of the progress toward an acceptable final result using a series of approximate results.

Hierarchy of needs theory. There is a hierarchy of five needs—physiological, safety, love, esteem, and self-actualization—and as each need is sequentially satisfied, the next need becomes dominant.

Hygiene factors. Those factors—such as company policy and administration, supervision, and salary—that, when present in a job, placate workers. When these factors are present, people will not be dissatisfied.

Hypothesis. A tentative explanation about the relationship between two or more variables.

I

Iceberg principle. Term for a warning to marketing managers on the hidden danger of relying on sales volume alone as the means of evaluating a marketing plan.

Injunction. A court order decreeing that a person or an organization either take or refrain from taking an action. Those who violate injunctions are held in contempt of court.

Internal auditing. Activities involved with evaluating the adequacy of the internal control system; verifying and safeguarding assets; reviewing the reliability of the accounting and reporting systems; and ascertaining compliance with established policies and procedures.

Intrinsic rewards. Rewards that are internal to the work itself. Common examples are a feeling of accomplishment, increased responsibility, and the opportunity to achieve.

J

Job analysis. Developing a detailed description of the tasks involved in a job, determining the relationship of a given job to other jobs, and ascertaining the knowledge, skills, and abilities necessary for an employee to perform the job successfully.

Job description. A written statement of what a job holder does, how it is done, and why it is done.

Job design. The way that tasks are combined to form complete jobs.

Job enlargement. The horizontal expansion of jobs.

Job enrichment. Refers to basic changes in the content and level of responsibility of a job so as to provide greater challenge to the worker.

Job rotation. An *on-the-job development* method that moves people through highly diversified and differentiated jobs to give them a variety of experience.

Job sharing. Involves the filling of a job by two or more part-time employees, each working part of a regular work week and sharing the benefits of one full-time worker.

Jury of executive opinion. A forecasting technique that relies on the executives' past experiences and on their intuition regarding the future.

L

Life cycle. A series of stages in which attitudes and behavioral tendencies change over time as a result of developing maturity, income, and status.

Line and staff organizations. Organizations that have direct, vertical relationships between different levels and also specialists responsible for advising and assisting other managers.

Linear programming. A technique of mathematics used in operations research for solving certain kinds of problems involving many variables where a beset value or set of best values is to be found.

Lobbyist. A professional who seeks to influence government through personal contacts, information-gathering techniques, political contributions, and letter-writing campaigns.

Lockout. A lockout occurs when the employer closes down operations in an effort to force the union to cease harassing activities or to accept certain work rules demanded by the employer.

M

Management. The activity of working with people to determine, interpret, and achieve organizational objectives by performing the functions of *planning, organizing, staffing, leading,* and *controlling;* not necessarily synonymous with *leadership, supervision,* or *entrepreneurship.*

Management by exception (MBE) or exception principle. A *control* technique that enables managers to direct attention to the most critical control areas and permit employees at lower levels of management to handle routine variations.

Managerial grid. A two-dimensional matrix developed by Robert Blake and Jane Mouton that shows concern for people on the vertical axis and concern for production on the horizontal axis.

Marketing. That area of business that directs the flow of goods and services from producer to consumer in order to satisfy customers and to achieve the organization's objectives.

Marketing mix. The unique blend of pricing, promotion, product offerings, and distribution system designed to reach a specific group of consumers.

Middle managers. Managers above the supervisory level but subordinate to the firm's most senior executives.

Motivation. The willingness to exert high levels of effort toward organizational goals, conditioned by the effort's ability to satisfy some individual needs.

Municipal bond. A bond entitling the owner to receive interest exempt from federal income taxes.

N

Needs assessment. A basic procedure for determining the quantitative and/or qualitative extent of the discrepancies between what is and what is not required.

Nominal group technique. A group decision method in which individual members meet face-to-face to pool their judgments in a systematic but independent fashion.

Nonprobability sample. Any sample in which there is little or no attempt to ensure that a representative cross-section of the population is obtained.

Nonverbal communication. Messages conveyed through body movements, the intonations or emphasis we give to words, facial expressions, and the physical distance between the sender and receiver.

O

Operations research. The use of analytic methods adapted from mathematics for solving operational or planning problems. The objective is to provide management with a logical, mathematical basis for making sound predictions and decisions. Among the common scientific techniques used in operations research are the following: linear programming, probability theory, information theory, game theory, Monte Carlo method, and queuing theory.

Organization. A consciously coordinated social unit, composed of two or more people, that functions on a relatively continuous basis to achieve a common goal or set of goals.

Organization culture. The pattern of values, beliefs, and expectations shared by organization members; the shared assumption about how the organization should go about its work and what should be evaluated and rewarded.

Organizing. One of the basic management functions; determining what resources and which activities are required to achieve the organization's objectives, combining these into a formal structure, assigning the responsibility for accomplishing the objectives to responsible subordinates, and then delegating to those individuals the authority necessary to carry out their assignments.

OSHA. Occupational Safety and Health Act.

P

Participative management. A process where subordinates share a significant degree of decision-making power with their immediate superiors.

Performance appraisal (evaluation review). The formal system by which managers evaluate and rate the quality of subordinates' performance over a given period of time; a quantifiable or nonquantitative method of *control.*

Planning. The management process by which a manager anticipates the future and designs a program of action to meet it.

Policy. An understanding by members of a group that makes the actions of each group member more predictable to other members.

Political behavior. Those activities that are not required as part of one's formal role in the organization but that influence, or attempt to influence, the distribution of advantages and disadvantages within the organization.

Politics. A network of interactions by which power is acquired, transferred, and exercised on others.

Product liability. The legal obligation of sellers to pay damages to individuals who are injured by defective products or unsafely designed products.

Productivity. A measure of the relationship between inputs (labor, capital, natural resources, energy, and so forth) and the quality and quantity of outputs (goods and services).

Program evaluation and review technique (PERT). A scheduling and control technique that aids in scheduling sophisticated, nonrepetitive technical projects by (1) focusing management's attention on key program steps, (2) pointing to potential problem areas, (3) evaluating progress, and (4) giving management a reporting device.

Promotion. The means by which organizations communicate with their target markets regarding the merits and characteristics of their offerings. Performs three basic tasks: (1) informing, (2) persuading, and (3) reminding.

Public relations. The part of the promotional mix that evaluates public attitudes, identifies the policies and procedures of a firm or the individuals in it, and executes a program of action to promote public understanding and acceptance.

Q

Quality circle. A voluntary work group of employees who meet regularly to discuss their quality problems, investigate causes, recommend solutions, and take corrective actions.

Quality of work life. A process by which an organization responds to employee needs by developing mechanisms to allow them to share fully in making the decisions that affect their lives at work.

R

Recruitment. The process of attracting individuals—in sufficient numbers, and with appropriate qualifications—and encouraging them to apply for jobs with the organization.

Role playing. A developmental technique using unrehearsed dramas. Participants become actors and actresses who are assigned specific roles to play. They are usually told about the circumstances that add to a given situation or problem they are to act out. Roles are often rotated to teach empathy and to broaden the participant's perspective on the given problems. Observers are generally used to feed back reactions and to analyze the behaviors of the participants.

Rules and regulations. *Standing plans* that specifically state what can and cannot be done under a given set of circumstances; the result of a *policy* being adhered to in *every* instance.

S

Scientific management. The name given to the principles and practices that grew out of the work of Frederick Taylor and his followers and that are characterized by concern for efficiency and systematization in management.

Scientific method. A formal way of doing research that comprises observation of events, hypothesis formulation, experimentation, and acceptance or rejection of hypotheses.

Selection. The process of identifying those recruited individuals who will best be able to assist the firm in achieving organizational goals.

Simple random sample. A type of sample in which every element of the population has an equal chance of being selected as part of the sample.

Sinking fund. Money which has been set aside or invested for the definite purpose of meeting payments on debt at some future time. It is usually a fund set up for the purpose of accumulating money over a period of years in order to have money available for the redemption of long-term obligations at the date of maturity.

Span of control. A principle of organization that states that there is a limit to the number of subordinates who should report to one superior, since a supervisor has only a certain amount of time, energy, and attention to devote to supervision.

Staffing. The formal process of ensuring that the organization has qualified workers available at all levels to meet its short- and long-term business objectives.

Strategic planning. A way in which management plans to cope with the ever-changing external environment over the long run. Outlines the long-run mission of the organization and determines how an organization will generally utilize its resources.

Stratified sample. Segment of the population that has one or more common characteristics but is heterogeneous with regard to the variable(s) under examination.

Stress. A condition characterized by emotional strain and/or physical discomfort which, if it goes unrelieved, can impair one's ability to cope with the environment.

Synthesis. The combination of parts or elements into an entity, requiring logical reasoning in advancing from principles and propositions to conclusions.

System analysis. The examination of an activity, procedure, method, or technique to determine what must be accomplished and how the necessary operations may best be accomplished.

Systems approach. The viewing of any organization or entity as an arrangement of interrelated parts that interact in ways that can be specified and to some extent predicted.

T

Team building. A conscious effort to develop effective work groups throughout the organization.

Theory. A set of systematically interrelated concepts or hypotheses that purport to explain and predict phenomena.

Theory X. The assumption that employees dislike work, are lazy, dislike responsibility, and must be coerced to perform.

Theory Y. A view of management by which a manager believes people are capable of being responsible and mature.

Theory Z. As defined by author William Ouchi, a managerial approach used in Japan that emphasizes long-range planning, consensus decision making, and strong, mutual worker-employer loyalty.

Tunnel vision. Occurs when people have mental blinders, such as individual biases, that can restrict the search for an adequate solution to a relatively narrow range of alternatives.

Type A behavior. Aggressive involvement in a chronic, incessant struggle to achieve more and more in less and less time and, if necessary, against the opposing efforts of other things or other persons.

Type A individuals. Individuals who tend to feel very competitive, are prompt for appointments, do things quickly, and always feel rushed.

Type B behavior. Rarely harried by the desire to obtain a wildly increasing number of things or participate in an endlessly growing series of events in an ever-decreasing amount of time.

Type B individuals. Individuals who tend to be more relaxed, take one thing at a time, and express their feelings.

U

Unemployment compensation. Benefits paid to workers when they are out of work through no fault of their own. Payment is provided through a tax collected by the U.S. government from both employees and employers; the amount and duration of benefits vary from state to state and with individual cases.

Unity of command. A principle of organization that states that no member of an organization should report to more than one superior.

Unprogrammed decisions. Decisions that occur infrequently and, because of differing variables, require a separate decision each time the decision situation occurs.

W

Work sampling. Creating a miniature replica of a job to evaluate the performance abilities of job candidates.

Z

Zero-base budgeting (ZBB). A *budgetary control* method that divides an organization's programs into "decision packages" consisting of goals, activities, and resources needed and then computes costs "from scratch," as if the program had never existed.

APPENDIX B

PROFESSIONAL ASSOCIATIONS AND SOCIETIES CONCERNED WITH PARK, RECREATION, AND LEISURE SERVICES

Alabama Recreation and Park Society
P.O. Box 4744
Montgomery, AL 36103

Alaska Recreation and Park Association
Box 102664
Anchorage, AK 99510

Arizona Parks and Recreation Association
3124 E. Roosevelt
Phoenix, AZ 85008

Arkansas Recreation and Park Association
Cooperative Extension Service
P.O. Box 391
Little Rock, AR 72203

California Association of Park and Recreation Commissioners and Board Members
P.O. Box 161118
Sacramento, CA 95816

California Parks and Recreation Society
P.O. Box 161118
Sacramento, CA 95816

Colorado Parks and Recreation Association
P.O. Box 1037
Wheat Ridge, CO 80034

Connecticut Recreation and Park Association, Inc.
15 Gilead St.
Hebron, CT 06248

Delaware Recreation and Park Society
DNREC Parks and Recreation Division
89 Kings Highway
Dover, DE 19901

District of Columbia Recreation and Park Society
2901 20th St. NE
Washington, DC 20018

Ethnic Minority Society
c/o Conejo Recreation and Park District
155 West Wilbur Rd.
Thousand Oaks, CA 91360

European Recreation Society
HQ USAFE/MWO PSC 2
Box 10545
APO AE 09012

Florida Recreation and Park Association
411 Office Plaza Drive
Tallahassee, FL 32301

Georgia Recreation and Park Association, Inc.
1285 Parker Road
Conyers, GA 30207

Hawaii Recreation and Park Association
P.O. Box 22214
Honolulu, HI 96822

Idaho Recreation and Park Association
2951 North Government Way
Coeur D'Alene, ID 83814

Illinois Association of Park Districts
211 East Monroe
Springfield, IL 62701

Illinois Park and Recreation Association
1N141 County Farm Rd.
Winfield, IL 60190

Indiana Park and Recreation Association
101 Hurricane Street
Franklin, IN 46131

Iowa Park and Recreation Association
University of Iowa
203 Field House
Iowa City, IA 52242

Kansas Recreation and Park Association
Jayhawk Tower
700 SW Jackson Street
Suite 705
Topeka, KS 66603

Kentucky Recreation and Park Society
P.O. Box 43414
Louisville, KY 40235

Louisiana Recreation and Park Association
P.O. Drawer 14589
Baton Rouge, LA 70803

Maine Recreation and Park Association
P.O. Box 907
Yarmouth, ME 04096

Maryland Recreation and Park Association
201 Gun Road
Baltimore, MD 21227

Massachusetts Recreation and Park Association
P.O. Box 5135
Cochituate, MA 01778

Michigan Recreation and Park Association
2722 E. Michigan, Suite 201
Lansing, MI 48912

Minnesota Recreation and Park Association
5005 W. 36th St.
Saint Louis Park, MN 55416

Mississippi Recreation and Park Association
P.O. Box 16451
Hattiesburg, MS 39402

Appendix B

Missouri Park and Recreation Association
1203 Missouri Blvd.
Jefferson City, MO 65109

Montana Recreation and Park Association
P.O. Box 1704
Helena, MT 59624

Nebraska Recreation and Park Association
2740 A Street
Lincoln, NE 68502

Nevada Recreation and Park Society
749 Veterans Memorial Drive
Las Vegas, NV 89101

New England Park Association
Wickham Park
1319 West Middle Turnpike
Manchester, CT 06040

New Hampshire Recreation and Park Association
58 Hanson St.
Rochester, NH 03867

New Jersey Recreation and Parks Association
2 Griggstown Causeway
Princeton, NJ 08540

New Mexico Recreation and Park Association
P.O. Box 4179
Albuquerque, NM 87106

New York State Recreation and Park Society
119 Washington Ave.
Albany, NY 12210

North Carolina Recreation and Park Society
883 Washington Street
Raleigh, NC 27605

North Dakota Parks and Recreation Association
420 East Front
Bismarck, ND 58504

Ohio Parks and Recreation Association
1069-A W. Main St.
Westerville, OH 43081

Oklahoma Recreation and Parks Society
P.O. Box 13116
Oklahoma City, OK 73113

Oregon Recreation and Park Association
P.O. Box 89
Astoria, OR 97103

Park Law Enforcement Association
Oklahoma State Parks
9620 Alameda Drive
Norman, OK 73071

Pennsylvania Recreation and Park Society
723 S. Atherton Street
State College, PA 16801

Rhode Island Park and Recreation Association
Warwick Recreation and Park Department
975 Sandy Lane
Warwick, RI 02886

South Carolina Recreation and Park Society
P.O. Box 8453
Columbia, SC 29202

South Dakota Park and Recreation Association
600 East 7th St.
Sioux Falls, SD 57102

Tennessee Recreation and Park Society
2704 12th Ave. South
Nashville, TN 37204

Texas Recreation and Park Society
508 W. 12th St.
Austin, TX 78701

Utah Recreation and Park Association
645 S. Guardsman Way
Salt Lake City, UT 84108

Vermont Recreation and Park Association
8 South, 103 South Main Street
Waterbury, VT 05676

Virginia Recreation and Park Society
Rt. 4, Box 155
Mechanicsville, VA 23111

Washington Recreation and Park Association
350 South 333rd St., Suite 103
Federal Way, WA 98003

West Virginia Recreation and Park Association
400 Hal Greer Blvd.
Marshall Univ./Div. of HPER
Huntington, WV 25755

Wisconsin Park and Recreation Association
7000 Greenway, Suite 201
Greendale, WI 53129

Wyoming Recreation and Park Association
240 Lincoln St.
Lander, WY 82520

APPENDIX C

UNITED STATES GOVERNMENT-INDEPENDENT AGENCIES CONCERNED WITH PARK, RECREATION, AND LEISURE SERVICES

America the Beautiful Fund
219 Shoreham Bldg.
Washington, DC 20005-9202

American Alliance for Health, Physical Education, and Recreation and Dance
1900 Association Dr.
Reston, VA 22091

American Association for Leisure and Recreation (AALR)
1900 Association Dr.
Reston, VA 22091-1599

American Association of Zookeepers, Inc.
Administrative Offices
635 S.W. Gage Blvd.
Topeka, KS 66606-2066

American Association of Zoological Parks and Aquarium
Executive Office/Conservation Center
7970-D Old Georgetown Rd.
Bethesda, MD 20814

American Camping Association, Inc.
5000 State Rd. 67N
Martinsville, IN 46151

American Conservation Association, Inc.
30 Rockefeller Plaza
Rm. 5402, New York, NY 10112

American Forest Council
1250 Connecticut Ave.
NW, Suite 320
Washington, DC 20036

American Forest Foundation
1250 Connecticut Ave.
NW, Suite 320
Washington, DC 20036

American Forests
1516 P St. NW
Washington, DC 20005

American Hiking Society
P.O. Box 20160
Washington, DC 20041-2160

American Recreation Coalition
1331 Pennsylvania Ave. NW
#726
Washington, DC 20004

Americans for the Environment
1400 16th St. NW
Box 24
Washington, DC 20036-2266

Association for Conservation Information, Inc.
South Carolina Wildlife and Marine Resources Department
Box 12559
Charleston, SC 29412

Boy Scouts of America
National Office
P.O. Box 152079
1325 West Walnut Hill Ln.
Irving, TX 75015-2079

Boy Scouts of America
Southern Region
P.O. Box 440728
Kennesaw, GA 30144

Boy Scouts of America
Central Region
P.O. Box 3085
Naperville, IL 60566-7085

Boy Scouts of America
Northeast Region
P.O. Box 350
Dayton, NJ 08810-0350

Boy Scouts of America
Western Region
P.O. Box 3464
Sunnyvale, CA 94088-3464

The Camp Fire Club of America
230 Camp Fire Road
Chappaqua, NY 10514

Camp Fire, Inc.
4601 Madison Ave.
Kansas City, MO 64112

Commission on National Parks and Protected Areas (CNPPA)
IUCN
28 Rue Mauverney
CH1196, Switzerland

The Conservation Fund
1800 North Kent St.
Suite 1120
Arlington, VA 22209

Ducks Unlimited, Inc.
One Waterfowl Way
Memphis, TN 38120-2351

Environmental Protection Agency
Administrator
401 M St., SW
Washington, DC 20460

Friends of the Earth
218 D St., SE
Washington, DC 20003

Girl Scouts of the United States of America
420 Fifth Ave.
New York, NY 10018

Institute for Conservation Leadership
2000 P St., NW
Suite 413
Washington, DC 20036

The Izaak Walton League of America, Inc.
1401 Wilson Blvd.
Level B
Arlington, VA 22209

Appendix C 459

Muscle Shoals Technical Library
National Fertilizer and Environmental Research Center
Manager
Muscle Shoals, AL 35660

National Association of State Outdoor Recreation Liaison Officers
Executive Director
Arizona State Park
800 West Washington St.
Suite 415
Phoenix, AZ 85007

National Association of State Park Directors
Assistant Director, Game and Parks Commission
P.O. Box 30370
Lincoln, NE 68503-0370

National Association of State Recreation Planners
205 Butler St. SE
Suite 1352
Atlanta, GA 30334-9404

National Park Association
1101 17th St. NW
Washington, DC 20036

National Parks and Conservation Association
1776 Massachusetts Ave. NW
Washington, DC 20036-9202

National Science Foundation
1800 G St., NW,
Washington, DC 20550

Partners in Parks
4916 Butterworth Pl. NW
Washington, DC 20016

Peace Corps of the United States
1990 K St., NW
Washington, DC 20526

Resources for the Future
1616 P St. NW
Washington, DC 20036

Sierra Club
730 Polk St.
San Francisco, CA 94109

Tennessee Valley Authority
400 West Summit Hill Dr.
Knoxville, TN 37902

YMCA Earth Service Corps
909 4th Ave.
Seattle, WA 98104

APPENDIX D

U.S. STATE AND TERRITORIAL AGENCIES AND CITIZENS' GROUPS CONCERNED WITH PARK, RECREATION, AND LEISURE SERVICES

ALABAMA

Alabama Department of Conservation and Natural Resources
64 N. Union St.
Montgomery, AL 36130

Alabama Association of Conservation Districts
275 Novtan Rd.
Mobile, AL 36608

ALASKA

Department of Environmental Conservation
410 Willoughby Ave.
Juneau, AK 99801-1795

Department of Natural Resources
400 Willoughby
5th Fl.
Juneau, AK 99801

ARIZONA

Outdoor Recreation Coordinating Commission
800 W. Washington
Suite 415
Phoenix, AZ 85007

Arizona Association of Conservation Districts
Box 1076
Springerville, AZ 85938

Arizona State Parks
800 W. Washington
Suite 415
Phoenix, AZ 85007

ARKANSAS

Department of Parks and Tourism
One Capitol Mall
Little Rock, AR 72201

Arkansas Association of Conservation Districts
Rt. 2
Piggott, AR 72454

CALIFORNIA

Department of Conservation
801 K St.
24th Fl.
Director's Office
Sacramento, CA 95814

Department of Parks and Recreation
1416 Ninth St.
P.O. Box 942896
Sacramento, CA 94296-0001

COLORADO

Department of Natural Resources
1313 Sherman
Rm., 718
Denver, CO 80203

State Forest Service
Colorado State University
Ft. Collins, CO 80523

CONNECTICUT

Department of Environmental Protection
State Office Bldg.
165 Capitol Ave.
Hartford, CT 06106

DELAWARE

Department of Natural Resources and Enviromental Control
89 Kings Highway
P.OP. Box 1401
Dover, DE 19903

Division of Fish and Wildlife
89 Kings Highway
P.OP. Box 1401
Dover, DE 19903

DISTRICT OF COLUMBIA

District of Columbia
Conservation District
2100 Martin Luther King Jr. Ave. SE
Suite 203
Washington, DC 20020

FLORIDA

Department of Agriculture and Consumer Services
3125 Conner Blvd.
Tallahassee, FL 32399-1650

Recreation & Parks
3125 Conner Blvd.
Tallahassee, FL 32399-1650

GEORGIA

Department of Natural Resources
205 Butler St., SE
East Towers
Atlanta, GA 30334

HAWAII

Department of Parks and Recreation
490 Naval Hospital Rd.
Agana Heights, HI 96910

IDAHO

Idaho State Parks and Recreation
Statehouse Mail
Boise, ID 83720-8000

ILLINOIS

Department of Conservation
Lincoln Tower Plaza
524 S. Second St.
Springfield, IL 62701-1787

Resource Marketing & Education
Lincoln Tower Plaza
524 S. Second St.
Springfield, IL 62701-1787

Planning and Development
Lincoln Tower Plaza
524 S. Second St.
Springfield, IL 62701-1787

Division of Planning
Lincoln Tower Plaza
524 S. Second St.
Springfield, IL 62701-1787

Resource Management
Lincoln Tower Plaza
524 S. Second St.
Springfield, IL 62701-1787

INDIANA

Indiana Department of Natural Resources
402 W. Washington St.
#C256
Indianapolis, IN 46204-2212

Natural Resources Commission
402 W. Washington St.
#C256
Indianapolis, IN 46204-2212

Lands and Cultural Resources Advisory Council
402 W. Washington St.
#C256
Indianapolis, IN 46204-2212

Division of Outdoor Recreation
402 W. Washington St.
#C256
Indianapolis, IN 46204-2212

Division of State Parks
402 W. Washington St.
#C256
Indianapolis, IN 46204-2212

IOWA

Department Natural Resources
E. Ninth and Grand Ave.
Wallace Bldg.
Des Moines, IA 50319-0034

KANSAS

Kansas Department of Wildlife and Parks
900 Jackson St.
Suite 502
Topeka, KS 66612-1220

Department of Fish and Wildlife Resources
#1 Game Farm Rd.
Topeka, KS 66612

KENTUCKY

Department of Parks
10th Fl. Capital Plaza Bldg.
Frankfort, KY 40601

Resort Parks
10th Fl. Capital Plaza Bldg.
Frankfort, KY 40601

Department for Natural Resources
107 Mero St.
Frankfort, KY 40601

Kentucky-Tennessee Society of American Foresters
P.O. Box 149
Sewanee, TN 37375

LOUISIANA

Office of State Parks, Department of Culture, Recreation, and Tourism
P.O. Box 44426
Baton Rouge, LA 70804

State Office of Conservation
P.O. Box 94275
Capitol Sta.
Baton Rouge, LA 70804-9275

MAINE

Department of Conservation
State House Station #22
Augusta, ME 04333

Bureau of Parks and Recreation
State House Station #22
Augusta, ME 04333

Natural Resources Council of Maine
271 State St.
Augusta, ME 04333

MARYLAND

Department of Natural Resources
580 Taylor St.
Tawes State Office Bldg.
Annapolis, MD 21401

Maryland-National Capital Park and Planning Commision
6609 Riggs Rd.
Hyattsville, MD 20782

Prince George's County Planning
6609 Riggs Rd.
Hyattsville, MD 20782

Montgomery County Parks
6609 Riggs Rd.
Hyattsville, MD 20782

Prince George's County Director
Parks and Recreation
6609 Riggs Rd.
Hyattsville, MD 20782

MASSACHUSETTS

Executive Office of Environmental Affairs
Leverett Saltonstaff Bldg.
100 Cambridge St.
Rm. 2000
Boston, MA 02202

Department of Environmental Management
Commissioner
100 Cambridge St.
Rm. 1905
Boston, MA 02202

Massachusetts Association of Conservation Commission (MACC)
10 Juniper Rd.
Belmont, MA 02178

Massachusetts Association of Conservation Districts
84 Churck St.
Gilbertville, MA 01031

MICHIGAN

Department of Agriculture
4th Fl., Ottawa Bldg.
P.O. Box 30017
Lansing, MI 48909

Michigan Association of Conservation Districts
16059 Pardee Rd.
Galien, MI 49113

Michigan Forest Association
1558 Barrington St.
Ann Arbor, MI 48103

MINNESOTA

Minnesota Conservation Federation
1036 Cleveland Ave., S.
Suite B
St. Paul, MN 55116-1887

MISSISSIPPI

Department of Wildlife, Fisheries, and Parks
2906 Building
P.O. Box 451
Jackson, MS 39205

Planning and Policy
2906 Building
P.O. Box 451
Jackson, MS 39205

Wildlife/Fisheries Division
2906 Building
P.O. Box 451
Jackson, MS 39205

Bureau of Marine Resources
2906 Building
P.O. Box 451
Jackson, MS 39205

Parks/Recreation Divison
2906 Building
P.O. Box 451
Jackson, MS 39205

Outdoor Recreation Grants
2906 Building
P.O. Box 451
Jackson, MS 39205

MISSOURI

Department of Conservation
P.O. Box 180
Jefferson City, MO 65102-0180

Department of Natural Resources
P.O. Box 176
Jefferson City, MO 65102

Division of Parks, Recreation and Historic Preservation
P.O. Box 176
Jefferson City, MO 65102

Parks
P.O. Box 176
Jefferson City, MO 65102

MONTANA

Department of Fish, Wildlife, and Parks
1420 East Sixth
Helena, MT 59620

Department of Natural Resources and Conservation
1520 East Sixth Ave.
Helena, MT 59620-2301

Board of Natural Resources and Conservation
1520 East Sixth Ave.
Helena, MT 59620-2301

NEBRASKA

Game and Parks Commission
2200 N. 33rd St.
P.O. Box 30370
Lincoln, NE 68503

NEVADA

Department of Conservation and Natural Resources
Capitol Complex
123 W. Nye Ln.
Carson City, NV 89710

Division of State Parks
Capitol Complex
123 W. Nye Ln.
Carson City, NV 89710

NEW HAMPSHIRE

Department of Resources and
Economic Development
P.O. Box 856
172 Pembroke Rd.
Concord, NH 03302-0856

Division of Parks
P.O. Box 856
172 Pembroke Rd.
Concord, NH 03302-0856

New Hampshire Association of
Conservation Commissions
54 Portsmouth St.
Concord, NH 03301

New Hampshire Association of
Conservation Districts
63 High Range Rd.
Londonderry, NH 03053

NEW JERSEY

Department of Environmental
Protection and Energy
401 E. State St.
CN 402
Trenton, NJ 08625-0402

Division of Parks and Forestry
CN 404
Trenton, NJ 08625-0404

Green Acres and Recreation Program
CN 412
Trenton, NJ 08625-0412

New Jersey Conservation Foundation
300 Mendham Rd.
Morristown, NJ 07960

NEW MEXICO

Energy, Minerals, and Natural
Resources Department
2040 Pacheco St.
Santa Fe, NM 87505

Parks and Recreation
Villagra Building
P.O. Box 1147
Santa Fe, NM 87504-1147

NEW YORK

Adirondack Park Agency
P.O. Box 99
Ray Brook, NY 12977

Department of Environmental
Conservation
50 Wolf Rd.
Albany, NY 12233

Division of Lands and Forests
50 Wolf Rd.
Albany, NY 12233

Forest Resource Management Bureau
50 Wolf Rd.
Albany, NY 12233

Land Resources Bureau
50 Wolf Rd.
Albany, NY 12233

Division of Managment Planning and
Information Systems Development
50 Wolf Rd.
Albany, NY 12233

State Office of Parks, Recreation and
Historic Preservation
Empire State Plaza
Albany, NY 12238

NORTH CAROLINA

Department of Environmental,
Health, and Natural Resources
P.O. Box 27687
Raleigh, NC 27611

NORTH DAKOTA

Parks and Recreation Department
1835 Bismarck Expressway
Bismarck, ND 58504

OHIO

Department of Natural Resources
Fountain Square
Columbus, OH 43224

Division of Forestry
Fountain Square
Columbus, OH 43224

Division of Natural Areas and
Preserves
Fountain Square
Columbus, OH 43224

Division of Parks and Recreation
Fountain Square
Columbus, OH 43224

Ohio Conservation and Outdoor
Education Association
121 Beckett Ct., St.
Clairsville, OH 43950

OKLAHOMA

Tourism and Recreation
Department
2401 North Lincoln
Suite 500
Oklahoma City, OK 73105-4492

Division of Travel & Tourism
2401 North Lincoln
Suite 500
Oklahoma City, OK 73105-4492

Division of Resorts
2401 North Lincoln
Suite 500
Oklahoma City, OK 73105-4492

OREGON

Oregon Department of Forestry
2600 State St.
Salem, OR 97310

Oregon Natural Resources Council
Yeon Bldg. 1050
522 SW Fifth Ave.
Portland, OR 97204

PENNSYLVANIA

State Conservation Commission
Department of Environmental
Resources
P.O. Box 8555
Harrisburg, PA 17105-8555

Pennsylvania Recreation and Park
Society, Inc.
723 South Atherton St.
State College, PA 16801-4629

PUERTO RICO

Department of Natural Resources
P.O. Box 5887
Puerta de Tierra Sta.
San Juan, Puerto Rico 00906

RHODE ISLAND

Department of Environmental Management
9 Hayes St.,
Providence, RI 02908

SOUTH CAROLINA

Department of Parks, Recreation, and Tourism
Edgar A. Brown Bldg.
1205 Pendleton St.
Columbia, SC 29201

State Parks Division
1205 Pendleton St.
Office of the Governor
Columbia, SC 29201

Tourism Division
1205 Pendleton St.
Office of the Governor
Columbia, SC 29201

Recreation Division
1205 Pendleton St.
Office of the Governor
Columbia, SC 29201

Division of Natural Resources
1205 Pendleton St.
Office of the Governor
Columbia, SC 29201

SOUTH DAKOTA

Game, Fish, and Parks Department
523 East Capitol
Pierre, SD 57501-3182

South Dakota Resources Coalition
P.O. Box 7020
Brookings, SD 57007

TENNESSEE

Department of Environment and Conservation
401 Church St.
Nashville, TN 37243

Division of Recreation Services
401 Church St.
Nashville, TN 37243

Tennessee Conservation League
300 Orlando Ave.
Nashville, TN 37209-3200

TEXAS

Parks and Wildlife Department
4200 Smith School Rd.
Austin, TX 78744

UTAH

State Department of Natural Resources
1636 W. North Temple
Salt Lake City, UT 84116-3156

VERMONT

Department of Forests, Parks, and Recreation
Waterbury Complex
10 South
Waterbury, VT 05671

State Natural Resources Conservation Council
Agency of Natural Resources
103 S. Main St.
Center Bldg.
Waterbury, VT 05676

VIRGINIA

Department of Conservation and Recreation
203 Governor St.
Suite 302
Richmond, VA 23219

Division of Planning/ Recreation Resources
203 Governor St.
Suite 326
Richmond, VA 23219

Division of State Parks
203 Governor St.
Suite 306
Richmond, VA 23219

VIRGIN ISLANDS

Department of Planning and Natural Resources
Suite 231
Nisky Center
St. Thomas, U.S. 00803

WASHINGTON

Department of Natural Resources
P.O. Box 47001
Olympia, WA 98504-7001

State Parks and Recreation Commission
7150 Cleanwater Ln KY-11
Olympia, WA 98504-5711

Washington Recreation and Parks Association
350 S. 33rd St.
Suite 103
Federal Way, WA 98003

WEST VIRGINIA

Division of Natural Resources
1900 Kanawha Blvd., East
Charleston, WV 25305

Wildlife Resources
1900 Kanawha Blvd., East
Charleston, WV 25305

WISCONSIN

Department of Natural Resources
Box 7921
Madison, WI 53707

WYOMING

Wyoming State Parks and Historic Sites
2301 Central Ave.
Barrett Bldg.
Cheyenne, WY 82002

APPENDIX E

CANADIAN GOVERNMENT AGENCIES AND CITIZENS' GROUPS CONCERNED WITH PARK, RECREATION AND LEISURE SERVICES

ALBERTA

Department of Forestry, Lands, and Wildlife
Main Fl., North Tower
Petroleum Plaza
9945-108 St.
Edmondton, Alberta
Canada T5K 2G6

The Alberta Fish and Game Association
6924-104 St.
Edmonton Alberta
Canada T6H 2L7

Alberta Wilderness Association
Box 6398, Station D.
Calgary, Alberta
Canada T2P 2E1

BRITISH COLUMBIA

Ministry of Tourism and Ministry Responsible for Culture
1117 Wharf St.
Victoria, Canada V8V 2Z2

Outdoor Recreation Council of British Columbia
334-1367 West Broadway
Vancouver, British Columbia
Canada V6H 4A9

MANITOBA

Department of Natural Resources
Rm. 314
Legislative Bldg.
Winnipeg,, Canada R3C 0V8

NEW BRUNSWICK

Department of Natural Resources and Energy
P.O. Box 6000
Fredericton, New Brunswick
Canada E3B 5H1

NEWFOUNDLAND

Department of Tourism and Culture
P.O. Box 8700
St., John's, Newfoundland
Canada A1B 4J6

NORTHWEST TERRITORIES

Department of Economic Development and Tourism, Government of the Northwest Territories
Box 1320, Yellowknife
Northwest Territories
Canada X1A 2L9

NOVA SCOTIA

Department of Natural Resources
P.O. Box 698
Halifax, Nova Scotia
Canada B3J 2T9

ONTARIO

Ministry of Natural Resources
Toronto, Ontario
Canada M7A 1W3

The Conservation Council of Ontario
Suite 506
489 College St.
Toronto, Ontario
Canada M6G 1A5

PRINCE EDWARD ISLAND

Department of the Environment
P.O. Box 2000
Charlottetown, Canada C1A 7N8

QUEBEC

Department of Recreation, Fish, and Game
Place de la Capitale 150
Boul. Rene-Levesque Est.
Quebec, Canada G1R 4Y1

SASKATCHEWAN

Saskatchewan Natural Resources
3211 Albert St.
Regina, Canada S4S 5W6

YUKON TERRITORY

Department of Renewable Resources
Box 2730
Whitehorse, Canada Y1A 2C6

APPENDIX F

UNITED STATES GOVERNMENT AGENCIES— EXECUTIVE BRANCH CONCERNED WITH PARK, RECREATION, AND LEISURE STUDIES

Denver Wildlife Research Center
Animal Damage Control Program,
U.S. Department of Agriculture
Animals Plant Health Inspection Service
P.O. Box 25266, Bldg. 16
Denver, CO 80225-0266

USDA FOREST SERVICE

Headquarters
P.O. Box 96090
Washington, DC 20090-6090

Region 1, Northern
P.O. Bx 7669
Missoula, MT 59807

Region 2, Rocky Mountain
740 Sims St.
Lakewood, CO 80401

Region 3, Southwestern
Federal Bldg.
517 Gold Ave., SW
Albuquerque, NM 87102

Region 4, Intermountain
Federal Office Bldg.
324 25th St.
Ogden UT 84401

Region 5, California
630 Sansome St.
San Francisco, CA 94111

Region 6, Pacific Northwest
333 SW 1st Ave.
Box 3623
Portland, OR 97208

Region 8, Southern
Suite 800
1720 Peachtree Rd., NW
Atlanta, GA 30367

Region 9, Eastern
310 W. Wisconsin Ave.
Room 500
Milwaukee, WI 53203

Region 10, Alaska
Federal Office Bldg.
Box 21628
Juneau, AK 99802-1628

National Oceanic and Atmospheric Administration
Herbert C. Hoover Bldg.
Rm. 5128
14th and Constitution Ave., NW
Washington, DC 20230

U.S. Army Corps of Engineers
20 Massachusetts Ave., NW
Washington, DC 20314-1000

NAVAL FACILITIES ENGINEERING COMMAND

Headquarters
200 Stovall St.
Alexandria, VA 22332-2300

Natural Resources Division
Code 143
200 Stovall St.
Alexandria, VA 22332-2300

Natural Resources Specialists
Headquarters
200 Stovall St.
Alexandria, VA 22332-2300

U.S. MARINE CORPS

Headquarters
2 Navy Annex
Washington, DC 20380-1775

Natural Resources Management Office
2 Navy Annex
Washington, DC 20380-1775

Bureau of Indian Affairs
Interior South Bldg.
1951 Constitution Ave., NW
Washington, DC 20245

Bureau of Land Management
U.S. Department of Interior
1849 C St., NW
Rm 5600
Washington, DC 20240

INDEX

A

Absenteeism, and stress, 400
Accountability, in goal setting, 62
Advertising, 83
Agitators, 274-275
American Society for Training and Development, 292
Americans with Disabilities Act, 212, 213, 215, 216
Authoritarian approach, 255
Autonomous group working, 235

B

Boards and commissions, 19
Brainstorming
    basic rules of, 4
    defined, 3
    for problem solving, 8, 126, 128
Brainstorming sessions
    establishing groups for, 4-5
    followup to, 7
    leading, 5-7
    pitfalls to avoid in, 6-7
    suggestions for improving, 7-8
Burnout, in work teams, 55
Business ethics
    code of ethics for, 14, 16
    common dilemmas in, 10, 11-12
    costs of downward trend in, 12
    ethicists' role in, 13, 15
    maintaining standards for, 13-15
    suggestions for improving, 15-16
    supervisors' role in, 13-14, 15
Business etiquette, 365-373
    cultural differences and, 371
    and dressing, 368
    foreign trade and, 371-372
    introductions in, 366-367
    meals and, 369
    meetings, proper etiquette for, 367. *See also* Meetings, planning of
    suggestions for, 372-373
    telephone conversations and, 367-368

C

Career coasters, 276-277
Career planning and development, 170-176
    accounting for change in, 169-170
    criteria for, 171-172
    problems facing dual-career couples, 173-175
    purpose of, 170
    role of management in, 172-173
    suggestions for improving, 176
Carrot and stick approach, 255
CEO-Board relations, 177-182
    accountability of CEO in, 182
    balance in, 177-178
    in co-designing policy, 178
    overlapping functions of 178-180
    policy functions of CEO, 178
Cluster chain, 71
Comment cards, 33
Committees, 17-25
    advantages and disadvantages of, 19-22
    chairpersons for, 22-23
    functions of, 17
    improving management of, 24-25
    selection of, 23-24

size considerations for, 24
types of, 18-19
Communication. *See* Organizational communication
Communication chains
formal, 70
in grapevine networks, 70-71
Communication materials, 148-149
Competitors, direct and indirect, 85
Compressed work week, 345
Consultants, steps for selecting, 301
Cover letter
for resume, functions of, 358, 359
rules for constructing, 358-359
Creative individuals, improved management of, 79-80
Creativity
improved management of, 79-80
individual, characteristics of, 77-78
management attitudes toward, 75, 76, 78-79
management of, 75-80
in the marketing process, 89
organizational, 78-80
reasons for lack of, 78-79
training for, 77
and the value of creative spirit, 79
Critical path method (CPM), 137
Customer expectations, perceptions of, 30
Customer information, obtaining, 32
Customer service
assessment of, 34
companies, characteristics of, 31
and customers' expectations, 30
effectiveness of factors in, 34
importance of, 28
inadequacy of, in service organizations, 28-30
maintaining, 31
measuring effectiveness of, 33-34
obstacles to, 29-30
obtaining customer information for, 32-33
purpose of, 27-28
research on, 27-28
suggestions for improving, 34-35
See also Marketing

D

Decision making
ability, suggestions for improving, 41-42
analysis, using "situation variables," 38-39
careless, causes of, 39
"Groupthink Syndrome" in, 39-41
nature of, 37-38
in project management, 138
role of computers in, 41
Delegation, 45-50
of creative freedom, 77
as an employee motivator, 258
guidelines for, 47-50
meaning of, 45-47
obstacles to, 46
responsibility and authority in, 46-47, 50
understanding concept of, 47
Demotivation, rules for avoiding, 258-259, 260
Department of Commerce, 141
Development goal, 65
Disabled workers, 211-219
ADA definition of, 212
approaches to hiring, 215-217
changing attitudes toward, 211-212, 217-218
employment issues involving, 212-219
interviewing of, 217, 219
job requirements and, 215-217, 218
legal issues involving, 212
myths about, 213
social acceptance of, 214-215
special accommodations for, 213-214
suggestions for hiring, 218-219
training for, 214
workplace attitudes involving, 214-215, 217-218
Dissatisfiers. *See* Motivators
Downsizing
as cause of stress, 408
employee termination and, 323, 331
problem employees and, 273
and use of temporary employees, 314
Drucker, Peter, 27

E

Education
of customers, 35
employee, about customer service, 35
public. *See* Public relations
Employees
    assistance to, following termination, 329
    termination of, 323-332
Employee appraisals, 191-196
    common flaws in, 192,193
    employee's responsibility in, 195-196
    flaws in systems for, 192, 193
to improve job satisfaction, 249
interviews, problems with, 193-194
    manager's role in, 191, 194-195
    objectives for, 194
as ongoing process, 194
suggestions for improving, 195-196
    systems for, 192, 193
    unsatisfactory performance and, 195
Employee education. *See* In-service training
Employee evaluation. *See* Employee appraisals
Employee fitness, 197-202
    corporate programs for, 198
    exercise required to achieve, 199, 201
    motivational programs to encourage, 259
    nutritional aspects of, 199-200, 201
    personal benefits of, 200
    statistics supporting, 197-198
    stress reduction and, 201
    suggestions for improving fitness and lifestyle, 201
    while traveling, 200
Employee grievance policy, 326
Employee Retirement Income Security Act, 326
Employee training needs assessment, 296-297
Employees
agitators as, 274-275
    categories of, 274-277
    characteristics of, 273
    identifying, 273-274
    perceived moods of, for training, 295-297
    problems of, 273-279
    temporary, 313-321
Employment interviewing, 203-210
    of disabled workers, 217, 219
    in employee appraisals, 193-194, 195-196
    importance of, 203,
    interviewer's role in, 205-206
    manager's approach to, 204
    process, sources of error in, 206-207
    process, stages of, 203-204
    questions to ask during, 207-208
    questions to avoid in, 208-209
    skills, suggestions for improving, 208-209
Energizers, 276-277
Esteem needs, 252
Equal Employment Opportunity Commission (EEOC), sexual harassment and, 284, 285, 286
Ethics
    defined, 10
    levels of, 10
    See also Business ethics
Etiquette. *See* Business etiquette
External stressors, 399

F

Feedback
    to customers, 35
    on goals, 68
    in work teams, 55, 59
Fitness. *See* Employee fitness
Flow charts, 137
Focus groups, 33

G

Goal setting
    attainable goals in, 66
    establishing a program for, 65-68
factors affecting, 61-62
    failure of, 66-67
    kinds of goals in, 64-65
    management participation in, 61, 63, 65, 68
    and performance evaluation, 63

purpose of, 61, 62
reasons for, 63
research on, 61
suggestions for, 390-391
training for, 67
Goals
   defined, 65
   in job-enrichment programs, 237-238
   for marketing, 86-87
   in participation management environment, 268
      seen as constraints, 238
   short- and long-term, 87
   types of, 64-65
   worker self-esteem and, 434, 435
Goals and objectives, 48
   and personnel policies, 183
   in project management, 135-136
   of public relations program, 146-147
   strategic planning for achievement of, 159
   team approach to, 52
Goldbricker, 275
Gossip chain, 71
Grapevine
   communication chains in, 70-71
   defined, 69
   and formal communications - networks, 70
   information, accuracy of, 73
   organizational response to, 72, 73
   participation by managers in, 72
   purposes of, 70
   reasons for activity on, 71-72
   rumors on, 73, 74
   as stress-reduction mechanism, 72-73
   successful use of, 74
Groupthink
   in project management, 138
   symptoms of, 40-41
   syndrome, 40
in work teams, 55
Groupthink, 39

H

Hawthorne studies, 234, 244
Herzberg's two-factor theory, 234, 235, 244
   applied to job satisfaction, 234, 235, 244
   in organizational profile, 253
How to Be a More Creative Executive, 75-76
Hygiene factors, 234

I

Identity materials, for organization, 148
Improvement teams, 263-264
Innovation. *See* Creativity
In-service training, 221-231
   corporate approaches to, 224
   defined, 222
   for delegation, 46, 49
   developing programs for, 225-229
   evaluation method for, 223
   evaluation of, 228, 230
   factors affecting success of, 221-222, 225-226
   general methods of, 223-224
   informal approach to, 227-228
   in job enrichment programs, 239
   management commitment to, 221-222
   measuring results of, 222-223
   needs, strategies to meet, 225
   organization as basis for, 224-225, 230
   and patterns of adult learning, 226, 230
   problem employees and, 277
   programs, successful development of, 230
   scope of, 221
   sexual harassment issues in, 287-288
   status and importance of, 229-230
   techniques for, 229, 230
   use of outside firms for, 228-229
   using in-house resources, 226-228, 230
   for volunteers, 338-340
Internal stressors, 399
Interpretive Guidelines on Sexual Harassment, 284
Interviews
   exit, 298
   personal, for customer service, 32

INDEX 477

J

Janis, Irving. *See* Groupthink
Job bidding, 349
Job descriptions
    and disabled workers, 214-217
    and organizational charts, 109
    using standardized questionnaires to prepare, 216-217
    of volunteers' role, 336
Job dissatisfaction, role of stress in, 400
Job enlargement, 234
Job enrichment, 233-241
    characteristics of, 236, 237
    compared to job satisfaction, 233-234, 240
    defined, 234, 235
    goal setting for, 237-238
    interviews in, 240
    Japanese-style management for, 236-237
    job characteristics model of, 235-236, 237
    job satisfaction and productivity as key to, 240
    pay and training issues in, 239
    pros and cons of, 238-239
    quality-of-worklife approach, 237
    strategies for, 235-237
    union attitudes toward, 239, 240
    work-design techniques for, 234-235
Job performance, 246-249
    benefit of improving, 247
    correlation with job satisfaction, 248-249
    defined, 246-247
    empowerment strategy for, 247
    as a factor in job enrichment, 246-247
Job rotation, 234
Job satisfaction, 243-249
    attitudes toward work and, 246
    creation of, 244-245
    defined, 243-244
    employee needs and, 245-246
    empowerment strategy for, 247
    formal needs theory and, 244-245
    importance of job performance to, 246-248
    management plan and, 249
    measurement of, 236
    objectives and goals, 249
    productivity and, 247
Job sharing, 345-347
Job standards, 47

L

Leaders. *See* Leadership
Leadership
    and charisma, 389
    and followers, 387, 388, 390
    defining, 387-388, 390
    future of, 390
    and leaders, characteristics of, 387, 388, 390
    participative, 80
    in problem solving, 126
    situational analysis of, 388
    suggestions for, 390-391
    traits, research on, 388, 389
Letters
    basic kinds of, 383-384
    checklist for writing, 384-385
Liaison communicators. *See* Grapevine, successful use of
Listening, effective attitudes toward speakers and, 393, 394, 395, 396
    barriers to, 394
    to customers, 31
    feedback from, 395
    hearing problems and, 397
    improving skills for, 396-397
    as negotiating skill, 100-101
    nonverbal communication and, 397
    positive results of, 394-395
    time spent in, 393
Location, 85

M

Mail surveys, 32
Management
    of creativity,
    problem employees and, 277-278
    role of, in strategic planning, 153-155
    top-down approach to, 51
Manipulation approach, 255
Market research, 85-86, 88, 89

Marketing
- activities available for, 87
- budget for, 87
- customer and benefits analysis for, 86
- customer orientation of, 81, 84, 85
- defined, 81-82, 83
- function, diagrammed, 82
- identifying opportunities for, 85-86
- improving the process of, 89
- key element of, 81, 88
- long-term benefits of, 88
- plan for, 83-89

Marketing plan
- benefits addressed in, 84-85
- content of, 83
- defining customers in, 84
- environmental analysis for, 85
- evaluating success of, 88
- as "how-to" manual, 89
- market segmentation for, 81
- opportunities identified in, 85-86
- product/service evaluation in, 84
- reason to develop, 83
- specific goals for, 86-87
- as timetable of activities, 87

Maslow, Abraham, 251
Maslow's Hierarchy of Needs, 234-235
- in study of job satisfaction, 234-235, 244
- as theory of motivation, 251-153

McGregor, Douglas
- organization theory of, 107, 108, 251, 253-254
- X and Y theory of, 253-254

"Medium is the message," 116

Meetings
- cost of, 91
- defined, 92
- improving effectiveness of, 93-94
- planning of, 91-95
- pricing and control of, 91
- reasons for failure of, 92-93
- suggestions for, 94-95
- types of, 92

Mission, defined, 65
Money approach, 255
Morale, downsizing and, 259
Motivation, 251-260
- by avoiding demotivation, 258-259
- classical theories of, 251-255
- defined, 251
- effects of downsizing on, 259
- through employee ownership, 256
- of employees, suggestions for, 260
- external factors influencing, 257-258
- by force, 254-255
- in Herzberg's two-factor theory, 253
- importance of leadership in, 258-259
- inspiration and, 258
- by manipulation, 255
- in Maslow's hierarchy of needs, 251-253
- material rewards and, 255-256
- McGregor's theory X / theory Y classifications and, 253-254
- methods of, 254-255
- organizational programs for, 256-259
- by persuasion, 255
- use of contests in, 255
- Wedgwood's seven approaches to, 255

Motivational programs, 256-257
- assumptions to consider, 256
- to encourage employee fitness, 259
- management's role in evaluating, 259
- motivation work design, 256
- Zero Defects program, 256

Motivators, 253
Multiple management committee, 19

N

NASA, 135
Negotiation, 97-104
- "best alternative" in, 104
- improving, 103-104
- power relationships in, 99-100
- preparation for, 99
- reducing conflict in, 102-103
- skill development for, 100-104

Negotiator
- characteristics of, 97-100
- self-esteem of, 100

News, about organization, 147
Nice guy approach, 255

INDEX 479

O

Objective
  defined, 65
  for training, 299
On-the-job training. *See* In-service training
Organization culture, work team enhancement of, 55
Organization theory, 45
  and organizational charts, 107-108
  perspective of, 161
Organizational chart
  authority relationships in, 106
  construction, false assumptions in, 107
  developing and using, 109
  difficulties with, 106-107
  as formal statement, 107
  meaningful groups in, 108
  purposes of, 106
Organizational communication
  in CEO-board relations, 180-181, 182
  defined, 111-112
  delivery of, 116-117
  "5-W and H" test of, 118
  goals of, 115-116
  horizontal communication in, 114-115
  improved by goal setting, 64
  message competition in, 117
  Murphy's law of, 111, 118
  in participation management environment, 268
  problem employees and, 278
  as stress reducer, 408
  structure of, 112-113, 115-116, 117-118
  suggestions for improving, 118-119
  vertical communication in, 112-114
  See also Grapevine; Public relations
Organizational surveys, 264
Organizations
  charts of, 105-110
  sociometric studies of, 107
Orientation and training of volunteers, 338-339
Participation management, 261-271
  benefits of, 261-262
  current status of, 261-262
  effectiveness of, 262, 266-267
  empowerment as, 264
  gain sharing and, 263
  improvement teams in, 263-264
  leadership role in, 267, 269
  organizational surveys for, 264
  participative goal-setting for, 264
  pitfalls of, 266
  quality circles and, 264
  self-managed teams as, 263
  spectrum of management approaches to, 267
  suggestion systems and, 263
  suggestions for implementation, 268-269
  ways to involve employees in, 262-264

P

Participative goal-setting, 274
Participative management. *See* Participation management
Part-time workers. *See* Temporary employees
Paternalistic approach, 255
Peak time, 348-349
Performance appraisals
  assessment of training needs, 298
  based on goals, 67
  building self-esteem with, 435
  and goal-setting concept, 63
  PIN technique for, 435
  in termination process, 325, 327
  See also Employee appraisals
Performance evaluations. *See* Performance appraisals
Personal interviews, 32
Personnel policies, 183-189
  beneficiaries of, 186
  and compliance with regulations, 185, 188, 189
  diversity recognized in, 187
  enforcement of, 185, 186
  issues to address in, 187-188
  negative attitudes towards, 183-184
  and perceptions of fairness, 186-187
  suggestions for developing, 188-189

written, importance of, 184-185, 186, 189
Persuasion, 411-421
    anticipating objections to, 418
    and audience point of view, 417, 418
    content and relationships levels of, 411
    credibility and salesmanship in, 412
    importance of confidence in, 420
    improving skills in, 420
    preparation for, 416-419
    "smoke screen" objections to, 418
    techniques for, 412-415
    timing of, 417
    tips for success in, 416
    understanding resistance to, 417
Peters, Tom, 3
Peterson, Donald, 292
Physiological needs, 251
Plan, 184
Planner, role of, 153-155
Planning. *See* Strategic planning
Policy, 184
POSDCORB, 153
Presentations, effective, 375-379
    analysis of audience for, 376-377, 378, 379
    delivery notes for, 377
    determining objectives of, 376, 377
    examples of, 378
    planning for, 376
    use of visual aids in, 377-378
    value of, in business, 375
Price, 85
Probability chain, 71
Problem employees, 273-279
    career coasters, 276-277
    energizers as, 275-276
    goldbrickers as, 275
    in-service training as deterrent to, 277
    management of, 277-278 *and*
Problem solving
    barriers to, 124-125
    brainstorming in, 8, 126, 128
    changes and conflicts in, 125-126
    determining problem in, 122-123
    final decisions in, 127
    goals in, 64
    leadership in, 126

objectives for, 123-124
selecting alternatives in, 126-127
solution implementation in, 127-128
step-by-step approach to, 121-128
suggestions for, 128
value of goal setting in, 64
Procedure, 184
Procrastination, 423-428
    approach to overcoming, 427
    negative effects of, 425, 426
    as "putting off," 426
    reasons for, 423, 425-426
    self-assessment exercise for, 424
    suggestions for eliminating, 427-428
Product, 84
Productivity
    improved, through goal setting, 61
    self-esteem and, 432-434
Program Evaluation and Review Technique (PERT), 137
Project management
    characteristics of, 132
    flow chart development in, 137
    goals and objectives in, 135-136
    groupthink in, 138
    NASA example of, 134
    outside the organization, 134
    scenarios for, 136
    structure, advantages of, 131-132
    suggestions for, 137-138
    team responsibilities in, 135
    time and resource estimates for, 136-137
    See also Committees
Project manager
    characteristics of 132, 134
    responsibilities of, 132, 134-135
Promotion, defined, 83
Pros and Cons of Participative Management, The, 265
Public relations, 141-150
    communication skills for, 145
    consultants in, 144
    contribution of, 147
    defined, 141
    evaluating success of, 150
    implementation of, 147-149
    materials supporting, 148-149
    methods of practice, 143-144

program, objectives of, 146-147
publics dealt with in, 143
as staff function, 144
suggestions for improving, 149-150
value of, 142
Public service activities, 147-148
Publicity. *See* Public relations
"Putting off." *See* Procrastination

Q

Quality circles, 263

R

Rational approach, 255
Resumes
chronological type of, 355-356
cover letter for, 358-359
defined, 354
factors causing rejection of, 361
functional type of, 356-357
importance of, 354
industry-related type of, 357
as key to job search, 353
reasons to write, 354-355
suggestions for improving, 360
things to avoid in, 358
types of, 355-357
useful words for, 359
Rewards, for good work, 48
Rule, 184
Rumors, 73, 74

S

Safety needs, 251
Satisfiers. *See* Motivators
Self-actualization needs, 252
Self-esteem
defined, 429-430
employee, suggestions for enhancing, 435-436
feedback to promote, 436
form of, 431
gender differences in, 430
management's role in fostering, 434-435
national preoccupation with, 429

of negotiators, 100
organization goals and, 434-435
productivity and, 432-434
stability of, and careers, 431-432
stress and, 433-434
of subordinates, 48
theories of, 430
workplace stress and, 433-434
Self-managed teams, 263
Semi-structured interview, 204
Service, 85
Sexual harassment, 281-290
cost to organization of, 285-286
court test for, 283
definition of, 282-284
education and training about, 287-288
elements of, 282-283
filing complaints about, 286
manager's responsibility regarding, 283-284
meaningful recourses for, 285
national focus on, 282
organization's liability for, 287
organization's policy on, 284-285
strategies for prevention of, 284-285
suggested guidelines for planning, 289
training topics related to, 287-288
valid claims of, 283
workplace behavior/discrimination issues and, 288-289
Single strand chain, 70
Situation variables, 38-39
Skills, negotiation, 100-104
Social needs, 252
Socrates, 141
Special events, 148
"Square peg in the round hole." *See* Energizers
Staff development, 291-304
assessment of training needs and, 295
equipment aids for, 300
evaluation of programs, 301-303
importance of, 291-292
management support of, 292-294
measuring results of programs for, 294

outside training resources for, 300-301
package training programs for, 300
programs, excuses not to evaluate, 302
programs, suggestions for improving, 303-304
programs and principles for planning, 294-295
purpose of, 291, 295
questions answered by evaluation, 303
training needs for, 295-299
training objectives for, 299
transfer to work site of, 303
See also In-service training
Staggered fixed schedule, 344
Standing committees, 18
Strategic long-term plan, 53
Strategic planning
 and budgeting, 157
 defined, 152
 failure, reasons for, 155-158
 function, suggestions for improving, 158-159
 fundamentals of, 152
 logical steps to, 151-152
 management role in, 153-155
 as organization function, 151-159
 planners, role of, compared to management, 153-155
 vital tasks of, 151
Stress
 attitude toward change and, 401, 406-407
 blaming for high turnover rates, 400
 causes of, 399, 400-401
 control of one's life and, 407-408
 coping with, 406-407, 409
 cost of, 400
 grapevine's role in reducing, 72-73
 involvement with others and, 407-409
 management of, 399-410
 organization's response to, 408
 perfectionism as cause of, 409
 potential, preparing employees for, 408
 quiz to determine level of, 402-404, 408
 research findings on, 400
 signs of, 400-401
 stressors and, 399-400, 408
 suggestions for improving management of, 409
 work-relatedness of, 405-406
Stress management interventions (SMI), 408
Stringer, Robert A., Jr., theory of motives of, 254
Structured interview, 204
Substance abuse, 305-312
 defined, 305
 employee assistance programs (EAP) for, 308, 311
 identification of in workplace, 307
 negative effects of, 305
 organization policy regarding, 306-307
 prevention of, through education, 308, 311
 suggestions for managing, 311
 supervisory training in, 309, 311
 table of substances and their effects, 310-311
Substance abuse policy
 developing or revising, 306-307, 311
 employee assistance programs in, 307
 essential elements of, 306, 307
 establishment of, in the workplace, 309
 value of communication to, 309
Suggestion systems, 263

T

Task, defined, 65
Task force committee, 18
Team building
 suggestions for, 58-59
 value of, 58
Team management, environment for, 51, 53
Telephone surveys, 32
Temporary employees, 313-321
 characteristics of, 315

estimating need for, 317, 318, 320
facilities and workload for, 319
informing permanent staff about, 318-320
instructions, successful use of, 317-381
outside services providers of, 315-317
pros and cons of using, 315-316
reasonable expectations of, 320
reasons for hiring, 314-315, 321
statistics on, 314
suggestions for improving use of, 320-321
supervision and orientation of, 318, 321
value of communication to, 309
Temporary help service, 315-316
questions to ask before using, 317, 320
steps in successful use of, 318-320
Terminating employees, 324-332
under "at will" employment, 324, 325
avoiding litigation from, 325
under contract, 326
effective management of process for, 330, 331
explanations of reasons for, 326, 329-339
importance of documentation in, 326-327, 329
interview in, 328
and management credibility, 327
organization policy for, 323
in progressive discipline system, 327
quick method of, 32-324
in supportive manner, 329
typical reasons for, 327-328
unacceptable reasons for, 324-325
Termination interview, 328
Time management
to avoid procrastination, 427, 428
defined, 162
hints for more productive, 162-164
as negotiating skill, 98
and organizational charts, 107-108
perspective of, 162
in project management, 136-137
suggestions for improving, 165

Title VII, U.S. Civil Rights Act of 1964, 282
Toll-free numbers, 33
Training
to accomplish delegated tasks, 46
for creativity, 77
for goal setting, 67
objectives of, 299
in teamwork skills, 54, 55

U

Unethical behavior, reasons for, 10, 12-13
Unstructured interview, 204
general types of, 204-205
preparation steps for, 205-206

V

Volunteer training, purpose of, 339-340
Volunteerism, 333-341
benefits of, 337
programs, suggestions for improving, 340-341
recent history of, 333
statistic regarding, 334
Volunteers
characteristics of, 334-335
in meetings, 95
myths about, 335
orientation and training of, 338-340
recognition for, 337-338
recruitment of, 333, 336-337, 340
specific need for, 336, 340
V-time, 348

W

Work goals, 64
Work sharing 347-348
Work teams
defined, 51-52
disadvantages of, 55
effective, criteria for, 52-53, 57
effective management of, 54
fostering cooperation in, 54-55
goal orientation in, 56, 59
groupthink in, 55

improved communication in, 55
institquting concept of, 53-55
manager's role in building, 57-58
members of, 53-54
patience and persistence in, 56
self-managing, concept of, 52, 53-55
synergistic effect of, 52
and the value of feedback, 55, 59
Work unit participation quiz, 265
answers to, 270-271
Work-design techniques, 234-235
Workplace behavior, guidelines for, 288-289
Work-time options, 343-351
compressed work week, 345
flextime as, 344
job bidding, 349
job sharing, 346-347
peak time, 348-349
staggered fixed schedule as, 344
suggestions for designing successful options for, 350-351
V-time, 348
work sharing, 347-348
Writing, 381-386
business letters, 383-385
elements of, 381-382
importance of, in business, 381-382
rules for improving, 382-383
suggestions for improving skills in, 385-386
Wrongful termination lawsuits, steps to reduce chances of, 325-326

Z

Zero D
effects program, 256-257